NATO ASI Series

Advanced Science Institutes Series

A series presenting the results of activities sponsored by the NATO Science Committee, which aims at the dissemination of advanced scientific and technological knowledge, with a view to strengthening links between scientific communities.

The Series is published by an international board of publishers in conjunction with the NATO Scientific Affairs Division

A Life Sciences B Physics	Plenum Publishing Corporation London and New York
C Mathematical and Physical Sciences D Behavioural and Social Sciences E Applied Sciences	Kluwer Academic Publishers Dordrecht, Boston and London
F Computer and Systems Sciences G Ecological Sciences H Cell Biology	Springer-Verlag Berlin Heidelberg New York London Paris Tokyo Hong Kong

Series H: Cell Biology Vol. 36

The ASI Series Books Published as a Result of Activities of the Special Programme on CELL TO CELL SIGNALS IN PLANTS AND ANIMALS

This book contains the proceedings of a NATO Advanched Research Workshop held within the activities of the NATO Special Programme on Cell to Cell Signals in Plants and Animals, running from 1984 to 1989 under the auspices of the NATO Science Committee.

The books published as a result of the activities of the Special Programme are:

Vol. 1: Biology and Molecular Biology of Plant-Pathogen Interactions. Edited by J.A. Baily. 1986.

Vol. 2: Glial-Neuronal Communication in Development and Regeneration. Edited by H.H. Althaus and W. Seifert. 1987.

Vol. 3: Nicotinic Acetylcholine Receptor: Structure and Function. Edited by A. Maelicke. 1986.

Vol. 4: Recognition in Microbe-Plant Symbiotic and Pathogenic Interactions. Edited by B. Lugtenberg. 1986.

Vol. 5: Mesenchymal-Epithelial Interactions in Neural Development. Edited by J.R. Wolff, J. Sievers, and M. Berry. 1987.

Vol. 6: Molecular Mechanisms of Desensitization to Signal Molecules. Edited by T.M. Konjin, P.J.M. Van Haastert, H. Van der Starre, H. Van der Wel, and M.D. Houslay. 1987.

Vol. 7: Gangliosides and Modulation of Neuronal Functions. Edited by H. Rahmann. 1987.

Vol. 9: Modification of Cell to Cell Signals During Normal and Pathological Aging. Edited By S. Govoni and F. Battaini. 1987.

Vol. 10: Plant Hormone Receptors. Edited by D. Klämbt. 1987.

Vol. 11: Host-Parasite Cellular and Molecular Interactions in Protozoal Infections. Edited by K.-P. Chang and D. Snary. 1987.

Vol. 12: The Cell Surface in Signal Transduction. Edited by E. Wagner, H. Greppin, and B. Millet. 1987.

Vol. 19: Modulation of Synaptic Transmission and Plasticity in Nervous Systems. Edited by G. Hertting and H.-C. Spatz. 1988.

Vol. 20: Amino Acid Availability and Brain Function in Health and Disease. Edited by G. Huether. 1988.

Vol. 21: Cellular and Molecular Basis of Synaptic Transmission. Edited by H. Zimmermann. 1988.

Vol. 23: The Semiotics of Cellular Communication in the Immune System. Edited by E.E. Sercarz, F. Celada, N.A. Mitchison, and T. Tada. 1988.

Vol. 24: Bacteria, Complement and the Phagocytic Cell. Edited by F.C. Cabello and C. Pruzzo. 1988.

Vol. 25: Nicotinic Acetylcholine Receptors in the Nervous System. Edited by F. Celementi, C. Gotti, and E. Sher. 1988.

Vol. 26: Cell to Cell Signals in Mammalian Development. Edited by S.W. de Laat, J.G. Bluemink, and C.L. Mummery. 1989.

Vol. 27: Phytotoxins and Plant Pathogenesis. Edited by A. Graniti, R.D. Durbin, and A. Ballio. 1989.

Vol. 31: Neurobiology of the Inner Retina. Edited by R. Weiler and N.N. Osborne. 1989.

Vol. 32: Molecular Biology of Neuroreceptors and Ion Channels. Edited by A. Maelicke. 1989.

Vol. 33: Regulatory Mechanisms of Neuron to Vessel Communication in the Brain. Edited by F. Battaini, S. Govoni, M.S. Magnoni, and M. Trabucchi. 1989.

Vol. 35: Cell Separation in Plants: Physiology, Biochemistry and Molecular Biology. Edited by D.J. Osborne and M.B. Jackson. 1989.

Vol. 36: Signal Molecules in Plants and Plant-Microbe Interactions. Edited by B.J.J. Lugtenberg. 1989.

Signal Molecules in Plants and Plant-Microbe Interactions

Edited by

Ben J. J. Lugtenberg

Department of Plant Molecular Biology
Leiden University, Leiden, The Netherlands

Springer-Verlag Berlin Heidelberg New York
London Paris Tokyo Hong Kong

Published in cooperation with NATO Scientific Affairs Division

Proceedings of the NATO Advanced Research Workshop on Molecular Signals in Microbe-Plant Symbiotic and Pathogenic Systems, held at Biddinghuizen, The Netherlands, May 21–26, 1989

ISBN-13: 978-3-642-74160-9 e-ISBN-13: 978-3-642-74158-6
DOI: 10.1007/978-3-642-74158-6

Library of Congress Cataloging-in-Publication Data. NATO Advanced Research Workshop on Molecular Signals in Microbe-Plant Symbiotic and Pathogenic Systems (1989 : Biddinghuizen, Netherlands) Signal molecules in plants and plant-microbe interactions / edited by Ben J. J. Lugtenberg. p. cm.—(NATO ASI series. Series H, Cell biology ; vol. 36) "Proceedings of the NATO Advanced Research Workshop on Molecular Signals in Microbe-Plant Symbiotic and Pathogenic Systems, held at Biddinghuizen, the Netherlands, May 21–26, 1989"—T.p. verso. "Published in cooperation with NATO Scientific Affairs Division."
ISBN-13: 978-3-642-74160-9 (U.S. : alk. paper)
1. Plant-microbe relationships—Molecular aspects—Congresses. 2 Cellular signal transduction—Congresses. I. Lugtenberg, Ben, 1941– . II. North Atlantic Treaty Organization. Scientific Affairs Division. III. Title. IV. Series. QR351.N38 1989 591.8'8—dc20 89-21996

Softcover reprint of the hardcover 1st edition 1989

2131/3140-543210 – Printed on acid-free-paper

PREFACE

This volume comprises the lectures of the participants at the NATO Advanced Research Workshop held at the Congress Centre The Flevohof at Biddinghuizen, The Netherlands, May 21-26, 1989. The purpose of the workshop was to bring together experts in plant molecular biology, plant pathology and symbiosis in order to discuss recent progress in the field of microbe-plant interactions at the molecular level, to promote integration of various disciplines, and to plan future research and applications.

Plants have developed a variety of sophisticated defence mechanisms to cope with an environment in which many different microbes live. Most microbes which colonize plant tissues are harmless. Some microbes do enter the plant but kill only one or a few plant cells because the plant's hypersensitive response prevents further damage. Other microbes have developed ways to attack plants more successfully, resulting in enormous losses of crop yields. Finally, specific microbes have reached an agreement with the host plant which is beneficial for both: they live in symbiosis with the plant and provide the host plant for example with substantial amounts of atmospheric nitrogen. Chemical protection of crops is a necessity in modern crop management and often depends on analogy with natural signal molecules. However, this treatment has sometimes negative effects as well. Therefore scientists are also looking for alternative, biological, ways to control crop pests. In order to capitalize the recent rapid progress made in the mentioned fields, a multidisciplinary approach is required. Plant molecular biologists, phytopathologists, molecular microbiologists, plant physiologists, soil biologists and agronomists should collaborate closely and design recommendations for future research on the molecular basis of microbe-plant interactions.

Against this background specialists from thirteen countries discussed the results of their most recent work on the molecular basis of microbe-plant interactions. It appeared that we are beginning to understand these interactions at the molecular level. Substantial progress has been made in identification of signal molecules and of their function and mechanism of action.

The atmoshpere of the workshop was excellent. Several lectures and discussions resulted in new collaborations and opened links to other disciplines. The scientific success of the workshop was to a major extent the result of the excellent, critical and friendly advice of my colleagues Prof. Peter Albersheim, Prof. Klaus Hahlbrock and Prof. Robbert Schilperoort, to whom I am most grateful. The efforts of all speakers and discussion leaders was remarkable. I thank all participants for a week of excellent science.

The meeting would not have been possible without the help of many people, most of whom I will not mention here. I want to express my admiration for Sonja Chang, who spent many days at the organization of the workshop. I am also in debt to her for her careful and efficient handling of the manuscripts and for improving and preparing many of the excellent typescripts from which this book was produced. The selfless handling of logistic and financial aspects of the meeting by many of my colleagues from the Department of Plant Molecular Biology of the Leiden University was warmly appreciated. Also the valuable advice of Dr. Alain Jubier at the NATO office in Brussels should be mentioned in particular.

The financing of the workshop was helped greatly by contributions of several bodies, from which the Scientific Affairs Division of NATO was our major sponsor. On behalf of all participants I thank our sponsors for the often generous grants which enabled us to finance this scientific meeting.

We hope that this volume will prove to be useful in sharpening our perspective of how microorganisms and plants communicate at the molecular level.

Leiden, August 1989 B. Lugtenberg

CONTENTS

I. SIGNAL COMMUNICATION IN ANIMALS AND THEIR POSSIBLE RELEVANCE FOR PLANTS

II. PLANT SIGNALS

V. SIGNALS IN THE INTERACTION OF PLANTS WITH (PATHOGENIC) BACTERIA

VI. SIGNALS IN PLANT-FUNGUS COMMUNICATION

VII. SIGNALS IN NODULATION AND NITROGEN FIXATION

Animal Receptors - Examples of Cellular Signal Perception Molecules

Michael G. Hahn

Complex Carbohydrate Research Center & Department of Botany, University of Georgia, 220 Riverbend Road, Athens, GA 30602, USA

INTRODUCTION

Considerable information has accumulated about the mechanisms by which animal cells perceive extracellular signals. In sharp contrast, very little is known about cellular signal perception in plant cells. The purpose of this article is to examine signal perception mechanisms in animal cells and to highlight general principles that may prove applicable to the study of plant systems. I will focus on the receptors in target cells that specifically bind the signal molecule(s). This is the initial event in the signal cascade that eventually leads to changes in cellular metabolism (*e.g.*, gene expression) in targeted cells. Secondary messengers, which are often involved in the signal transduction process subsequent to the binding of signal molecules to receptors, are the focus of another review (see article by C. A. West, this volume).

One concept that is clearly emerging from studies of mammalian receptors is that receptors can be grouped into families that utilize common mechanisms for signal transmission. Analysis of the proteins that make up these families is revealing structural and functional similarities among members of a given receptor family. Examples from three families of animal receptors will be discussed here. Steroid hormone receptors are examples of receptors, located within the cell, that directly regulate gene expression after binding signal molecules. The other two families represent plasma membrane-localized receptors that cause the release of second messengers that relay the signal to the cell nucleus. The insulin receptor is an example of a receptor that has intrinsic protein kinase activity and transports its ligand to the cell interior *via* receptor-mediated endocytosis. Finally, neurotransmitter receptors, specifically muscarinic cholinergic receptors and adrenergic receptors, are examples of receptors that are involved in the control of ion flow across the membrane and are linked to second messenger production through guanine nucleotide binding proteins. Using these examples, experimental approaches and general principles of receptor structure and function that might be applicable to the study of plant receptors will be discussed.

NATO ASI Series, Vol. H 36
Signal Molecules in Plants and Plant-Microbe Interactions
Edited by B. J. J. Lugtenberg

STEROID RECEPTORS

Steroid hormones are metabolites of cholesterol that are ubiquitous in their distribution, being found in virtually all eukaryotic organisms. These hormones have been shown to trigger and control a number of physiological and developmental processes in many organisms. The effects of different steroid hormones on cellular metabolism are mediated by intracellular receptors that are specific for each steroid. Upon association of the steroid with its receptor, the complex acquires an increased affinity for chromosomal binding sites in the nucleus. Association of the receptor:hormone complex with specific sites in the genome results in increased transcription of specific genes. This ability of the steroid receptors to act as transcriptional regulatory proteins has focused attention on their structure and mode of action. Review articles on steroid hormone receptor structure [*e.g.*, (Sherman & Stevens 1984)] and function [*e.g.*, (Ringold 1985; Yamamoto 1985; Gustafsson *et al* 1987; Evans 1988; Beato 1989; Burnstein & Cidlowski 1989)], as well as complete books on this subject [*e.g.*, (Eriksson & Gustafsson 1983; Sheridan *et al* 1988)] have appeared regularly.

Purification and characterization of steroid hormone receptors was complicated by their low abundance (10^3 to 10^4 per cell). The use of synthetic steroids and/or analogs radiolabeled to high specific radioactivity (50-100 Ci/mmole), photo-affinity labeling techniques, and affinity chromatography methods have allowed the purification of steroid receptors [see chapters 5-9 in Sheridan *et al* (1988)]. Another approach that has been used recently to identify an aldosterone receptor is the generation of an anti-idiotypic antibody (Lombes *et al* 1989) that recognizes the receptor.

The development of purification methods and ligand-binding assays for steroid receptors permitted studies of their subcellular localization and their biochemical properties. In the absence of steroids, the receptors are soluble intracellular proteins that are commonly found in the cytoplasmic fraction of cell extracts. Upon steroid hormone binding, the receptors undergo a process termed "activation" or "transformation" that greatly increases the affinity of the receptors for binding sites in the cell nucleus where they are found upon disruption of steroid-treated cells. These observations led to the proposal of a two-step mechanism for the interaction of steroids with their target cells (Jensen *et al* 1968). In this model, steroids passively diffuse into cells where they combine with free receptors localized in the cytoplasm. After hormone binding and activation, the receptors migrate to the cell nucleus and bind with high affinity to chromosomal DNA. This model was called into question by two studies that reported localization of most of the free receptors in cell nuclei using either receptor-specific monoclonal antibodies (King & Greene 1984) or cell enucleation techniques (Welshons *et al* 1984). There is now general agreement that activated steroid receptors reside in the nucleus. However, the location of unoccupied receptors is still a matter of debate [see chapters 1,3,4 of

Sheridan *et al* (1988)], although most workers seem to agree that at least a majority of unoccupied receptors are localized in the cell nucleus.

Both occupied and unoccupied steroid receptors have been isolated from several target tissues and their properties examined and compared [reviewed by Sherman & Stevens (1984)]. Activation of receptors following steroid hormone binding is accompanied by decreased sedimentation rates (indicating a change in size and/or conformation), altered chromatographic properties, increased affinity for chromatin and DNA, and changes in receptor phosphorylation. It is still unclear whether the changes in receptor phosphorylation are a cause or result of activation of the receptor. Unactivated receptors are multimeric proteins, possibly consisting of one or more receptor monomers and other proteins, most prominently, a 90 kD heat shock protein (Denis *et al* 1988; Bresnick *et al* 1989). Binding of steroid to free receptors leads to dissociation of the multimer to give a 90-100 kD monomer that is the activated form of the receptor (Fig. 1). The activated receptor binds to specific DNA sequences in the chromosome, thereby modulating the transcription of genes located near the receptor binding site (see below). The steroid receptor can be viewed, therefore, as a hormone-dependent transcription factor. Indeed, recent evidence from recombinant DNA studies demonstrates that steroid receptors are structurally related to known transcription factors (see below).

Studies on purified activated steroid receptors suggested that the polypeptides consist of discrete, linearly arranged, functional domains. Proteolytic cleavage of activated steroid

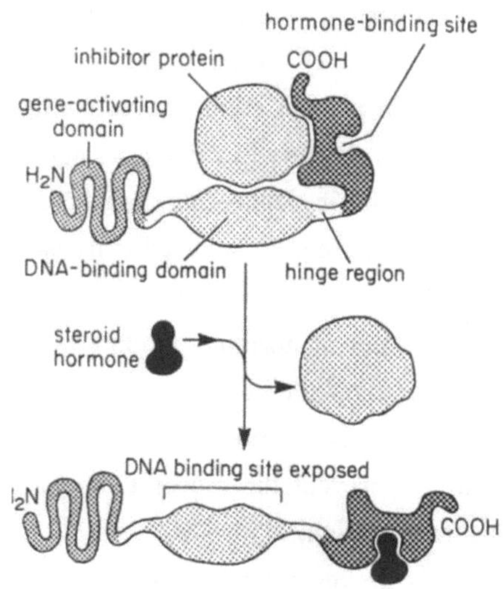

Figure 1. Schematic diagram of the activation of steroid hormone receptors upon hormone binding. Several proteins together with one or more steroid receptor monomers may constitute the unactivated receptor multimer. [Redrawn from Alberts *et al* (1989), p. 691, with permission, copyright Garland Publishing, Inc.]

receptor proteins and examination of naturally occurring truncated receptor mutants suggested that the binding sites for steroid and DNA are located in separate domains on the polypeptide. The single steroid binding site appears to be located in the C-terminal third of the polypeptide. The DNA-binding domain appears to be adjacent to the steroid-binding domain.

Receptor-specific polyclonal antibodies and monoclonal antibodies generated against purified steroid receptor proteins were used to screen expression libraries for cDNA clones encoding the human glucocorticoid (Miesfeld *et al* 1984; Hollenberg *et al* 1985) and estrogen (Walter *et al* 1985; Green *et al* 1986; Krust *et al* 1986) receptors. Subsequently, cDNAs encoding receptors for mineralocorticoid [*e.g.*, (Arriza *et al* 1987)], progesterone [*e.g.*, (Loosfelt *et al* 1986; Gronemeyer *et al* 1987)], and testosterone [*e.g.*, (Chang *et al* 1988; Lubahn *et al* 1988)] have been cloned and sequenced. Comparative analysis of these sequences has shown that steroid receptors belong to a receptor superfamily (Evans 1988), which also includes receptors for thyroid hormone (Weinberger *et al* 1986a; Thompson *et al* 1987), retinoic acid (Giguere *et al* 1987; Petkovich *et al* 1987), vitamin D (McDonnell *et al* 1987), as well as an avian oncogene product v-*erb A* (Green *et al* 1986; Krust *et al* 1986; Weinberger *et al* 1986b). Additional sequence comparison studies are leading to the identification of other members of this nuclear receptor superfamily [*e.g.*, (Hazel *et al* 1988; Trapman *et al* 1988)] that also appear to be ligand-dependent transcriptional regulatory proteins.

Analysis of cloned steroid receptor genes [reviewed by Evans (1988), Beato (1989), Burnstein & Cidlowski (1989)] has confirmed and extended the results of studies on purified steroid receptor proteins described earlier. The sequences of the cloned receptor genes show two regions of sequence similarity (Fig. 2) corresponding to the DNA-binding and steroid-binding domains identified earlier in protein studies, thus confirming the idea that steroid receptor polypeptides are made up of discrete functional domains. Construction of receptor deletion mutants has proved very useful in delineating domains involved in hormone binding and DNA binding (discussed below) and regions involved in transcriptional activation of target genes [*e.g.*, (Dobson *et al* 1989)]. Deletion mutations at the C-terminal end of the glucocorticoid receptor gene led to loss of hormone binding by the mutant receptor. More significantly, loss of the hormone binding site yielded a constitutively active receptor, *i.e.*, the altered receptor protein activated gene transcription in the absence of the hormone. Thus, the hormone-binding region appears to prevent the receptor, in the absence of hormone, from binding to DNA and activating transcription.

The DNA-binding domain is the most highly conserved region among the members of this receptor superfamily (Fig. 2). Sequence analysis of this region also suggests that the DNA-binding domain contains a structural motif similar to that found in a transcription factor from *Xenopus laevis* that is rich in cysteine residues (Evans & Hollenberg 1988). These cysteine-rich regions are thought to be able to adopt conformations coordinated around zinc atoms,

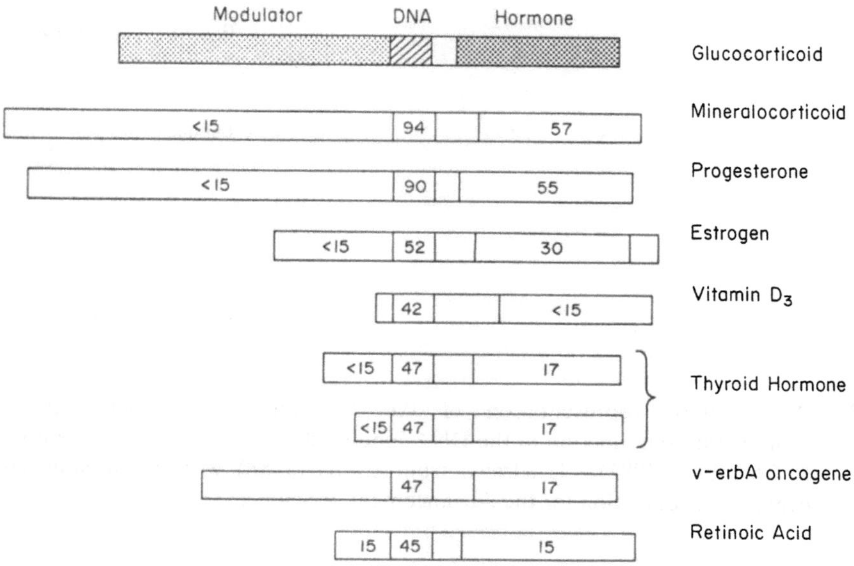

Figure 2. Schematic amino acid sequence comparisons of members of the steroid receptor superfamily showing the domain structure of the receptors. The identified domains are the hypervariable, amino terminal **Modulator** domain important in trans-activation of hormone-regulated genes, and the more conserved **DNA-** and **Hormone-**binding domains. The percentages of amino acid identity of each region are shown in relation to the human glucocorticoid receptor. [Adapted from Evans (1988) with permission, copyright American Association for the Advancement of Science]

thus creating "fingers" that can bind to DNA and/or RNA. While it is not known whether this "zinc-finger" configuration has any functional significance in DNA binding by steroid receptors *in vivo*, deletions or insertions in this region abolish DNA binding *in vitro* (Giguére *et al* 1986; Gronemeyer *et al* 1987). These mutant receptors retain the ability to bind steroid hormone, suggesting that the structure of the remainder of the receptor protein is intact. An elegant experiment that confirmed the function of the DNA-binding domain was the substitution of the putative DNA-binding domain of the human estrogen receptor with that of the human glucocorticoid receptor (Green & Chambon 1987) (Fig. 3). The resulting hybrid receptor activated transcription, in an estrogen-dependent manner, of a glucocorticoid responsive gene. The hybrid receptor did not activate transcription of an estrogen responsive gene. Refinement of this technique (Green *et al* 1988), together with amino acid substitution studies (Mader *et al* 1989), has shown that the N-terminal "zinc finger" largely determines target gene specificity. Recent functional expression, in high yield, of the DNA-binding domain of the human glucocorticoid receptor in *Escherichia coli* (Dahlman *et al* 1989) should permit detailed structural and conformational analysis of this domain.

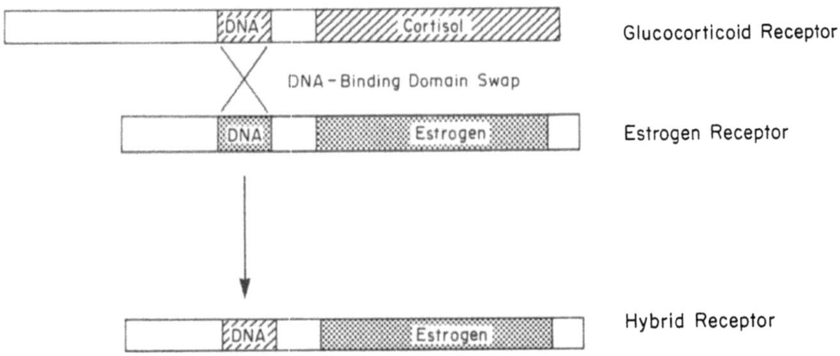

Figure 3. Diagrammatic representation of the "domain swap" experiment used to demonstrate the function of the DNA-binding domain of a steroid receptor (Green & Chambon, 1987). [Redrawn from Evans (1988) with permission, copyright American Association for the Advancement of Science]

Recombinant DNA techniques have also proved very useful in elucidating the nature of the steroid receptor binding site(s) in chromosomal DNA [reviewed by Evans (1988), Beato (1989), Burnstein & Cidlowski (1989)]. The glucocorticoid-inducible mouse mammary tumor virus system [reviewed by Ringold (1985)] has proved particularly useful in these studies. Application of glucocorticoids to mouse mammary carcinoma cells stimulates the production of mouse mammary tumor virus, a typical retrovirus. Reverse transcription of the viral RNA to give proviral DNA results in the generation of long terminal repeats that contain the regulatory signals involved in controlling viral gene transcription. In particular, signals that are recognized by the glucocorticoid receptor are encoded in this region. These signal sequences have properties reminiscent of enhancer sequences [reviewed by Khoury & Gruss (1983)] in that they will function in both forward and reverse orientations and at variable distances relative to a heterologous promoter (Chandler *et al* 1983). Analysis of chimeric genes generated by fusion of the long terminal repeat, or deletion mutants thereof, to a reporter gene resulted in the identification of a family of octanucleotides related to the degenerate sequence **5' AGA(A/T)CAG(A/T) 3'** that are present at least once in several putative glucocorticoid receptor binding sites in the viral terminal repeat. Very similar sequences (termed glucocorticoid response elements or, more generally, **H**ormone **R**esponse **E**lements) have been found in several other glucocorticoid-regulated genes using analogous techniques. Related but not identical sequences have also been identified proximal to genes that are transcriptionally activated by other steroid hormone receptors.

Detailed knowledge of steroid receptor structure and the identification of the hormone response elements in nuclear DNA are allowing experiments to determine how transcriptional activation of target genes occurs. The apparent dyad symmetry of the hormone response elements suggests that they interact with receptor dimers. Evidence supporting such a model has been obtained from studies of activated estrogen and glucocorticoid receptor proteins (Sabbah *et al* 1989; Wrange *et al* 1989) and from studies of the interaction between the DNA binding domain of the progesterone receptor and its hormone response element (Tsai *et al* 1988). Other recent studies demonstrate that several hormone response elements, or a combination of hormone response elements and binding sites for other transcription factors, must be clustered in order to obtain maximum transcriptional activation (Jantzen *et al* 1987; Strähle *et al* 1988; Ankenbauer *et al* 1988). These studies suggest that cooperative binding of two or more transcriptional regulatory proteins to their target sequences is important for transcriptional activation of genes.

INSULIN RECEPTOR

Insulin is a peptide hormone secreted by the β cells of the pancreas that elicits a wide variety of biological responses in cells capable of responding to the hormone [reviewed by Czech (1977), Rosen (1987)]. These biological responses include activation of transport systems for metabolites (*e.g.*, sugars, amino acids), activation of enzymes (*e.g.*, glycogen synthase), inhibition of enzymes, changes in gene expression, and promotion of cell growth. Some of the effects of insulin occur within minutes, *e.g.*, stimulation of hexose transport, while other insulin-mediated effects require hours, *e.g.*, cell growth. The plethora of insulin-mediated changes in cellular metabolism has focused attention on the study of insulin's mode of action, in particular, on the first step of the signal cascade, the binding of insulin to its receptor.

Receptors for insulin were first demonstrated almost twenty years ago by Cuatrecasas using iodinated insulin (Cuatrecasas 1971). Since that time, advances in techniques for protein purification and the application of recombinant DNA techniques have significantly advanced our knowledge of the structure and function of the insulin receptor [reviewed by Czech (1985), Rosen (1987)]. However, much remains to be learned about how the occupied receptor triggers the observed physiological changes. The insulin receptor is present in most, if not all, mammalian cells at concentrations varying between 10^2 and 10^5 receptors per cell, depending on the cell type. Successful detergent solubilization of the insulin receptor in a form that retains its ability to bind the hormone permitted the purification of the receptor by ligand affinity chromatography. The insulin receptor is an integral membrane glycoprotein consisting of two subunits (α - $M_r \approx 130$ kD; β - $M_r \approx 95$ kD) with a subunit composition $\alpha_2\beta_2$. The α-subunits are linked to each other and to the β-subunits by disulfide bonds (Fig. 4). These two subunits are cleaved from a larger precursor protein (containing one α- and one β-subunit) as demonstrated by pulse-chase experiments. The receptors present on the cell surface are in a constant state of turnover that appears to occur via receptor-mediated endocytosis [*e.g.*, (Backer *et al* 1989; Trischitta *et al* 1989; Bottaro *et al* 1989)], a process that is stimulated by insulin binding. In view of the intrinsic enzymatic activity of the receptor (see below), the functional role, if any, of receptor-mediated endocytosis in signal transduction remains to be elucidated.

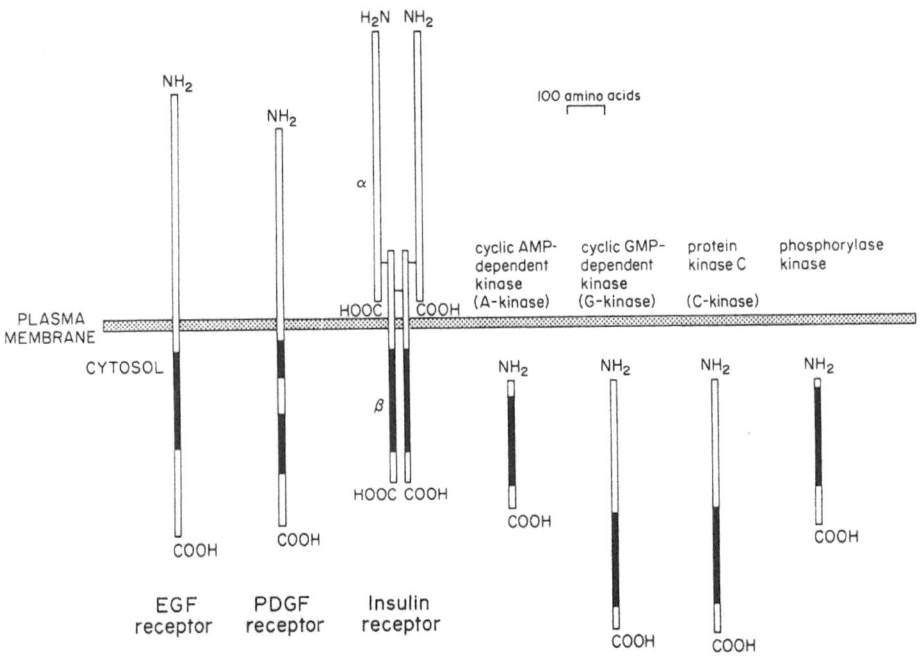

Figure 4. Diagrammatic representation of the structures of the insulin receptor and the membrane receptors for epidermal growth factor (EGF), platelet-derived growth factor (PDGF), as well as several cytoplasmic protein kinases, showing the conserved protein kinase domains. [Redrawn from Alberts *et al* (1989), p. 707, with permission, copyright Garland Publishing, Inc.]

Different functional domains have been localized to different subunits of the insulin receptor. Affinity cross-linking and photo-affinity-labeling experiments demonstrated that the insulin binding site was located on the α-subunit. Binding of insulin to the receptor in intact cells was shown to rapidly activate protein kinases [reviewed by Czech *et al* (1988)] resulting in increased phosphorylation of the β-subunit of the receptor on seryl, threonyl and tyrosyl residues, as well as increased phosphorylation of other polypeptides in treated cells. An insulin-dependent protein tyrosine kinase activity co-purified precisely with the insulin receptor, demonstrating that the insulin receptor itself was a ligand-dependent protein-tyrosine kinase (Kasuga *et al* 1983; Petruzzelli *et al* 1984) that autophosphorylates the β-subunit of the receptor on specific tyrosine residues. Affinity labeling studies demonstrated that the β-subunit has an ATP-binding site, suggesting that the protein kinase domain was localized to the β-subunit. Other plasma membrane receptors for peptide hormones and growth factors, such as the epidermal growth factor receptor [reviewed by Carpenter (1987), Staros *et al* (1989)] and the platelet-derived growth factor [reviewed by Heldin & Ronnstrand (1989), Williams (1989)], have also been shown to be tyrosine-specific protein kinases suggesting a functional and structural relationship among these receptors (Fig. 4).

The role of phosphorylation of the insulin receptor in receptor function has not been completely elucidated [reviewed by Rosen (1987), White & Kahn (1989)]. Autophosphorylation of receptor tyrosine residues occurs within seconds resulting in an activation of the protein tyrosine kinase activity of the receptor. In contrast, phosphorylation of receptor seryl and threonyl residues occurs more slowly (within minutes), and appears to be carried out by as yet unidentified protein kinases. Seryl/threonyl phosphorylation decreases the tyrosine kinase activity of the insulin receptor, and thus may play a role in modulating the signal transduction pathway triggered by insulin.

The insulin receptor tyrosine kinase phosphorylates not only itself but also a number of other soluble and membrane-bound proteins [*e.g.*, (Bernier *et al* 1987; Chou *et al* 1987; Perrotti *et al* 1987; Margolis *et al* 1988)]. For the most part, the functions of these proteins, as well as their possible role in the signal transduction pathway connecting insulin with metabolic changes in target cells, remain to be established. Evidence has been presented suggesting a role in intracellular signaling for a 15-kD cytosolic protein that is phosphorylated on tyrosine upon binding of insulin to the insulin receptor (Bernier *et al* 1987; 1988; Hresko *et al* 1988). This protein is thought to play a role as an intermediary in insulin-activated glucose uptake in fat cells.

The recent application of recombinant DNA techniques has confirmed and extended much of the work carried out at the protein level on the insulin receptor. Partial amino acid sequences of tryptic peptides generated from the purified receptor were used to design synthetic oligonucleotide probes for use in isolating cDNA clones for the human insulin proreceptor (Ullrich *et al* 1985; Ebina *et al* 1985b). The complete amino acid sequence of the receptor was deduced from the nucleotide sequence. More recently, the intron-exon structure of the human insulin proreceptor gene (Seino *et al* 1989) and the nucleotide sequence of its promoter region were reported (Araki *et al* 1987; Seino *et al* 1989). Analysis of the deduced amino acid sequence of the receptor confirmed that the two subunits of the receptor arise from a common precursor protein. The α-subunit was found to contain most of the potential glycosylation sites and appeared to lack a sequence of hydrophobic amino acids sufficiently long to serve as a membrane anchor. The β-subunit contained a 23-amino acid transmembrane region, and a protein kinase domain homologous to tyrosine kinases encoded by oncogenes. In fact, comparisons of various protein kinase domains for which sequence data are available demonstrated that the kinase domains of a wide variety of receptor tyrosine kinases, cytoplasmic protein kinases, and oncogene products are highly conserved [reviewed by Hanks *et al* (1988)]. Identification of additional protein kinases, including kinases that also function as receptors, using this conserved domain is likely (Hunter 1987).

Availability of cDNA clones encoding the complete sequence of the insulin proreceptor permitted the development of experimental systems to study the structure:function relationships of the insulin receptor in molecular detail. Expression of transfected insulin receptor cDNAs has been achieved in *Xenopus* oocytes (Ebina *et al* 1985a) and several mammalian cell lines (Ebina *et al* 1985a; Whittaker *et al* 1987). Functionality of the transfected insulin receptor was demonstrated by several criteria including binding of insulin, autophosphorylation, and enhanced insulin-induced stimulation of glucose uptake and glycogen synthesis. These experiments provide conclusive evidence that the cloned gene encodes the insulin receptor.

Application of recombinant DNA techniques has resulted in the construction of a number of mutant insulin receptor cDNAs. Expression of these mutant receptor cDNAs in transfected cells has helped delineate domains involved in insulin binding, tyrosine kinase activity, and coupling to insulin-sensitive cellular responses. One approach (already described above for steroid receptors) is the construction of hybrid receptor proteins using domains from different receptor proteins. Thus, fusion of DNA encoding the ligand-binding domain of the insulin receptor to DNA encoding the transmembrane and kinase domains of the epidermal growth factor receptor, followed by expression of the chimeric gene in transfected cells yielded a hybrid receptor that underwent insulin-stimulated autophosphorylation of the epidermal growth factor kinase domain (Riedel *et al* 1986). However, it was not reported whether the hybrid receptor mediated any insulin-dependent biological response characteristic of epidermal growth factor. Another hybrid receptor construct, consisting of the extracellular domain of the human insulin receptor and the transmembrane and kinase domains of a protein encoded by the chicken v-*ros* oncogene, had insulin-stimulatable kinase activity, but did not mediate insulin-activated deoxyglucose uptake in transfected cells (Ellis *et al* 1987a). This hybrid receptor also did not cause insulin-dependent transformation (*i.e.*, oncogene activity) of the transfected cells as measured by cell growth characteristics.

Single amino acid substitutions, created by site-directed mutagenesis, at a lysine residue located in the putative ATP-binding site of the β-subunit abolished the tyrosine kinase activity of the mutated receptor expressed in transfected cells (Chou *et al* 1987; Ebina *et al* 1987; McClain *et al* 1987). The mutant receptors were properly processed and inserted into the plasma membrane, and retained the capacity to bind insulin normally. However, insulin-mediated effects on deoxyglucose uptake, protein phosphorylation, and glycogen synthesis were abolished in cells containing a mutant receptor. Furthermore, the mutant receptor did not undergo receptor-mediated endocytosis (McClain *et al* 1987), which occurs after insulin binds to the wild-type receptor. Additional site-directed mutagenesis experiments in which phenylalanine was substituted for tyrosyl residues, whose phosphorylation is believed to be important for kinase activation, yielded mutant receptors that had greatly reduced insulin-inducible tyrosine kinase activity (Ellis *et al* 1986). Mutation of another tyrosyl residue thought to be important for receptor function in signal transduction yielded a receptor that had the same insulin-binding properties and kinase activity, and underwent autophosphorylation characteristic of the wild-type receptor (White *et al* 1988). However, binding of insulin to the mutant receptor expressed in transfected cells did not activate cellular responses characteristic of the wild-type insulin receptor, including tyrosyl phosphorylation of other proteins. Expression of a cDNA encoding a truncated insulin receptor protein lacking all of the α-subunit (insulin-binding domain) and the membrane-spanning domain of the β-subunit yielded a soluble tyrosine kinase activity, but basal uptake of deoxyglucose did not increase in the transfected cells (Ellis *et al* 1987b). Deletion of the C-terminal 43 amino acids of the proreceptor (*i.e.*, of the β-subunit) yielded a mutant receptor that was synthesized and processed normally, was able to bind insulin, appeared to have normal kinase activity both with respect to autophosphorylation of the receptor and phosphorylation of cellular proteins, and underwent receptor-mediated endocytosis, yet did not mediate insulin-sensitive deoxyglucose uptake or activation of glycogen synthetase (McClain *et al* 1988; Maegawa *et al* 1988).

The studies with mutant and hybrid insulin receptors demonstrate that the tyrosine kinase activity of the receptor is essential but not sufficient for the receptor to mediate the biological effects of insulin. Other aspects of insulin receptor function remain to be elucidated at the molecular level. How the binding of insulin to the α-subunit of the receptor on the extra-cellular surface triggers activation of tyrosine kinase on the β-subunit on the cytoplasmic side of the plasma membrane remains unknown. The role of the insulin receptor in mediating the effects of insulin leading to altered cellular metabolism also remains to be clarified. For example, what role do the phospho-oligosaccharides, recently identified as possible "second messengers" in the insulin-activated signal cascade (Saltiel & Cuatrecasas 1986; Kelly *et al* 1987; Mato *et al* 1987; Alemany *et al* 1988), play, and how does insulin binding to its receptor lead to the release of these second messengers?

NEUROTRANSMITTER RECEPTORS

Signal transduction in the mammalian nervous system has been intensively studied during the past two decades. Propagation of a nerve impulse requires signal transduction from one cell to the next and involves the release of a signal, the neurotransmitter, from the end of a neuron. The neurotransmitter then binds to specific receptors on an adjacent neuron or on a cell in the target tissue (*e.g.*, muscle) causing a localized depolarization of the cell membrane by opening an ion channel, thereby propagating the nerve impulse. The two most prevalent neurotransmitters are acetylcholine (utilized by the cholinergic nerve fibers) and norepinephrine (used by the adrenergic nerve fibers). Receptors for both neurotransmitters have been intensively studied, and significant information has accrued about the structure and function of these receptors. The discussion here will focus primarily on the receptors for acetylcholine (cholinergic receptors).

Receptors for acetylcholine have been subdivided into two major classes based upon the ability of the alkaloids nicotine and muscarine to mimic some or all of the effects of acetylcholine on receptive cells (*i.e.*, nicotine and muscarine are acetylcholine agonists) (Mayer 1985). The two different receptor classes are found in different cell types of the nervous system. Nicotinic receptors are found in neuromuscular junctions in striated muscles, sympathetic ganglion cells, adrenal medullary cells, and in the central nervous system. Muscarinic receptors are found at neuromuscular junctions in smooth muscles, the heart, secretory organs, the gastrointestinal tract, and the central nervous system. Characteristics of the muscular nicotinic cholinergic receptor will be briefly described first. This receptor is the prototype of an integral membrane receptor controlling an ion channel. Recent progress toward characterization of muscarinic cholinergic receptor structure and function will then be discussed in somewhat greater detail.

Nicotinic cholinergic receptors were the first to be studied in detail because of their availability in large quantities from fish electric organs. The nicotinic receptor has been purified to homogeneity from a number of tissues and extensively characterized [reviewed by Conti-Tronconi & Raftery (1982), Stroud & Finer-Moore (1985), Maelicke (1988)]. Briefly, the muscle and electric organ receptors are integral membrane glycoproteins consisting of four subunits (α - $M_r \approx 40$ kD, β - $M_r \approx 50$ kD, γ - $M_r \approx 60$ kD, δ - $M_r \approx 65$ kD) in a molar ratio of 2:1:1:1. All five subunits span the membrane and form a rosette when viewed in the electron microscope (Kistler *et al* 1982; Brisson & Unwin 1985). The central hole of this rosette is presumed to be the ion channel, which opens upon binding of acetylcholine. Affinity-purified nicotinic cholinergic receptors have been inserted into artificial lipid bilayers in a functional form, demonstrating that the ion channel and the structures controlling its opening and closing are intrinsic parts of the receptor. Furthermore, receptor subunits isolated from different organisms can be combined *in vitro* to form functional receptors, indicating functional homology between different nicotinic cholinergic receptors (see below).

The primary sequences of each of the subunits of nicotinic cholinergic receptors from several different sources have been deduced from the sequences of the corresponding cDNAs [reviewed by Maelicke (1987)] and agree with partial amino acid sequences determined previously. There is strong sequence similarity between sequences of receptor subunits isolated from different organisms. Furthermore, the primary sequences of the different subunits revealed extensive sequence similarities, suggesting their evolution from a single ancestral subunit. More recently, evidence has been obtained for different structural and functional subclasses of nicotinic cholinergic receptors, particularly in neuronal tissues [*e.g.*, (Schoepfer *et al* 1988; Wada *et al* 1988)]. The molecular characterization of these new receptor subtypes has not yet progressed as far as has analogous research on the muscarinic cholinergic receptor subtypes described below.

Availability of the genes for each of the subunits has made possible transgenic experiments to elucidate receptor function [*e.g.*, (Mishina *et al* 1984; Boulter *et al* 1987)]. *Xenopus* oocytes, which lack acetylcholine-responsive ion channels, are commonly used for these experiments. Expression of the genes for all four subunits is required in order to obtain a functional receptor. Site-directed mutagenesis studies have proved very useful in identifying sequences in the various subunits that are important for ligand binding (both agonist and antagonist) and proper formation of the transmembrane channel [*e.g.*, (Mishina *et al* 1985; Leonard *et al* 1988)]. Reconstitution and transgenic expression studies provide unambiguous evidence that the proteins originally isolated as acetylcholine binding-proteins are in fact the nicotinic acetylcholine receptors.

Muscarinic cholinergic receptors [reviewed by Järv & Bartfai (1988), Nathanson (1987)] have very different properties from those described above for nicotinic cholinergic receptors. Binding of agonists to muscarinic cholinergic receptors results in activation of one or more cellular responses, including inhibition of adenylate cyclase, stimulation of cGMP synthesis, increased turnover of phosphatidylinositol, in addition to causing changes in ion conductance of the plasma membrane. The induction of these different responses is brought about through the action of guanine nucleotide-binding regulatory proteins [G proteins, reviewed by Stryer & Bourne (1986), Gilman (1987)]. G proteins function as intermediaries between receptor(s) and effector molecules (such as adenylate cyclase) that form the next link in the signal cascade. The responses elicited upon agonist binding vary depending on the tissue being studied. In addition, different tissues show different affinities for both agonists and antagonists (Hammer *et al* 1980; Birdsall & Hulme 1983; Cohen & Sokolovsky 1987). These results led to the hypothesis that different subtypes of muscarinic cholinergic receptors exist in different tissues. However, the molecular basis for the observed differences in ligand binding and cellular responses remained unknown. The recent application of molecular cloning techniques has achieved remarkable progress toward explaining muscarinic receptor function at the molecular level, and demonstrated that muscarinic cholinergic receptors belong to a larger family of receptors, all of which exert their effects through interactions with G proteins.

Muscarinic cholinergic receptors have been purified from several sources after detergent solubilization followed by affinity chromatography [*e.g.*, (Haga & Haga 1985; Repke 1987)]. The purified receptor is an integral membrane protein composed of a single polypeptide having a molecular mass of about 70 kD as determined by SDS-PAGE. It is not yet clear whether the receptor is a glycoprotein (Repke 1987), though the amino acid sequences deduced from cDNA sequences (see below) contain consensus glycosylation sites. The purified solubilized receptor could be reconstituted in phospholipid vesicles together with purified G protein to give a functional receptor that showed high affinity agonist binding with concomitant agonist-dependent GTPase activity (Haga *et al* 1985; 1986). Reconstitution experiments in phospholipid vesicles also have been used to study the interaction of muscarinic cholinergic receptors with specific G proteins [*e.g.*, (Florio & Sternweis 1989)].

Partial amino acid sequences of tryptic peptides generated from the purified receptor were used to design synthetic oligonucleotide probes used to isolate cDNA clones for cerebral (Kubo *et al* 1986b) and atrial (Kubo *et al* 1986a; Peralta *et al* 1987a) muscarinic cholinergic receptors. The complete amino acid sequences of these muscarinic receptors was deduced from the nucleotide sequences. Analysis of these amino acid sequences revealed significant sequence similarities with the visual opsins [reviewed by Nathans (1987)], as well as with another mammalian neurotransmitter receptor, the β-adrenergic receptor (Frielle *et al* 1987; Kobilka *et al* 1987a). More recently, this receptor superfamily has expanded to include the α-adrenergic receptors (Cotecchia *et al* 1988; Regan *et al* 1988), serotonin receptors (Kobilka *et al* 1987c; Fargin *et al* 1988; Julius *et al* 1988;), substance K (a tachykinin neuropeptide) receptor (Masu *et al* 1987), an oncogenic angiotensin receptor (Jackson *et al* 1988), and yeast mating type receptors [reviewed by Herskowitz & Marsh (1987)].

The amino acid sequence similarities among the members of this receptor family are localized to seven regions of the proteins. These regions are rich in hydrophobic amino acids and are presumed to form seven transmembrane α-helices (Fig. 5). With the exception of bacteriorhodopsin (Nathans 1987), there is little experimental evidence to support this topography for these receptors. Some experimental evidence has accumulated confirming aspects of this model for the β_2-adrenergic receptor (Dohlman *et al* 1987; 1988; Wong *et al* 1988). The remaining regions of the amino acid sequences of these receptors show little, if any, similarities, presumably reflecting the different cellular responses elicited by the different receptors or receptor subclasses.

The isolation of distinct cDNA clones for different receptor subtypes, for example, for the muscarinic cholinergic receptors (Kubo *et al* 1986a; 1986b; Bonner *et al* 1987; Fukuda *et al* 1987; Peralta *et al* 1987b; Shapiro *et al* 1988), clearly demonstrates that the pharmacologically different subtypes are in fact different proteins. That is, the receptor subtypes are not distinguished by different modifications or different local environments of the

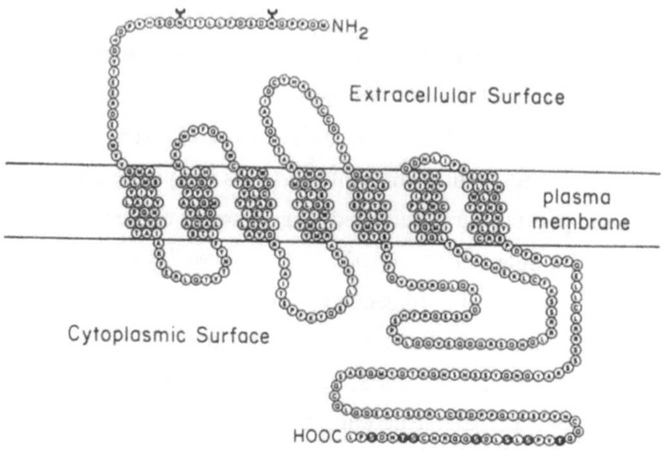

Figure 5. Proposed structural organization of the β_2-adrenergic receptor in the plasma membrane thought to be typical of the topography of the superfamily of G protein coupled receptors. [Reproduced from Sibley *et al* (1988) with permission, copyright The Endocrine Society]

same receptor protein, or the interaction of the same receptor protein with different G proteins. This conclusion was substantiated by studies on the expression of recombinant receptor proteins either via injection of translatable mRNA into *Xenopus* oocytes or by stable transfection of receptor genes into cell lines lacking the receptor. Injection of porcine cerebral muscarinic receptor (Kubo *et al* 1986b; Fukuda *et al* 1987) or porcine cardiac muscarinic receptor (Fukuda *et al* 1987) mRNAs into *Xenopus* oocytes resulted in expression of functional receptors having the ligand-binding properties (both agonist and antagonist) and ion channel responses characteristic of these two receptor subtypes *in vivo*. Genes for a number of different muscarinic cholinergic receptor subtypes have been transfected into mammalian cell lines lacking muscarinic receptors and expressed (Bonner *et al* 1987; Peralta *et al* 1987a; 1987b; 1988; Neher *et al* 1988; Shapiro *et al* 1988; Stein *et al* 1988). The transfected cells exhibited the physiological responses (*e.g.*, ion channel responses, ligand-binding properties, effects on adenylate cyclase and phosphatidylinositol turnover) characteristic of the receptor whose gene had been introduced into the cell line. Similar experiments have been carried out with different adrenergic receptor subtypes with analogous results (Frielle *et al* 1987; Kobilka *et al* 1987b; Bouvier *et al* 1988; Cotecchia *et al* 1988). Transfection experiments have also been used to demonstrate that cells can contain multiple G proteins that selectively couple to different muscarinic receptors (Ashkenazi *et al* 1989). The data from the transfection experiments further support the hypothesis that the physiological response of a particular cell to the binding of acetylcholine is determined in large part by the structure of the receptor protein itself and by the nature of the G protein(s) that interact with that receptor.

The availability of full-length cDNA clones has also allowed studies to identify functional domains on individual receptors belonging to the superfamily of G protein-coupled receptors. For example, sequential deletions of different parts of the β-adrenergic receptor followed by studies of the expression of the resulting mutant receptors has yielded information about which parts of the receptor protein are required for proper expression, plasma membrane localization and ligand binding (Dixon *et al* 1987), as well as for receptor sequestration (Strader *et al* 1987). Site-directed mutagenesis of the β-adrenergic receptor has revealed the importance of particular amino acid residues or sequences in the receptor protein in its interactions with ligands (Strader *et al* 1988), G-protein (O'Dowd *et al* 1988), and adenylate cyclase (Fraser *et al* 1988). Construction of chimeric adrenergic receptors combining different parts of the α_2- and β_2-adrenergic receptor subtypes has provided information regarding the structural domains that confer specificity for agonist and antagonist binding, as well as those involved in G protein coupling (Kobilka *et al* 1988).

SUMMARY AND PROSPECTS FOR PLANT RECEPTOR STUDIES

Remarkable progress has been made during the past twenty years in research on the structure and function of receptors in animals. A wide variety of methods have been developed to identify, purify, and characterize animal receptors, techniques that for the most part can readily be applied to the study of plant receptors.

Receptor isolation can be divided into three stages: 1) Demonstration of ligand-specific binding that correlates with a measurable biological activity of the ligand; 2) Purification of the molecule(s) responsible for the ligand-specific binding; 3) Demonstration that a purified ligand-binding molecule is a physiological receptor, *i.e.*, mediates the physiological response(s) of cells to the ligand. A prerequisite for receptor studies is the availability of homogeneous ligand in order to perform meaningful ligand-binding studies essential for receptor identification. Since receptors generally are present in low abundance in cells and have high affinity for their ligands, the ligand must be labeled to permit sensitive measurement of ligand binding to its receptor at concentrations well below receptor saturation. Radiolabeling of the ligand to a high specific radioactivity (> 50 Ci/mmole) is a common approach to this problem. If derivatization of the ligand is necessary to achieve such suitable labeling, then the derivatized ligand must be shown to have the same biological properties as the underivatized ligand. Ligand agonists and antagonists, and biologically less active molecules structurally related to the ligand, are necessary tools to demonstrate the specificity of ligand binding. Such a range of molecular tools are available for the isolation of plant receptors for some plant ligands [*e.g.*, auxin (Libbenga *et al* 1986), gibberellins (Stoddart 1986), fusicoccin (Meyer *et al* 1989), glucan elicitors (see Hahn *et al*, this volume)]. The preparation of a suitable range of molecules is difficult but achievable for some ligands of interest to plant scientists, *e.g.*, oligosaccharins (Albersheim & Darvill 1985).

Purification of the molecules that specifically bind the ligand represents a difficult problem, mainly because of the low abundance of binding sites in cells. Ligand affinity chromatography, a powerful method to achieve purification of binding proteins, has been used in the purification of most animal receptors. Additional approaches that proved successful are covalently tagging binding sites using photo- or chemical-affinity labeling techniques. The generation and use of anti-idiotypic antibodies that specifically recognize ligand-binding sites has also been used successfully for several mammalian receptors [*e.g.*, (Sege & Peterson 1978; Cleveland *et al* 1983; Schreiber *et al* 1980; Lombes *et al* 1989)], although this approach is experimentally difficult. Recently, a novel strategy has been described that may prove broadly applicable for directly cloning binding proteins from an expression library using a ligand probe [reviewed by Singh *et al* (1989)]. These techniques are only beginning to be applied to the study of plant receptors. Affinity chromatography has been applied to the purification of binding proteins for plant hormones [*e.g.*, (Löbler & Klämbt 1985; Shimomura *et al* 1986)]. Derivatives of several plant hormones have been prepared that are suitable for photo-affinity labeling of binding proteins (Hornberg & Weiler 1984; Melhado & Brodsky 1988; Brinegar *et al* 1988; Cooper *et al* 1988). Several ligand-binding proteins have been identified using radiolabeled photo-affinity reactive ligands (Hornberg & Weiler 1984; Meyer *et al* 1989). Further development and exploitation of these powerful techniques for plant receptor identification and purification is clearly necessary.

The most difficult aspect of receptor research in any system is the necessity to differentiate between a receptor and an acceptor for a given ligand. Both may bind the ligand, but only the receptor-ligand complex can produce biological effects (*e.g.*, changes in gene expression) in targeted cells. Techniques for proving the physiological function of purified ligand-binding proteins are well established in a number of animal receptor systems as exemplified by the three receptor families described in this paper. In particular, the animal receptor studies demonstrate the power of recombinant DNA techniques in elucidating the structure and function of receptors. The development of these methods relied on substantial knowledge of the biochemistry of the cellular response to a given ligand (*e.g.*, the genes whose transcription is activated by a given steroid, or the opening of ion channels in response to acetylcholine). Such a knowledge base is still lacking for many of the ligands of interest to plant scientists. For example, the primary effects of most plant hormones remain to be conclusively established (Guern 1987). The situation is somewhat better for oligosaccharide elicitors of plant defense responses. The complete structures of active ligands are known [reviewed by Darvill & Albersheim (1984), Ryan (1987)], and, more significantly, considerable information has been obtained at the molecular level about the induction of phenylpropanoid metabolism [reviewed by Dixon (1986), Ebel (1986), Hahlbrock & Scheel (1989)], one rapid cellular response to elicitors.

What is lacking in plant systems is the availability of receptor reconstitution systems such as those employed with great success in animal receptor studies (*e.g.*, a plant equivalent for the *Xenopus* oocyte system). In part, the development of such reconstitution systems in plants will require additional research on the mechanisms of signal transduction in plant cells (see collection of recent reviews (Boss & Morré 1989) on the current state of knowledge). Examples of signal transduction mechanisms that have received little attention from plant biochemists include the roles of protein phosphorylation, endocytosis, or G proteins in mediating the effects of biologically active ligands. Protein phosphorylation, which plays a major role in receptor activation and deactivation in mammalian systems (Sibley *et al* 1988), is just beginning to be studied in plant systems (Ranjeva & Boudet 1987). While endocytosis has been shown in plant cells [reviewed by Steer (1988)], receptor-mediated endocytosis, such as that exhibited by the insulin receptor, has yet to be demonstrated in plant cells. A recent study using antibodies generated against a highly conserved amino acid sequence from animal G proteins suggests that plants also contain GTP-binding proteins (Blum *et al* 1988), although their structure and function remain to be elucidated. Plant cells may also have other unique mechanisms of signal transduction that should not be overlooked simply because they do not coincide with existing animal models. The development and characterization of mutant plant cell lines defective in receptor function [see Guern (1987) for examples of such mutants] would provide the target cells for receptor reconstitution studies. Transformation techniques for introducing putative receptor genes into such cell lines are already well developed in plant systems (Klee *et al* 1987). The examples from the animal receptor literature should provide ample incentive for plant scientists to develop the necessary knowledge base and molecular tools to elucidate the structure and function of plant receptors.

The animal receptor literature also provides impetus for research directions that might prove fruitful in studies to identify and characterize plant receptors. One of the themes highlighted in this article is that animal receptor proteins have been found to consist of linear arrays of functional domains. Thus, a search for protein kinases, which play such a prominent role in signal transduction in animal systems, using nucleic acid probes derived from the highly conserved kinase domain (Hanks 1987), might lead to the identification of plant protein kinases, some of which may have receptor functions (*e.g.*, the insulin receptor in animals). Alternatively, the use of antibodies against phosphorylated amino acid residues, in particular tyrosine, to screen expression libraries (Lindberg *et al* 1988) might pick up candidate receptor proteins that undergo autophosphorylation. Analogous rationales could be used to identify transcriptional regulatory proteins in plants using conserved features of the DNA binding domain (*e.g.*, "zinc finger" motif). This is a possible route to identify plant receptors that are analogous to animal steroid receptors. These types of approaches will ultimately rely on functional assays (such as reconstitution or transfection) to prove receptor function for proteins identified in such screens. Whether any of the animal receptors characterized thus far can be used as direct models for plant receptors must await the isolation and characterization of plant receptors.

The second major theme highlighted here is that animal receptors can be grouped into families based upon structural and/or functional similarities. For example, the study of the superfamily of G protein-coupled animal receptors has revealed that structurally related receptor types differ in the specific ligands that are bound (*e.g.*, acetylcholine, norepinephrine, serotonin), while related receptor subtypes differ in the binding properties of agonists and antagonists, and in the nature of the cellular response to binding of the ligand to the receptor. Thus, a given ligand can have different effects on different cells depending on the receptor subtype present in those cells and on the signal transduction mechanisms coupled to those receptors. In view of the pleiotropic effects of many plant hormones, it appears likely that the concept of receptor families for a given ligand, so prominent in animal receptors, will apply to plant systems as well.

ACKNOWLEDGEMENTS

I would like to thank my colleagues at the Complex Carbohydrate Research Center for stimulating discussions, particularly D. Mohnen, A. Darvill, R. Carlson, K. Brady, and S. Doares for their critical comments on this manuscript. I am grateful to R. Nuri for editorial assistance and to C. L. Gubbins Hahn for drawing the figures. Research in my laboratory is supported by a grant from the National Science Foundation (DMB 87-04022). The Complex Carbohydrate Research Center is supported in part by Department of Energy grant DE-FG09-87ER13810 as part of the USDA/DOE/NSF Plant Science Centers Program.

REFERENCES

Albersheim P, Darvill AG (1985) Oligosaccharins. *Sci Am* 253:58-64

Alberts B, Bray D, Lewis J, Raff M, Roberts K, Watson JD (1989) Molecular Biology of the Cell. Garland Publishing, Inc., New York, NY

Alemany S, Mato JM, Strålfors P (1988) Phospho-dephospho-control by insulin is mimicked by a phospho-oligosaccharide in adipocytes. *Nature* 330:77-79

Ankenbauer W, Strähle U, Schütz G (1988) Synergistic action of glucocorticoid and estradiol responsive elements. *Proc Natl Acad Sci USA* 85:7526-7530

Araki E, Shimada F, Uzawa H, Mori M, Ebina Y (1987) Characterization of the promoter region of the human insulin receptor gene. Evidence for promoter activity. *J Biol Chem* 262:16186-16191

Arriza JL, Weinberger C, Cerelli G, Glaser TM, Handelin BL, Housman DE, Evans RM (1987) Cloning of human mineralocorticoid receptor complementary DNA: Structural and functional kinship with the glucocorticoid receptor. *Science* 237:268-275

Ashkenazi A, Peralta EG, Winslow JW, Ramachandran J, Capon DJ (1989) Functionally distinct G proteins selectively couple different receptors to PI hydrolysis in the same cell. *Cell* 56:487-493

Backer JM, Kahn CR, White MF (1989) Tyrosine phosphorylation of the insulin receptor during insulin-stimulated internalization in rat hepatoma cells. *J Biol Chem* 264:1694-1701

Beato M (1989) Gene regulation by steroid hormones. *Cell* 56:335-344

Bernier M, Laird DM, Lane MD (1987) Insulin-activated tyrosine phosphorylation of a 15-kilodalton protein in intact 3T3-L1 adipocytes. *Proc Natl Acad Sci USA* 84:1844-1848

Bernier M, Laird DM, Lane MD (1988) Effect of vanadate on the cellular accumulation of pp15, an apparent product of insulin receptor tyrosine kinase action. *J Biol Chem* 263:13626-13634

Birdsall NJM, Hulme EC (1983) Muscarinic receptor subclasses. *Trends Pharmacol Sci* 4:459-463

Blum W, Hinsch K-D, Schultz G, Weiler EW (1988) Identification of GTP-binding proteins in the plasma membrane of higher plants. *Biochem Biophys Res Commun* 156:954-959

Bonner TI, Buckley NJ, Young AC, Brann MR (1987) Identification of a family of muscarinic acetylcholine receptor genes. *Science* 237:527-532

Boss WF, Morré DJ (eds) (1989) Second Messengers in Plant Growth and Development. Alan R. Liss, Inc., New York, NY

Bottaro DP, Bonner-Weir S, King GL (1989) Insulin receptor recycling in vascular endothelial cells. Regulation by insulin and phorbol ester. *J Biol Chem* 264:5916-5923

Boulter J, Connolly J, Deneris E, Goldman D, Heinemann S, Patrick J (1987) Functional expression of two neuronal nicotinic acetylcholine receptors from cDNA clones identifies a gene family. *Proc Natl Acad Sci USA* 84:7763-7767

Bouvier M, Hnatowich M, Collins S, Kobilka BK, Deblasi A, Lefkowitz RJ, Caron MG (1988) Expression of a human cDNA encoding the β_2-adrenergic receptor in Chinese hamster fibroblasts (CHW): Functionality and regulation of the expressed receptors. *Molec Pharmacol* 33:133-139

Bresnick EH, Dalman FC, Sanchez ER, Pratt WB (1989) Evidence that the 90-kDa heat shock protein is necessary for the steroid binding conformation of the L cell glucocorticoid receptor. *J Biol Chem* 264:4992-4997

Brinegar AC, Cooper G, Stevens A, Hauer CR, Shabanowitz J, Hunt DF, Fox JE (1988) Characterization of a benzyladenine binding-site peptide isolated from a wheat cytokinin-binding protein: Sequence analysis and identification of a single affinity-labeled histidine residue by mass spectrometry. *Proc Natl Acad Sci USA* 85:5927-5931

Brisson A, Unwin PNT (1985) Quaternary structure of the acetylcholine receptor. *Nature* 315:474-477

Burnstein KL, Cidlowski JA (1989) Regulation of gene expression by glucocorticoids. *Ann Rev Physiol* 51:683-699

Carpenter G (1987) Receptors for epidermal growth factor and other polypeptide mitogens. *Ann Rev Biochem* 56:881-914

Chandler VL, Maler BA, Yamamoto KR (1983) DNA sequences bound specifically by glucocorticoid receptor in vitro render a heterologous promoter hormone responsive in vivo. *Cell* 33:489-499

Chang C, Kokontis J, Liao S (1988) Molecular cloning of human and rat complementary DNA encoding androgen receptors. *Science* 240:324-326

Chou CK, Dull TJ, Russell DS, Gherzi R, Lebwohl D, Ullrich A, Rosen OM (1987) Human insulin receptors mutated at the ATP-binding site lack protein tyrosine kinase activity and fail to mediate postreceptor effects of insulin. *J Biol Chem* 262:1842-1847

Cleveland WL, Wassermann NH, Sarangarajan R, Penn AS, Erlanger BF (1983) Monoclonal antibodies to the acetylcholine receptor by a normally functioning auto-anti-idiotypic mechanism. *Nature* 305:56-57

Cohen S, Sokolovsky M (1987) Complexity apparent in muscarinic mechanisms. *Trends Pharmacol Sci* 8:41-44

Conti-Tronconi BM, Raftery MA (1982) The nicotinic cholinergic receptor: Correlation of molecular structure with functional properties. *Ann Rev Biochem* 51:491-530

Cooper G, Bourell J, Kaminek M, Fox JE (1988) Method for synthesis of 2-azido-N(6)-m-tritiobenzyl-aminopurine, a photoaffinity label for cytokinin-binding proteins in plants. *J Label Comp Radiopharm* 25:957-962

Cotecchia S, Schwinn DA, Randall RR, Lefkowitz R, Caron MG, Kobilka BK (1988) Molecular cloning and expression of the cDNA for the hamster α_1-adrenergic receptor. *Proc Natl Acad Sci USA* 85:7159-7163

Cuatrecasas P (1971) Insulin-receptor interactions in adipose cells: direct measurement and properties. *Proc Natl Acad Sci USA* 68:1264-1268

Czech MP (1977) Molecular basis of insulin action. *Ann Rev Biochem* 46:359-384

Czech MP (1985) The nature and regulation of the insulin receptor: Structure and function. *Ann Rev Physiol* 47:357-381

Czech MP, Klarlund JK, Yagaloff KA, Bradford AP, Lewis RE (1988) Insulin receptor signaling. Activation of multiple serine kinases. *J Biol Chem* 263:11017-11020

Dahlman K, Strömstedt P-E, Rae C, Jörnvall H, Flock J-I, Carlstedt-Duke J, Gustafsson J-Å (1989) High level expression in *Escherichia coli* of the DNA-binding domain of the glucocorticoid receptor in a functional form utilizing domain-specific cleavage of a fusion protein. *J Biol Chem* 264:804-809

Darvill AG, Albersheim P (1984) Phytoalexins and their elicitors - A defense against microbial infection in plants. *Ann Rev Plant Physiol* 35:243-275

Denis M, Gustafsson J-Å, Wikström A-C (1988) Interaction of the M_r=90,000 heat shock protein with the steroid-binding domain of the glucocorticoid receptor. *J Biol Chem* 263:18520-18523

Dixon RA (1986) The phytoalexin response: elicitation, signalling and control of host gene expression. *Biol Rev* 61:239-291

Dixon RAF, Sigal IS, Candelore MR, Register RB, Scattergood W, Rands E, Strader CD (1987) Structural features required for ligand binding to the β-adrenergic receptor. *EMBO J* 6:3269-3275

Dobson ADW, Conneely OM, Beattie W, Maxwell BL, Mak P, Tsai M-J, Schrader WT, O'Malley BW (1989) Mutational analysis of the chicken progesterone receptor. *J Biol Chem* 264:4207-4211

Dohlman HG, Bouvier M, Benovic JL, Caron MG, Lefkowitz RJ (1987) The multiple membrane spanning topography of the β_2-adrenergic receptor. Localization of the sites of binding, glycosylation, and regulatory phosphorylation by limited proteolysis. *J Biol Chem* 262:14282-14288

Dohlman HG, Caron MG, Strader CD, Amlaiky N, Lefkowitz RJ (1988) Identification and sequence of a binding site peptide of the β_2-adrenergic receptor. *Biochem* 27:1813-1817

Ebel J (1986) Phytoalexin synthesis: The biochemical analysis of the induction process. *Ann Rev Phytopathol* 24:235-264

Ebina Y, Ellis L, Jarnagin K, Edery M, Graf L, Clauser E, Ou J-H, Masiarz F, Kan YW, Goldfine ID, Roth RA, Rutter WJ (1985a) The human insulin receptor cDNA: The structural basis for hormone-activated transmembrane signalling. *Cell* 40:747-758

Ebina Y, Edery M, Ellis L, Standring D, Beaudoin J, Roth RA, Rutter WJ (1985b) Expression of a functional human insulin receptor from a cloned cDNA in Chinese hamster ovary cells. *Proc Natl Acad Sci USA* 82:8014-8018

Ebina Y, Araki E, Taira M, Shimada F, Mori M, Craik CS, Siddle K, Pierce SB, Roth RA, Rutter WJ (1987) Replacement of lysine residue 1030 in the putative ATP-binding region of the insulin receptor abolishes insulin- and antibody-stimulated glucose uptake and receptor kinase activity. *Proc Natl Acad Sci USA* 84:704-708

Ellis L, Clauser E, Morgan DO, Edery M, Roth RA, Rutter WJ (1986) Replacement of insulin receptor tyrosine residues 1162 and 1163 compromises insulin-stimulated kinase activity and uptake of 2-deoxyglucose. *Cell* 45:721-732

Ellis L, Morgan DO, Jong S-M, Wang L-H, Roth RA, Rutter WJ (1987a) Heterologous transmembrane signaling by a human insulin receptor-v-ros hybrid in Chinese hamster ovary cells. *Proc Natl Acad Sci USA* 84:5101-5105

Ellis L, Morgan DO, Clauser E, Roth RA, Rutter WJ (1987b) A membrane-anchored cytoplasmic domain of the human insulin receptor mediates a constitutively elevated insulin-independent uptake of 2-deoxyglucose. *Mol Endocrinol* 1:15-24

Eriksson H, Gustafsson J-Å (eds) (1983) Steroid Hormone Receptors: Structure and Function. Elsevier Science Publishers B.V., Amsterdam, The Netherlands

Evans RM (1988) The steroid and thyroid hormone receptor superfamily. *Science* 240:889-895

Evans RM, Hollenberg SM (1988) Zinc fingers: Gilt by association. *Cell* 52:1-3

Fargin A, Raymond JR, Lohse MJ, Kobilka BK, Caron MG, Lefkowitz RJ (1988) The genomic clone G-21 which resembles a β-adrenergic receptor sequence encodes the 5-HT_{1A} receptor. *Nature* 335:358-360

Florio VA, Sternweis PC (1989) Mechanisms of muscarinic receptor action on G_o in reconstituted phospholipid vesicles. *J Biol Chem* 264:3909-3915

Fraser CM, Chung F-Z, Wang C-D, Venter JC (1988) Site-directed mutagenesis of human β-adrenergic receptors: Substitution of aspartic acid-130 by asparagine produces a receptor with high-affinity agonist binding that is uncoupled from adenylate cyclase. *Proc Natl Acad Sci USA* 85:5478-5482

Frielle T, Collins S, Daniel KW, Caron MG, Lefkowitz RJ, Kobilka BK (1987) Cloning of the cDNA for the human β_1-adrenergic receptor. *Proc Natl Acad Sci USA* 84:7920-7924

Fukuda K, Kubo T, Akiba I, Maeda A, Mishina M, Numa S (1987) Molecular distinction between muscarinic acetylcholine receptor subtypes. *Nature* 327:623-625

Giguere V, Ong ES, Segui P, Evans RM (1987) Identification of a receptor for the morphogen retinoic acid. *Nature* 330:624-629

Giguére V, Hollenberg SM, Rosenfeld MG, Evans RM (1986) Functional domains of the human glucocorticoid receptor. *Cell* 46:645-652

Gilman AG (1987) G proteins: Transducers of receptor-generated signals. *Ann Rev Biochem* 56:615-649

Green S, Walter P, Kumar V, Krust A, Bornert J-M, Argos P, Chambon P (1986) Human oestrogen receptor cDNA: Sequence, expression and homology to v-*erb-A*. *Nature* 320:134-139

Green S, Chambon P (1987) Oestradiol induction of a glucocorticoid-responsive gene by a chimaeric receptor. *Nature* 325:75-78

Green S, Kumar V, Theulaz I, Wahli W, Chambon P (1988) The N-terminal DNA-binding 'zinc finger' of the oestrogen and glucocorticoid receptors determines target gene specificity. *EMBO J* 7:3037-3044

Gronemeyer H, Turcotte B, Quirin-Stricker C, Bocquel MT, Meyer ME, Krozowski Z, Jeltsch JM, Lerouge T, Garnier JM, Chambon P (1987) The chicken progesterone receptor: Sequence, expression and functional analysis. *EMBO J* 6:3985-3994

Guern J (1987) Regulation from within: The hormone dilemma. *Ann Bot* 60:75-102

Gustafsson J-Å, Carlstedt-Duke J, Poellinger L, Okret S, Wikström A-C, Brönnegård M, Gillner M, Dong Y, Fuxe K, Cintra A, Härfstrand A, Agnati L (1987) Biochemistry, molecular biology, and physiology of the glucocorticoid receptor. *Endoc Rev* 8:185-234

Haga K, Haga T (1985) Purification of the muscarinic acetylcholine receptor from porcine brain. *J Biol Chem* 260:7927-7935

Haga K, Haga T, Ichiyama A, Katada T, Kurose H, Ui M (1985) Functional reconstitution of purified muscarinic receptors and inhibitory guanine nucleotide regulatory protein. *Nature* 316:731-733

Haga K, Haga T, Ichiyama A (1986) Reconstitution of the muscarinic acetylcholine receptor. Guanine nucleotide-sensitive high affinity binding of agonists to purified muscarinic receptors reconstituted with GTP-binding proteins (G_i and G_o). *J Biol Chem* 261:10133-10140

Hahlbrock K, Scheel D (1989) Physiology and molecular biology of phenylpropanoid metabolism. *Ann Rev Plant Physiol Plant Mol Biol* 40:347-369

Hammer R, Berrie CP, Birdsall NJM, Burgen ASV, Hulme EC (1980) Pirenzepine distinguishes between different subclasses of muscarinic receptors. *Nature* 283:90-92

Hanks SK (1987) Homology probing: Identification of cDNA clones encoding members of the protein-serine kinase family. *Proc Natl Acad Sci USA* 84:388-392

Hanks SK, Quinn AM, Hunter T (1988) The protein kinase family: conserved features and deduced phylogeny of the catalytic domains. *Science* 241:42-52

Hazel TG, Nathans D, Lau LF (1988) A gene inducible by serum growth factors encodes a member of the steroid and thyroid hormone receptor superfamily. *Proc Natl Acad Sci USA* 85:8444-8448

Heldin C-H, Ronnstrand L (1989) The platelet-derived growth factor receptor. In: Moudgil VK (ed) Receptor Phosphorylation. CRC Press, Inc., Boca Raton, FL, p 149-162

Herskowitz I, Marsh L (1987) Conservation of a receptor/signal transduction system. *Cell* 50:995-996

Hollenberg SM, Weinberger C, Ong ES, Cerelli G, Oro A, Lebo R, Thompson EB, Rosenfeld MG, Evans RM (1985) Primary structure and expression of a functional human glucocorticoid receptor cDNA. *Nature* 318:635-641

Hornberg C, Weiler EW (1984) High-affinity binding sites for abscisic acid on the plasmalemma of *Vicia faba* guard cells. *Nature* 310:321-324

Hresko RC, Bernier M, Hoffman RD, Flores-Riveros JR, Liao K, Laird DM, Lane MD (1988) Identification of phosphorylated 422(aP2) protein as pp15, the 15-kilodalton target of the insulin receptor tyrosine kinase in 3T3-L1 adipocytes. *Proc Natl Acad Sci USA* 85:8835-8839

Hunter T (1987) A thousand and one protein kinases. *Cell* 50:823-829

Jackson TR, Blair LAC, Marshall J, Goedert M, Hanley MR (1988) The *mas* oncogene encodes an angiotensin receptor. *Nature* 335:437-440

Jantzen H-M, Strähle U, Gloss B, Stewart F, Schmid W, Boshart M, Miksicek R, Schütz G (1987) Cooperativity of glucocorticoid response elements located far upstream of the tyrosine aminotransferase gene. *Cell* 49:29-38

Jensen EV, Suzuki T, Kawashima T, Stumpf WE, Jungblut PW, DeSombre ER (1968) A two-step mechanism for the interaction of estradiol with rat uterus. *Proc Natl Acad Sci USA* 59:632-638

Julius D, MacDermott AB, Axel R, Jessell TM (1988) Molecular characterization of a functional cDNA encoding the serotonin 1c receptor. *Science* 241:558-564

Järv J, Bartfai T (1988) Muscarinic acetycholine receptors. In: Whittaker VP (ed) Handbook of Experimental Pharmacology, Vol. 86. The Cholinergic Synapse. Springer Verlag, Berlin, Germany, p 315-345

Kasuga M, Fujita-Yamaguchi Y, Blithe DL, White MF, Kahn CR (1983) Characterization of the insulin receptor kinase purified from human placental membranes. *J Biol Chem* 258:10973-10980

Kelly KL, Merida I, Wong EHA, DiCenzo D, Mato JM (1987) A phospho-oligosaccharide mimics the effect of insulin to inhibit isoproterenol-dependent phosphorylation of phospholipid methyltransferase in isolated adipocytes. *J Biol Chem* 262:15285-15290

Khoury G, Gruss P (1983) Enhancer elements. *Cell* 33:313-314

King WJ, Greene GL (1984) Monoclonal antibodies localize oestrogen receptor in the nuclei of target cells. *Nature* 307:745-747

Kistler J, Stroud RM, Klymkowsky MW, Lalancette RA, Fairclough RH (1982) Structure and function of an acetylcholine receptor. *Biophys J* 37:371-383

Klee H, Horsch R, Rogers S (1987) *Agrobacterium*-mediated plant transformation and its further applications to plant biology. *Ann Rev Plant Physiol* 38:467-486

Kobilka BK, Frielle T, Collins S, Yang-Feng T, Kobilka TS, Francke U, Lefkowitz RJ, Caron MG (1987a) An intronless gene encoding a potential member of the family of receptors coupled to guanine nucleotide regulatory proteins. *Nature* 329:75-79

Kobilka BK, MacGregor C, Daniel K, Kobilka TS, Caron MG, Lefkowitz RJ (1987b) Functional activity and regulation of human β_2-adrenergic receptors expressed in *Xenopus* oocytes. *J Biol Chem* 262:15796-15802

Kobilka BK, Dixon RAF, Frielle T, Dohlman HG, Bolanowski MA, Sigal IS, Yang-Feng TL, Francke U, Caron MG, Lefkowitz RJ (1987c) cDNA for the human β_2-adrenergic receptor: A protein with multiple membrane-spanning domains and encoded by a gene whose chromosomal location is shared with that of the receptor for platelet-derived growth factor. *Proc Natl Acad Sci USA* 84:46-50

Kobilka BK, Kobilka TS, Daniel K, Regan JW, Caron MG, Lefkowitz RJ (1988) Chimeric α_2-,β_2-adrenergic receptors: Delineation of domains involved in effector coupling and ligand binding specificity. *Science* 240:1310-1316

Krust A, Green S, Argos P, Kumar V, Walter P, Bornert J-M, Chambon P (1986) The chicken oestrogen receptor sequence: Homology with v-*erbA* and the human oestrogen and glucocorticoid receptors. *EMBO J* 5:891-897

Kubo T, Fukuda K, Mikami A, Maeda A, Takahashi H, Mishina M, Haga T, Haga K, Ichiyama A, Kangawa K, Kojima M, Matsuo H, Hirose T, Numa S (1986a) Cloning, sequencing and expression of complementary DNA encoding the muscarinic acetylcholine receptor. *Nature* 323:411-416

Kubo T, Maeda A, Sugimoto K, Akiba I, Mikami A, Takahashi H, Haga T, Haga K, Ichiyama A, Kangawa K, Matsuo H, Hirose T, Numa S (1986b) Primary structure of porcine cardiac muscarinic acetylcholine receptor deduced from the cDNA sequence. *FEBS Lett* 209:367-372

Leonard RJ, Labarca CG, Charnet P, Davidson N, Lester HA (1988) Evidence that the M2 membrane-spanning region lines the ion channel pore of the nicotinic receptor. *Science* 242:1578-1581

Libbenga KR, Maan AC, Van der Linde PCG, Mennes AM (1986) Auxin receptors. In: Chadwick CM, Garrod DR (eds) Hormones, Receptors and Cellular Interactions in Plants. Cambridge University Press, Cambridge, UK, p 1-68

Lindberg RA, Thompson DP, Hunter T (1988) Identification of cDNA clones that code for protein-tyrosine kinases by screening expression libraries with antibodies against phosphotyrosine. *Oncogene* 3:629-633

Lombes M, Edelman IS, Erlanger BF (1989) Internal image properties of a monoclonal auto-anti-idiotypic antibody and its binding to aldosterone receptors. *J Biol Chem* 264:2528-2536

Loosfelt H, Atger M, Misrahi M, Guiochon-Mantel A, Meriel C, Logeat F, Benarous R, Milgrom E (1986) Cloning and sequence analysis of rabbit progesterone-receptor complementary DNA. *Proc Natl Acad Sci USA* 83:9045-9049

Lubahn DB, Joseph DR, Sullivan PM, Willard HF, French FS, Wilson EM (1988) Cloning of human androgen receptor complementary DNA and localization to the X chromosome. *Science* 240:327-330

Löbler M, Klämbt D (1985) Auxin-binding protein from coleoptile membranes of corn (*Zea mays* L.). I. Purification by immunological methods and characterization. *J Biol Chem* 260:9848-9853

Mader S, Kumar V, De Verneuill H, Chambon P (1989) Three amino acids of the oestrogen receptor are essential to its ability to distinguish an oestrogen from a glucocorticoid- responsive element. *Nature* 338:271-274

Maegawa H, McClain DA, Freidenberg G, Olefsky JM, Napier M, Lipari T, Dull DJ, Lee J, Ullrich A (1988) Properties of a human insulin receptor with a COOH-terminal truncation. II. Truncated receptors have normal kinase activity but are defective in signaling metabolic effects. *J Biol Chem* 263:8912-8917

Maelicke A (1988) Structure and function of the nicotinic acetylcholine receptor. In: Whittaker VP (ed) Handbook of Experimental Pharmacology, Vol. 86. The Cholinergic Synapse. Springer Verlag, Berlin, Germany, p 267-313

Margolis RN, Taylor SI, Seminara D, Hubbard AL (1988) Identification of pp120, an endogenous substrate for the hepatocyte insulin receptor tyrosine kinase, as an integral membrane glycoprotein of the bile canalicular domain. *Proc Natl Acad Sci USA* 85:7256-7259

Masu Y, Nakayama K, Tamaki H, Harada Y, Kuno M, Nakanishi S (1987) cDNA cloning of bovine substance-K receptor through oocyte expression system. *Nature* 329:836-839

Mato JM, Kelly KL, Abler A, Jarett L (1987) Identification of a novel insulin-sensitive glycophospholipid from H35 Hepatoma cells. *J Biol Chem* 262:2131-2137

Mayer SE (1985) Neurohumoral transmission and the autonomic nervous system. In: Goodman Gilman A, Goodman LS, Gilman A (eds) The Pharmacological Basis of Therapeutics. Macmillan Publishing Co., Inc., New York,NY, p 56-90

McClain DA, Maegawa H, Lee J, Dull TJ, Ulrich A, Olefsky JM (1987) A mutant insulin receptor with defective tyrosine kinase displays no biologic activity and does not undergo endocytosis. *J Biol Chem* 262:14663-14671

McClain DA, Maegawa H, Levy J, Huecksteadt T, Dull TJ, Lee J, Ullrich A, Olefsky JM (1988) Properties of a human insulin receptor with a COOH-terminal truncation. I. Insulin binding, autophosphylation, endocytosis. *J Biol Chem* 263:8904-8911

McDonnell DP, Mangelsdorf DJ, Pike JW, Haussler MR, O'Malley BW (1987) Molecular cloning of complementary DNA encoding the avian receptor for vitamin D. *Science* 235:1214-1217

Melhado LL, Brodsky JL (1988) Synthesis of 4-azidoindole-3-acetic acid, a photoprobe causing sustained auxin activity. *J Org Chem* 53:3852-3855

Meyer C, Feyerabend M, Weiler EW (1989) Fusicoccin-binding proteins in *Arabidopsis thaliana* (L.) Heynh. *Plant Physiol* 89:692-699

Miesfeld R, Okret S, Wikström A-C, Wrange Ö, Gustafsson J-Å, Yamamoto KR (1984) Characterization of a steroid hormone receptor gene and mRNA in wild-type and mutant cells. *Nature* 312:779-781

Mishina M, Kurosaki T, Tobimatsu T, Morimoto Y, Noda M, Yamamoto T, Terao M, Lindstrom J, Takahashi T, Kuno M, Numa S (1984) Expression of functional acetylcholine receptor from cloned cDNAs. *Nature* 307:604-608

Mishina M, Tobimatsu T, Imoto K, Tanaka M, Fujita Y, Fukuda K, Kurasaki M, Takahashi H, Morimoto Y, Hirose T, Inayama S, Takahashi T, Kuno M, Numa S (1985) Location of functional regions of acetylcholine receptor α-subunit by site-directed mutagenesis. *Nature* 313:364-369

Nathans J (1987) Molecular biology of visual pigments. *Ann Rev Neurosci* 10:163-194

Nathanson NM (1987) Molecular properties of the muscarinic acetylcholine receptor. *Ann Rev Neurosci* 10:195-236

Neher E, Marty A, Fukuda K, Kubo T, Numa S (1988) Intracellular calcium release mediated by two muscarinic receptor subtypes. *FEBS Lett* 240:88-94

O'Dowd BF, Hnatowich M, Regan JW, Leader WM, Caron MG, Lefkowitz RJ (1988) Site-directed mutagenesis of the cytoplasmic domains of the human β_2-adrenergic receptor. Localization of regions involved in G protein-receptor coupling. *J Biol Chem* 263:15985-15992

Peralta EG, Ashkenazi A, Winslow JW, Smith DH, Ramachandran J, Capon DJ (1987a) Distinct primary structures, ligand-binding properties and tissue-specific expression of four human muscarinic acetylcholine receptors. *EMBO J* 6:3923-3929

Peralta EG, Winslow JW, Peterson GL, Smith DH, Ashkenazi A, Ramachandran J, Schimerlik MI, Capon DJ (1987b) Primary structure and biochemical properties of an M_2 muscarinic receptor. *Science* 236:600-605

Peralta EG, Ashkenazi A, Winslow JW, Ramachandran J, Capon DJ (1988) Differential regulation of PI hydrolysis and adenylyl cyclase by muscarinic receptor subtypes. *Nature* 334:434-437

Perrotti N, Accili D, Marcus-Samuels B, Rees-Jones RW, Taylor SI (1987) Insulin stimulates phosphorylation of a 120-kDa glycoprotein substrate (pp120) for the receptor-associated protein kinase in intact H-35 hepatoma cells. *Proc Natl Acad Sci USA* 84:3137-3140

Petkovich M, Brand NJ, Krust A, Chambon P (1987) A human retinoic acid receptor which belongs to the family of nuclear receptors. *Nature* 330:444-450

Petruzzelli L, Herrera R, Rosen OM (1984) Insulin receptor is an insulin-dependent tyrosine protein kinase: Copurification of insulin-binding activity and protein kinase activity to homogeneity from human placenta. *Proc Natl Acad Sci USA* 81:3327-3331

Ranjeva R, Boudet AM (1987) Phosphorylation of proteins in plants: Regulatory effects and potential involvement in stimulus/reponse coupling. *Ann Rev Plant Physiol* 38:73-93

Regan JW, Kobilka TS, Yang-Feng TL, Caron MG, Lefkowitz RJ, Kobilka BK (1988) Cloning and expression of a human kidney cDNA for an α_2-adrenergic receptor subtype. *Proc Natl Acad Sci USA* 85:6301-6305

Repke H (1987) Muscarinic receptor-detergent complexes with different biochemical properties: Selective solubilization, lectin affinity chromatography and ligand binding studies. *Biochim Biophys Acta* 929:47-61

Riedel H, Dull TJ, Schlessinger J, Ullrich A (1986) A chimaeric receptor allows insulin to stimulate tyrosine kinase activity of epidermal growth factor receptor. *Nature* 324:68-70

Ringold GM (1985) Steroid hormone regulation of gene expression. *Ann Rev Pharmacol Toxicol* 25:529-566

Rosen OM (1987) After insulin binds. *Science* 237:1452-1458

Ryan CA (1987) Oligosaccharide signaling in plants. *Ann Rev Cell Biol* 3:295-317

Sabbah M, Redeuilh G, Baulieu E-E (1989) Subunit composition of the estrogen receptor. Involvement of the hormone-binding domain in the dimeric state. *J Biol Chem* 264:2397-2400

Saltiel AR, Cuatrecasas P (1986) Insulin stimulates the generation from hepatic plasma membranes of modulators derived from an inositol glycolipid. *Proc Natl Acad Sci USA* 83:5793-5797

Schoepfer R, Whiting P, Esch F, Blacher R, Shimasaki S, Lindstrom J (1988) cDNA clones coding for the structural subunit of a chicken brain nicotinic acetylcholine receptor. *Neuron* 1:241-248

Schreiber AB, Couraud PO, Andre C, Vray B, Strosberg AD (1980) Anti-alprenolol anti-idiotypic antibodies bind to β-adreneric receptors and modulate catecholamine-sensitive adenylate cyclase. *Proc Natl Acad Sci USA* 77:7385-7389

Sege K, Peterson PA (1978) Use of anti-idiotypic antibodies as cell-surface receptor probes. *Proc Natl Acad Sci USA* 75:2443-2447

Seino S, Seino M, Nishi S, Bell GI (1989) Structure of the human insulin receptor gene and characterization of its promoter. *Proc Natl Acad Sci USA* 86:114-118

Shapiro RA, Scherer NM, Habecker BA, Subers EM, Nathanson NM (1988) Isolation, sequence, and functional expression of the mouse M1 muscarinic acetylcholine receptor gene. *J Biol Chem* 263:18397-18403

Sheridan PJ, Blum K, Trachtenberg MC (eds) (1988) Steroid Receptors and Disease: Cancer, Autoimmune, Bone, and Circulatory Disorders. Marcel Dekker, Inc., New York, NY

Sherman MR, Stevens J (1984) Structure of mammalian steroid receptors: Evolving concepts and methodological developments. *Ann Rev Physiol* 46:83-105

Shimomura S, Sotobaya T, Futai M, Fukui T (1986) Purification and properties of an auxin-binding protein from maize shoot membranes. *J Biochem* 99:1513-1524

Sibley DR, Benovic JL, Caron MG, Lefkowitz RJ (1988) Phosphorylation of cell surface receptors: A mechanism for regulating signal transduction pathways. *Endoc Rev* 9:38-56

Singh H, Clerc RG, LeBowitz JH (1989) Molecular cloning of sequence-specific DNA binding proteins using recognition site probes. *BioTechniques* 7:252-261

Staros JV, Fanger BO, Faulkner LA, Palaszewski PP, Russo MW (1989) Mechanism of transmembrane signaling by the epidermal growth factor receptor/kinase. In: Moudgil VK (ed) Receptor Phosphorylation. CRC Press, Inc., Boca Raton, FL, p 227-242

Steer MW (1988) Plasma membrane turnover in plant cells. *J Exp Bot* 39:987-996

Stein R, Pinkas-Kramarski R, Sokolovsky M (1988) Cloned M1 muscarinic receptors mediate both adenylate cyclase inhibition and phosphoinositide turnover. *EMBO J* 7:3031-3035

Stoddart JL (1986) Gibberellin receptors. In: Chadwick CM, Garrod DR (eds) Hormones, Receptors and Cellular Interactions in Plants. Cambridge University Press, Cambridge, UK, p 91-114

Strader CD, Sigal IS, Blake AD, Cheung AH, Register RB, Rands E, Zemcik BA, Candelore MR, Dixon RAF (1987) The carboxyl terminus of the hamster β-adrenergic receptor expressed in mouse L cells is not required for receptor sequestration. *Cell* 49:855-863

Strader CD, Sigal IS, Candelore MR, Rands E, Hill WS, Dixon RAF (1988) Conserved aspartic acid residues 79 and 113 of the β-adrenergic receptor have different roles in receptor function. *J Biol Chem* 263:10267-10271

Stroud RM, Finer-Moore J (1985) Acetylcholine receptor structure, function, and evolution. *Ann Rev Cell Biol* 1:317-351

Stryer L, Bourne HR (1986) G proteins: A family of signal transducers. *Ann Rev Cell Biol* 2:391-419

Strähle U, Schmid W, Schütz G (1988) Synergistic action of the glucocorticoid receptor with transcription factors. *EMBO J* 7:3389-3395

Thompson CC, Weinberger C, Lebo R, Evans RM (1987) Identification of a novel thyroid hormone receptor expressed in the mammalian central nervous system. *Science* 237:1610-1614

Trapman J, Klaassen P, Kuiper GGJM, van der Korput JAGM, Faber PW, van Rooij HCJ, Geurts van Kessel A, Voorhorst MM, Mulder E, Brinkmann AO (1988) Cloning, structure and expression of a cDNA encoding the human androgen receptor. *Biochem Biophys Res Commun* 153:241-248

Trischitta V, Wong K-Y, Brunetti A, Scalisi R, Vigneri R, Goldfine ID (1989) Endocytosis, recycling, and degradation of the insulin receptor. Studies with monoclonal antireceptor antibodies that do not activate receptor kinase. *J Biol Chem* 264:5041-5046

Tsai SY, Carlstedt-Duke J, Weigel NL, Dahlman K, Gustafsson J-Å, Tsai M-J, O'Malley BW (1988) Molecular interactions of steroid hormone receptor with its enhancer element: Evidence for receptor dimer formation. *Cell* 55:361-369

Ullrich A, Bell JR, Chen EY, Herrera R, Petruzzelli LM, Dull TJ, Gray A, Coussens L, Liao Y-C, Tsubokawa M, Mason A, Seeburg PH, Grunfeld C, Rosen OM, Ramachandran J (1985) Human insulin receptor and its relationship to the tyrosine kinase family of oncogenes. *Nature* 313:756-761

Wada K, Ballivet M, Boulter J, Connolly J, Wada E, Deneris ES, Swanson LW, Heinemann S, Patrick J (1988) Functional expression of a new pharmacological subtype of brain nicotinic acetylcholine receptor. *Science* 240:330-334

Walter P, Green S, Greene G, Krust A, Bornert J-M, Jeltsch J-M, Staub A, Jensen E, Scrace G, Waterfield M, Chambon P (1985) Cloning of the human estrogen receptor cDNA. *Proc Natl Acad Sci USA* 82:7889-7893

Weinberger C, Thompson CC, Ong ES, Lebo R, Gruol DJ, Evans RM (1986a) The c-*erb-A* gene encodes a thyroid hormone receptor. *Nature* 324:641-646

Weinberger C, Hollenberg SM, Rosenfeld MG, Evans RM (1986b) Domain structure of human glucocorticoid receptor and its relationship to the v-*erb-A* oncogene product. *Nature* 318:670-672

Welshons WV, Lieberman ME, Gorski J (1984) Nuclear localization of unoccupied oestrogen receptors. *Nature* 307:747-749

White MF, Kahn CR (1989) Function of the insulin receptor kinase in the intact cell. In: Moudgil VK (ed) Receptor Phosphorylation. CRC Press, Inc., Boca Raton, FL, p 115-134

White MF, Livingston JN, Backer JM, Lauris V, Dull TJ, Ullrich A, Kahn CR (1988) Mutation of the insulin receptor at tyrosine 960 inhibits signal transmission but does not affect its tyrosine kinase activity. *Cell* 54:641-649

Whittaker J, Okamoto AK, Thys R, Bell GI, Steiner DF, Hofmann CA (1987) High-level expression of human insulin receptor cDNA in mouse NIH 3T3 cells. *Proc Natl Acad Sci USA* 84:5237-5241

Williams LT (1989) Signal transduction by the platelet-derived growth factor receptor. *Science* 243:1564-1570

Wong SK-F, Slaughter C, Ruoho AE, Ross EM (1988) The catecholamine binding site of the β-adrenergic receptor is formed by juxtaposed membrane-spanning domains. *J Biol Chem* 263:7925-7928

Wrange Ö, Eriksson P, Perlmann T (1989) The purified activated glucocorticoid receptor is a homodimer. *J Biol Chem* 264:5253-5259

Yamamoto KR (1985) Steroid receptor regulated transcription of specific genes and gene networks. *Ann Rev Genet* 19:209-252

SECOND MESSENGERS IN ANIMALS AND THEIR POSSIBLE RELEVANCE FOR PLANTS

Charles A. West, Robert Bruce and Yue-Ying Ren
Department of Chemistry and Biochemistry
University of California, Los Angeles
Los Angeles, California 90024-1569
USA

INTRODUCTION

My task is to review the subject of second messengers in animals and its possible relevance for plants. Implicit in this title is the hope that useful guidance will be found in the results from the relatively well-studied animal regulatory systems as we begin to probe transmembrane signalling systems that operate in plants during pathogenic and symbiotic plant-microbe interactions. Some support for this position may be taken from arguments recently set forth by Janssens (1987). He has concluded that similarities are evident in signal transduction mechanisms in vertebrates, invertebrates and eukaryotic microbes, whereas signalling mechanisms in prokaryotes appear to be significantly different. Although much of the evidence on which these comparisons are based is still very preliminary, Janssens suggests that eukaryotic transduction mechanisms, along with many other properties of the eukaryotic cell, may well have originated with eukaryotic microbes. An extension of this reasoning suggests that higher plants would share these attributes as well. On the other hand, it may be well to remember that the signal transduction pathways we are considering in animal cells are mostly concerned with the regulation of basic cellular processes and the integration of metabolic activities within multicellular organisms, whereas we are considering in this Workshop signalling processes that probably evolved at a later stage during the interaction of two different organisms. We should not be surprised to see special features of the signalling pathways that operate in these cases.

The concept of a second messenger was developed with respect to the actions of cAMP in mediating the response of animal cells to hormones (Sutherland 1972). For the purposes of this discussion, I shall consider second messengers to be low molecular weight, diffusible molecules (1) that are generated at the inner surface of the plasma membrane or in the cytoplasm in response to an extracellular signal and (2) that participate in a pathway leading

NATO ASI Series, Vol. H 36
Signal Molecules in Plants and Plant-Microbe Interactions
Edited by B. J. J. Lugtenberg

to changes in the activities of selected intracellular proteins. The consideration of second messengers is most meaningful in the context of the signal transduction pathways in which they participate. Therefore, we will first review the general characteristics of some of the best understood signal transduction pathways involving second messengers in animals and then will consider for each the question of whether an analogous pathway may operate in plants. Detailed consideration of the molecular and physiological aspects of these regulatory systems will not be considered here because such details would probably not be very helpful at this stage in considering the relevance of these animal models for plants.

CYCLIC NUCLEOTIDES AS SECOND MESSENGERS

Figure 1 summarizes the relationships among the components that participate in the 3',5'-cyclic AMP (cAMP)-dependent signal transduction pathway in animal cells. The activity of adenylate cyclase is governed by two membrane bound protein complexes that may coexist in one membrane. One of these complexes leads to activation in the presence of suitable extracellular agonists and the other leads to inactivation by suitable extracellular antagonists. Agonists triggering the activation pathway include hormones such as glucagon, catecholamines, follicle stimulating hormone and corticotropin. The binding of an agonist to the exposed binding site of a β-adrenergic receptor (R_s) leads to the formation of an activated complex of three protein components, the receptor, the transducing $G_{s\alpha}$ with bound GTP and the effector, adenylate cyclase. Antagonists that promote the inactivation pathway include hormones such as angiotensin and catecholamines, some neurotransmitters and other substances such as opiates and adenosine. The binding of an antagonist to the α_2 receptor leads to formation of an active $G_{i\alpha}$ • GTP complex and the inactivation of associated adenylate cyclase. The resultant activity of adenylate cyclase and the rate of synthesis of intracellular cAMP is therefore determined by the balance of the activation and inactivation pathways, and ultimately by the relative levels of agonist and antagonist external to the cell. The intracellular pool size of cAMP is determined by the balance between synthesis catalyzed by adenylate cyclase and hydrolytic cleavage through the action of phosphodiesterase.

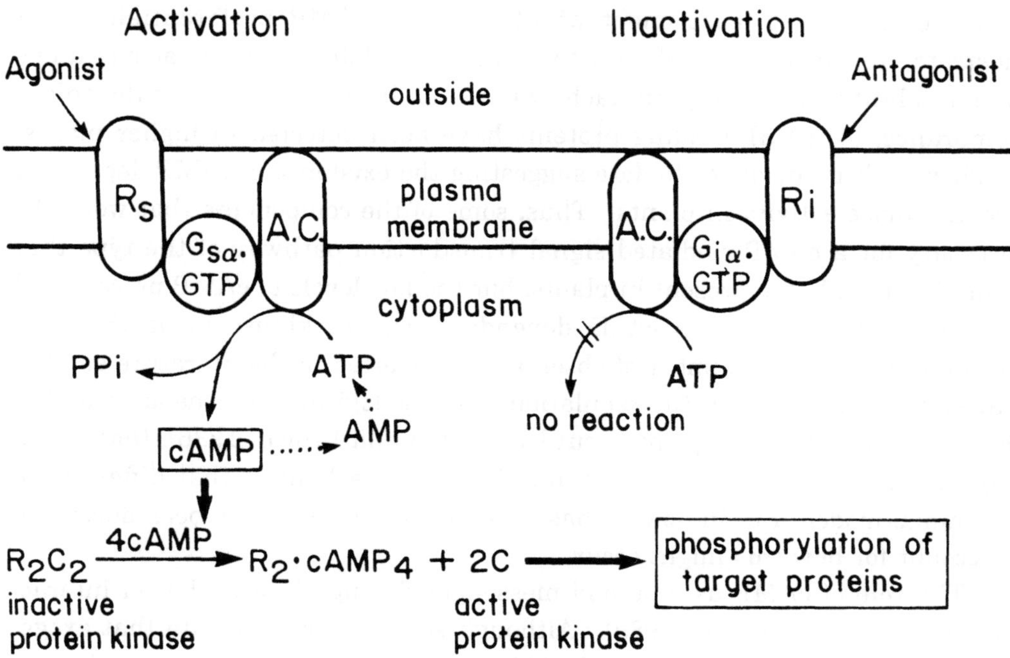

Figure 1. Signal induction pathways involving cAMP as a second messenger. R_s, R_i = receptors; $G_{s\alpha}$•GTP and $G_{i\alpha}$•GTP = α subunits of specific G proteins with bound GTP; A.C. = adenylate cyclase.

The dissociation of the regulatory subunit of c-AMP-dependent protein kinase to release active catalytic subunits is a function of the intracellular concentration of cAMP. The primary end-result of this cascade is the phosphorylation of serine and threonine residues of target enzymes and other regulatory proteins that serve as the substrates for this protein kinase. More specific secondary kinases also participate in this regulatory cascade. The net consequence is the coordinated activation or inactivation of catabolic and anabolic pathways of carbohydrate and lipid metabolism and, in some cases, the transcriptional activation of selected genes.

A considerable effort has been made to determine whether cAMP has a similar role in plant cells. Newton and Brown (1986) have reviewed the literature on the subject and have attempted to evaluate the status of the concept. Progress in this problem has been slowed by the apparently low levels of cAMP present in plant tissues (a few pmol per g dry weight) and resulting analytical difficulties. The accumulated evidence supports the following conclusions: (1) cAMP is present at low concentrations in a range of higher plants. (2) Among the phosphodiesterase activities for cyclic nucleotides in higher plants is at least one that utilizes 3',5'-cAMP preferentially as a substrate. (3) Adenylate cyclase activity has been detected in extracts of a few higher plants including spinach where it is localized in chloroplast membranes. (4) cAMP-binding proteins have been detected in higher plants. (5) There is little evidence to date suggesting the existence of cAMP-dependent protein kinase activity in plants. Thus, some of the components that would be necessary for a cAMP-mediated signal transduction pathway of the type that occurs in animals are present in plants, but the low levels of cAMP present and the lack of evidence for a cAMP-dependent protein kinase bring this into question. However, a number of observations reported in the literature suggest that cAMP is involved in the regulation of physiological phenomena in higher plants. It is possible, as pointed out by Newton and Brown (1986), that cAMP may function as a messenger in plants by some mechanism that differs from the animal model. But in no case has a satisfactory explanation been developed to account for how this might occur.

The role of cAMP as a second messenger during plant-pathogen interactions is no clearer (Boller 1989). Although some reports indicate that exogenously supplied cAMP or dibutyryl cAMP acted as elicitors of phytoalexin production, Hahn and Grisebach (1983) could detect no change in the low levels of cAMP in elicitor-treated soybean hypocotyls or cultured soybean suspension cells in comparison with untreated controls. On the other hand, Kurosaki *et al* (1987), reported a several-fold increase in cAMP within 30 min after treatment of carrot cultures with a phytoalexin elicitor preparation. The role of cAMP as a second messenger in plants, if any, remains uncertain.

3',5'-cyclic GMP (c-GMP) has been implicated as a second messenger in animal cells in a number of physiological contexts (Atlas 1986; Chinkers *et al* 1989), but relatively little is known about the signal transduction pathways in which it participates. There is no evidence to date for a second messenger role for c-GMP in plants.

PRODUCTS OF PHOSPHOLIPID METABOLISM AS SECOND MESSENGERS

Two substances that have well-established roles as second messengers in animal cells, inositol-1,4,5-trisphosphate (IP_3) and diacylglycerol (DAG), are produced simultaneously by the action of phospholipase C on a membrane lipid, phosphatidylinositol-4,5-bisphosphate ($PtdIP_2$). The activity of the phospholipase C that acts specifically on this substrate is governed by transmembrane signalling. Figure 2 summarizes the relationships among components

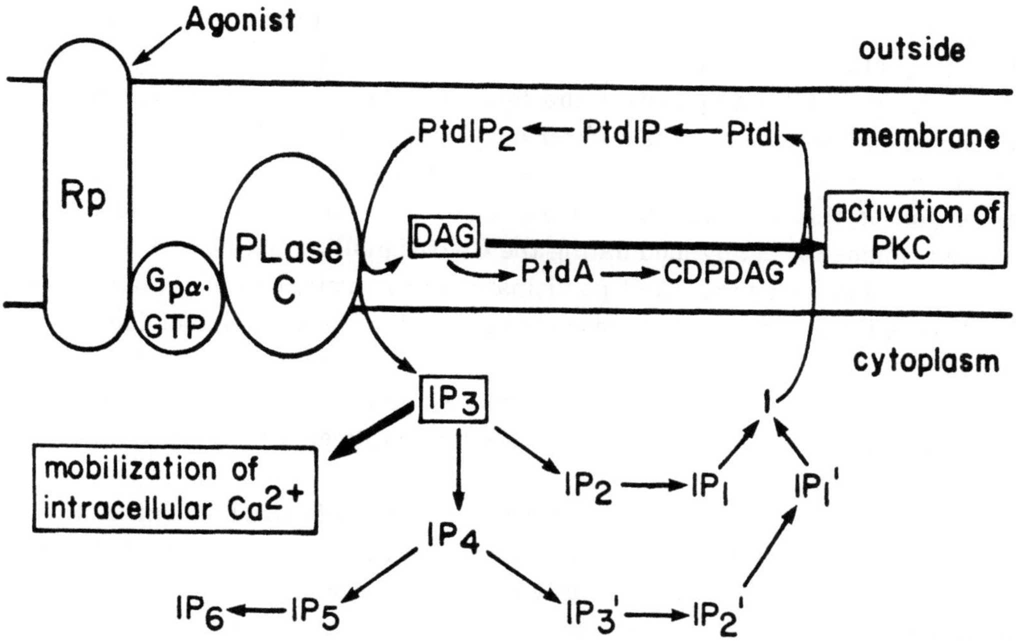

Figure 2. Phosphatidylinositol signal transduction pathway involving inositol-1,4,5-trisphosphate and diacylglycerol as second messengers. R_p = receptor; $G_{p\alpha}$•GTP = α-subunit of a specific G-protein with bound GTP; PLase C = phospholipase C; DAG = diacylglycerol; IP_3 = inositol-1,4,5-trisphosphate; PtdA = phosphatidic acid; PtdI = phosphatidyl inositol; PtdIP = phosphatidylinositol-4-phosphate; and $PtdIP_2$ = phosphatidyl inositol-4,5-bisphosphate.

of this regulatory cascade. A recent review by Berridge (1987) provides a more detailed account of the pathway. Agonists for this pathway, which include neurotransmitters, growth factors and hormones, bind to a plasma membrane receptor (Rp) to generate an activated form that in turn can activate the effector protein, phospholipase C, with the intermediate participation of a transducing Gp protein. Thus, this transducing complex is closely analogous to the adenylate cylase activation complex. The activated phospholipase C catalyzes the hydrolysis of a specific pool of Ptd IP_2 to release IP_3 to the cytoplasm and DAG to the membrane. The substrate Ptd IP_2 for this reaction is generated from Ptd I in the membrane by two successive kinase-catalyzed reactions.

The primary function of IP_3 as a second messenger in the cytoplasm is to stimulate the rapid release of Ca^{2+} from the endoplasmic reticulum, thereby elevating the concentration of free Ca^{2+} in the cytoplasm. The target molecules for Ca^{2+} action will be considered later. DAG, on the other hand, remains associated with the plasma membrane where it acts to stimulate protein kinase C. Protein kinase C acts as serine/threonine kinase to phosphorylate a diverse array of proteins; however, the physiological consequences of phosphorylation of target proteins by this kinase *in vivo* remain uncertain.

Berridge (1987) emphasizes the fact that these two second messengers generated by the action of one effector enzyme interact with one another in exerting their effects. IP_3 is thought to play a more direct role in initiating cellular responses through the elevated free Ca^{2+} pool, whereas DAG and protein kinase C appear to act by modulating the Ca^{2+}-signalling pathway and those of other second messengers. Ca^{2+} participates in the activation of protein kinase C, whereas DAG and protein kinase C contribute to the elevation of Ca^{2+} levels by inhibiting IP_3 hydrolysis, by stimulating a Ca^{2+}-pump and by stimulating the synthesis of PtdI. Thus, each messenger contributes to the action of the other, a situation prompting Berridge to refer to this as a dual signal pathway for cell activation.

The current status of the case for phosphoinositide involvement in second messenger production in plants has been reviewed recently by Boss (1988). The evidence is suggestive but incomplete, and a number of questions remain unanswered. The major points can be summarized as follows: (1) Plants are capable of the active biosynthesis of myoinositol (inositol) from glucose 6-P and its incorporation into a number of substances. (2) Among the phosphoinositide lipids identified in plants are PtdI, PtdIP, $PtdIP_2$ and LPtdIP (lysophosphoinositide phosphate). LPtdIP may conceivably act itself as a positive effector of membrane cation transport. (3) Careful analytical work has demonstrated that the plasma membranes of carrot cells are enriched in PtdIP and $PtdIP_2$ along with some LPtdIP. However, the substrate for phospholipase C action to produce second messengers in animal cells ($PtdIP_2$) is at a much lower concentration in plants than animals. (4) Phospholipase C is also enriched in plant plasma membranes, and, although the phospholipase C of wheat seedlings is stimulated by Ca^{2+}, it is not activated by GTP or nonhydrolyzable analogs of GTP. Thus, there is no indication of G-protein involvement in phosphoinositide breakdown by phospholipase C in plants to date. (5) IP_3 has been difficult to detect in plants, and there is no evidence for increases in cytoplasmic IP_3 following treatment with a stimulus. Administered IP_3 stimulates Ca^{2+} efflux from the tonoplast (Boudet and Ranjeva 1989). This, among other considerations has prompted these authors to propose that the vacuole serves as a storage site and source of second messengers in plants. (6) Although protein kinase C-type activity has been detected in plants, there is no convincing evidence to date of DAG activation of this enzyme. (7) Strasser *et al* (1986) detected no changes in phosphoinositides or inositol phosphates after treatment of cultured soybean cells with phytoalexin-inducing elicitors. They concluded that phosphoinositide metabolism is not involved unless it is very rapid and of short duration.

Other plasma membrane lipids of animal cells that may serve as precursors to second messengers include phosphatidylcholine as an alternative source of DAG (Exton 1988; Pelich and Vance 1989) and sphingolipids as a source of the possible second messengers sphinogosine and/or lysosphingolipids (Hannum and Bell 1989). The evidence to support these cases is not as well developed as for the phosphoinositides.

FREE CALCIUM IONS AS A SECOND MESSENGER

Ca^{2+} is an important second messenger in animal cells. A large number of agonists are known to increase the cytoplasmic concentration of unbound Ca^{2+}, while a few act as antagonists to reduce it. Thus, the intracellular Ca^{2+} levels are closely regulated by signal transduction systems. It seems likely on the basis of recent work that there are several regulatory signals that interact to govern changes in cytoplasmic Ca^{2+} levels (Exton 1988). These regulatory mechanisms are complex and not fully understood. A simplistic view is illustrated in Figure 3 to illustrate the general features. One important input is from the $PtdIP_2$ pathway described previously. The major effect of IP_3 derived from the stimulated breakdown of this lipid is to mobilize Ca^{2+} sequestered in an organelle, most likely the endoplasmic reticulum, to the cytosol. A second input involves the opening of a gated Ca^{2+} channel in the plasma membrane, thus permitting the uptake of extracellular Ca^{2+} into the cell. This process, which is agonist stimulated, is believed to operate with the participation of an agonist receptor and a transducing G-protein, but the mechanism for coupling these components and their precise natures remain uncertain. This process appears to be IP_3-independent. However, a cAMP-dependent protein kinase activated Ca^{2+}channel participates under appropriate circumstances (Schulman and Lou 1989).

The effector actions of Ca^{2+} are due to free Ca^{2+} and to a Ca^{2+}-calmodulin (CaM) complex. Some target proteins respond to Ca^{2+} concentrations directly and others to Ca^{2+}-CaM concentrations directly. Another large group of target regulatory proteins respond indirectly to Ca^{2+}-CaM after they are phosphorylated by Ca^{2+}CaM dependent protein kinases. These Ca•CaM-dependent protein kinases are of two types—one that accepts a specific protein, or a narrow range of proteins, as its substrate and another multifunctional protein kinase that has a broad specificity for protein substrates. This latter multifunctional-

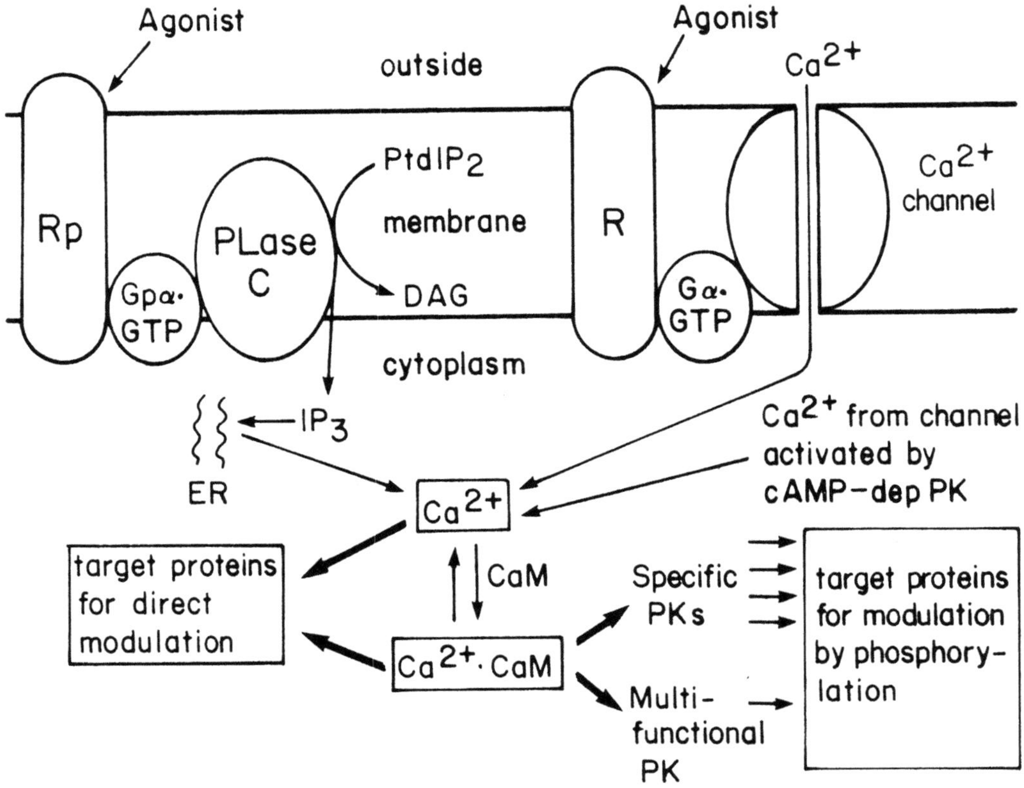

Figure 3. Signal transduction pathways involving Ca^{2+} as a second messenger. R_p and R = receptors; $G_{p\alpha}$ and G_α = specific G-protein α-subunits with bound GTP; PLase C = phospholipase C; IP_3 = inositol-1,4,5-trisphosphate; CaM = calmodulin; PK = protein kinase; and ER = endoplasmic reticulum.

type may participate in coordinated changes of the activities of regulatory enzymes in metabolic pathways that are under the influence of hormones or neurotransmitters as agonists (Schulman and Lou 1989).

Ca^{2+} has been implicated in the regulation of a number of basic processes associated with plant growth and development including cell elongation, cell division, hormone action and turgor regulation, for example. A general model for Ca^{2+} action as a second messenger has been proposed (Marmé 1989) based on the maintenance of free Ca^{2+} concentrations in the cytoplasm at low levels that can be modulated by the effects of external stimuli to change the balance between rates of efflux from the intracellular pool and influx from extracellular sources, and the storage or recall of Ca^{2+} from intracellular organelles. The activity of Ca^{2+}-responsive or Ca^{2+}-CaM (or other calcium binding protein)-

responsive enzymes is then modulated in turn as a function of the intracellular free Ca^{2+} concentration in the cytoplasm. There is considerable experimental support for this general model for cases where the external stimulus is a factor like light or a hormone that influences physiological processes central to normal growth and development. Major points from the review of Marmé (1989) and references contained therein are summarized below.

(1) Intracellular free Ca^{2+} concentrations are maintained in the sub-micromolar range in the cytoplasm of plant cells even though the extracellular concentration is many-fold higher. (2) Calmodulin is present in the cytoplasm of a number of plants that have been examined. (3) Cellular Ca^{2+} transport systems that could act to modulate cytoplasmic free Ca^{2+} concentration are known. Especially important in this regard is a CaM-dependent Ca^{2+}-Mg^{2+} ATPase in the plasma membrane that is responsible for the energy-dependent efflux of Ca^{2+} from the cytoplasm—a crucial factor in maintaining low intracellular Ca^{2+} levels. (4) The activities of a number of intracellular enzymes are influenced by Ca^{2+} or the Ca^{2+}/CaM complex, including the membrane Ca^{2+}-Mg^{2+} ATPase, NAD kinase and quinate:NAD^{+} oxido-reductase. Of special interest in connection with signal transduction pathways are reports from several laboratories of Ca^{2+} or Ca^{2+}/CaM or Ca^{2+}/phospholipid-dependent protein kinases. However, little is known of the regulation of these enzymes by Ca^{2+} *in vivo*, or the physiological functions of their phosphorylatable substrates.

These observations provide the framework for a model of how Ca^{2+} could serve as an intracellular second messenger and suggest that a signal transduction mechanism of this type must exist in plants. Gaps remain, however, in our knowledge of how the extracellular signals are coupled to changes in cytoplasmic Ca^{2+} concentration, and of the molecular events that relate the Ca^{2+} concentration to a specific physiological response.

Kauss and his associates have provided the most substantial evidence for Ca^{2+} involvement in the activation of components that participate in plant defense during a plant-pathogen interaction (Kauss 1987). Chitosan, a putative elicitor, and other polycations were shown to increase the permeability of soybean cells to ions and to stimulate callose deposition in the cells. Since extracellular Ca^{2+} was required for callose deposition and Ca^{2+} was shown *in vitro* to stimulate the activity of the β1,3-glucan synthase responsible for callose synthesis, it was hypothesized that increased concentrations of cytoplasmic Ca^{2+} after elicitor treatment mediate this response. Cytoplasmic Ca^{2+} levels are also implicated in the induction of phytoalexin biosynthesis by elicitors in soybean cells (Kauss 1987; Ebel 1986) and in carrot cells (Kurosaki *et al* 1987). These results make it evident that Ca^{2+} deserves attention as a second messenger in hypersensitive resistance responses.

OTHER SECOND MESSENGER CANDIDATES IN PLANTS

A number of other substances have been proposed as second messengers for plants, although generally these have been less extensively investigated in comparison with those already discussed. Two of the more attractive possibilities are the proton-transport activating phospholipid (PAP) described by Scherer (1989) and acetylcholine (Hartmann and Gupta 1989). Other candidates include lipoxygenase-generated oxygenated octadecanoid and dodecanoid metabolites derived from unsaturated fatty acids (Anderson 1989), pH changes in the cytoplasm (Felle 1989) and phytohormones (Boller 1989). These factors have not been systematically investigated in the context of a second messenger role in transmembrane signalling, so the evidence for their participation is incomplete. However these preliminary indications suggest they are worth further investigation.

GENERAL CHARACTERISTICS OF SIGNAL TRANSDUCTION PATHWAYS IN ANIMALS AND THEIR RELEVANCE FOR PLANTS

Three types of membrane-associated proteins are generally characteristic of transmembrane signalling systems in animals—a receptor, a transducing G-protein and an effector enzyme or ion-channel transport protein. It seems likely that analogous components are present in plant plasma membranes, although the evidence for existence of these components in plants is still very preliminary and incomplete, and the evidence for their functional roles in signal transduction systems in plants is virtually non-existent. Well-established second messengers in animals include cAMP, cGMP, inositol 1,4,5-trisphosphate, diacylglycerol and Ca^{2+}; of this group, reasonable evidence indicates the participation of Ca^{2+} as a second messenger in plants and circumstantial evidence favors the possibility that inositol 1,4,5-trisphosphate is involved, whereas there is little at the present time to support the view that cAMP, cGMP and diacylglycerol act as second messengers in plants. A number of other second messengers have been proposed, and it seems likely that the lists for both plant and animals will be extended. How much overlap there will be remains to be seen.

Protein kinases whose activities depend on the concentration of a second messenger (*e.g.* cAMP-dependent protein kinase, cGMP-dependent protein kinase, Ca^{2+}/calmodulin-dependent protein kinase, and Ca^{2+} and DAG-stimulated protein kinase C) frequently mediate the response of target proteins to changing concentrations of second messengers in animal cells. It follows that phosphorylation of target proteins to change their activities is the means for effecting regulation in these cases. This appears a likely possibility in plants as well, although the evidence to support this view is still very fragmentary. Blowers and Trevavas (1989) have reviewed the mounting evidence for soluble Ca^{2+}-dependent and H^{+}-dependent protein kinases that may be candidates for signal transduction components in plants. Ranjeva and Boudet (1987) have reviewed the regulation of protein phosphorylation in plants and its potential involvement in stimulus/response coupling.

A recent review by Shenolikar (1988) emphasizes an aspect of the animal signal transduction pathways that is not adequately represented by the cartoons in Figures 1, 2 and 3. As more is learned about these pathways, the more likely it seems that they are integrated into a coordinated, interactive regulatory network that can be fine-tuned in its response to a wide variety of stimuli. Consider the following features that appear to characterize these pathways as a group. (1) A second messenger in one pathway has the capacity to modulate the production and/or the responses to changes in levels of a second messenger of another pathway. Diacylglycerol, for example, modulates the responses to elevated Ca^{2+} and cAMP can stimulate the influx of extracellular Ca^{2+}. (2) Receptors may exist in families whose individual members may differ with respect to agonist and antagonist binding specificities. (3) Feedback or cross-over phosphorylation of receptors, G-proteins, ion channels and effector enzymes with resultant changes in the activities of these components is commonly seen. (4) The protein kinases and protein phosphatases that mediate many of the regulatory responses are a complex group of enzymes that are themselves subject to regulation. The interpretations of their actions is complicated by the large number of enzymes with broad and sometimes overlapping substrate specificities and the availability of many potential sites of phosphorylation and dephosphorylation on target proteins. A coherent picture of how these different signal transduction pathways are integrated into a network has not yet emerged, but there is enough known to indicate that they must be coordinated in some fashion and must possess a large number of regulatory sites to provide the fine tuning required.

Several observations made in connection with plant responses to elicitors suggest that interactive signal transduction pathways may be present in plants as well. Cell wall oligosaccharides from the fungal pathogen *Phytophthora infestans* were shown to enhance greatly the elicitation of sesquiterpenoid stress metabolites in potato by arachidonic acid even though they had little elicitor activity by themselves (Maniara *et al* 1984). Combinations of *P. megasperma* f. sp. *glycinea* glucan elicitor and plant-derived pectic oligosaccharide elicitor were found to give substantial elicitation of phytoalexins at concentrations where neither alone was very effective in soybean cotyledons (Davis *et al* 1986) and cultured parsley cells (Davis and Hahlbrock 1987).

Unpublished observations from our laboratory add to this list of synergistic interactions between two factors. R. Bruce has demonstrated that castor bean suspension cells respond to treatment with pectic fragment elicitors by accumulating substantial amounts of lignin-like material. Chitin, a fungal cell wall fragment, was not effective when added alone as a potential elicitor. However, the addition of chitin along with pectic fragment elicitors increased both the sensitivity to pectic fragments and the maximal response in lignin accumulation attainable in comparison with pectic fragments alone. In another set of experiments, Y.-Y. Ren has observed that chitin is an effective elicitor of diterpene hydrocarbon production in cultured rice cells. These diterpenes include two putative precursors of the momilactone and oryzalexin families of phytoalexins of rice. Cartwright *et al* (1980) reported that pretreatment of rice seedlings with the fungicide, 1,1-dichloro-2,2-dimethylcyclopropane carboxylic acid (WL 28325), enhanced the production of momilactones when the plants were subsequently challenged with the rice blast fungus, *Pyricularia oryzae.* Therefore, the influence of treatment of rice cell suspensions with WL 28325 on the subsequent production of diterpene hydrocarbons after treatment with chitin has been examined. Fungicide-treated cells give approximately a two-fold increase in diterpene production after chitin treatment in comparison with controls treated with chitin alone. The fungicide alone did not act as an elicitor at the concentrations employed.

None of these synergistic interactions between two factors described above has been explained in molecular terms. However, it seems reasonable to suggest that these are instances where two signal transduction pathways that interact positively with one another are stimulated by the two factors supplied. These observations point to the possibility of an interactive regulatory network in plants that can respond to a variety of signals generated during a plant-pathogen interaction.

In conclusion, I feel it is instructive to consider the general characteristics of signal transduction pathways in animals and it may be profitable to use animal models in designing experiments to elucidate signal transduction pathways in plants. However, not enough is yet known about signal transduction in plants to say how similar they may be at the molecular level to the animal models. One should be prepared for the unexpected, especially when considering specialized regulatory mechanisms such as those involved in plant-microbe interactions.

REFERENCES

Anderson JM (1989) Membrane-derived fatty acids as precursors to second messengers. In Boss WF, Morré DJ (eds) Second messengers in plant growth and development. Alan R Liss New York pp 181-212

Atlas SA (1986) Atrial natriuretic factor: a new hormone of cardiac origin. Rec Prog Hormone Res 42:207-249

Berridge MJ (1987) Inositol trisphosphate and diacylglycerol: two interacting second messengers. Ann Rev Biochem 56:159-193

Blowers, DP and Trewavas AJ (1989) Second messengers: their existence and relationship to protein kinases. In Boss WF, Morré DJ (eds) Second messengers in plant growth and development. Alan R Liss New York pp 1-28

Boller T (1989) Primary signals and second messengers in the reaction of plants to pathogens. In Boss WF, Morré DJ (eds) Second messengers in plant growth and development. Alan R Liss New York pp 227-255

Boss WF (1989) Phosphoinositide metabolism: Its relation to signal transduction in plants. In Boss WF, Morré DJ (eds) Second messengers in plant growth and development. Alan R Liss New York pp 29-56

Boudet AM, Ranjera R (1989) The vacuole: a potential store for second messengers in plants. In Boss WF, Morré DJ (eds) Second messengers in plant growth and development. Alan R Liss New York pp 213-225

Cartwright DW, Langcake P, Ride JP (1980) Phytoalexin production in rice and its enhancement by a dichlorocyclopropane fungicide. Physiol Plant Pathol 17:259-267

Chinkers M, Garbers DL, Chang M-S, Lowe DG, Chin H, Goedell DV, Schulz S (1989) A membrane form of guanylate cyclase is an atrial natriuretic peptide receptor. Nature 338:78-83

Davis KR, Darvill AG, Albersheim P (1986) Host pathogen interactions. XXXI. Several biotic and abiotic elicitors act synergistically in the induction of phytoalexin accumulation in soybean. Plant Mol Biol 6:23-32

Davis KR, Hahlbrock K (1987) Induction of defense responses in cultured parsley cells by plant cell wall fragments. Plant Physiol 85:1286-1290

Ebel J (1986) Phytoalexin biosynthesis: the biochemical analysis of the induction process. Annu Rev Phytopathol 24:235-264

Exton JH (1988) Mechanisms of action of calcium-mobilizing agonists: some variations on a young theme. FASEB J 2:2670-2676

Felle H (1989) pH as a second messenger in plants. In Boss WF, Morré DJ (eds) Second messengers in plant growth and development. Alan R Liss New York pp 145-166

Hahn MG, Grisebach H (1983) Cyclic AMP is not involved as a second messenger in the response of soybean to infection by Phytophthora megasperma f. sp.glycinea. Z Naturforsch 38c:578-582

Hannum YA, Bell RM (1989) Functions of sphingolipids and sphingolipid breakdown products in cellular regulation. Science 243:500-507

Hartmann E, Gupta R (1989) Acetylcholine as a signalling system in plants. In Boss WF Morré DJ (eds) Second messengers in plant growth and development. Alan R Liss New York pp 257-287

Janssens PMW (1987) Did vertebrate signal transduction mechanisms originate in eukaryotic microbes? TIBS 12:456-459

Kauss, H (1987) Some aspects of calcium-dependent regulation in plant metabolism. Annu Rev Plant Physiol 38:47-72

Kurosaki F, Tsurusawa Y, Nishi A (1987) The elicitation of phytoalexins by Ca^{2+} and cyclic AMP in carrot cells. Phytochemistry 26:1919-1923

Maniara G, Laine R, Kuć J (1984) Oligosaccharides from Phytophthora infestans enhance the elicitation of sesquiterpenoid stress metabolites by arachidonic acid in potato. Physiol Plant Pathol 24:177-186

Marmé D (1989) The role of calcium and calmodulin in signal transduction. In Boss WF and Morré DJ (eds) Second messengers in plant growth and development. Alan R Liss, New York, pp 57-80

Newton RP, Brown EG (1986) The biochemistry and physiology of cyclic AMP in higher plants. In Chadwick CM, Garrod DR (eds) Hormones, receptors and cellular interactions in higher plants. Cambridge Cambridge University Press pp 115-153

Pelich SL, Vance DE (1989) Signal transduction via phosphatidylcholine cycles. TIBS 14:28-31

Ranjeva R, Boudet AM (1987) Phosphorylation of proteins in plants: regulatory effects and potential involvement in stimulus/response coupling. Annu Rev Plant Physiol 38:73-93

Scherer GFE (1989) Ether phospholipid platelet activating factor (PAF) and a proton transport-activating phospholipid (PAP): potential new signal transduction constituents for plants. In Boss WF, Morré DJ (eds) Second messengers in plant growth and development. Alan R Liss New York pp 167-179

Schulman H and Lou L (1989) Multifunctional Ca^{2+}/calmodulin-dependent protein kinase: domain structure and regulation. TIBS 14:62-66

Shenolikar S (1988) Protein phosphorylation: hormones, drugs, and bioregulation. FASEB J 2:2753-2764

Strasser H, Hoffman C, Grisebach H, Matern U (1986) Are polyphosphoinositides involved in signal transduction of elicitor-induced phytoalexin synthesis in cultured plant cells? Z Naturforsch 41c:222-227

Sutherland EW (1972) Studies on the mechanism of hormone action. Science 177:401-408

OLIGOSACCHARINS - PLANT REGULATORY MOLECULES

A.G. Darvill, P. Albersheim, P. Bucheli, S. Doares, N. Doubrava, S. Eberhard, D.J. Gollin, M.G. Hahn, V. Marfà-Riera, W.S. York and D. Mohnen

Complex Carbohydrate Research Center and Department of Biochemistry
The University of Georgia
220 Riverbend Road
Athens, GA 30602
USA

The discoveries that complex carbohydrates are tissue-specific cell-surface antigens and the receptors for hormones, toxins, bacteria, and viruses have created considerable interest in complex carbohydrates. Furthermore, the complex carbohydrate portions of some glycoprotein hormones are required for their biological activity, and the complex carbohydrates of some glycoprotein enzymes keep the enzymes stable, while the carbohydrates of some glycoproteins direct the glycoproteins to their proper locations. Moreover, the immune response to glycoproteins is often directed to their carbohydrate side chains, and the carbohydrate chains have been shown to affect the activity and residence time of glycoprotein pharmaceuticals. Still another discovery that has generated considerable interest in complex carbohydrate research is that these molecules perform regulatory functions in plants and animals, often at the level of controlling gene expression.

Our laboratory has long pursued two seemingly unrelated lines of research -- one, elucidating the structure and function of the matrix polysaccharides of the walls of growing plant cells, and the other, studying how plants defend themselves against disease. The combined results from these projects led to our discovery that structurally defined fragments of plant cell wall polysaccharides are able to function in plants as chemical messages with specific regulatory properties. We call oligosaccharides with regulatory activity *oligosaccharins*. We have determined that oligosaccharins not only can trigger plant defenses against pathogens and other types of stress, but also are able to regulate the rate of growth and differentiation of plant tissues into roots, flowers, and vegetative buds.

The first characterized oligosaccharins were those that are commonly referred to as elicitors of defense responses in plants. The best

NATO ASI Series, Vol. H 36
Signal Molecules in Plants and Plant-Microbe Interactions
Edited by B. J. J. Lugtenberg

characterized elicitor isolated from fungal cell walls is a branched hepta-β-glucoside. This elicitor was purified to homogeneity as a hexaglucosyl glucitol from a partial acid hydrolyzate of isolated mycelial walls of the phytopathogen *Phytophthera megasperma* f.sp. *glycinea* [Sharp *et al* 1984a,b]. The biological activities of the hepta-β-glucoside and its reduced form, the hexaglucosyl glucitol, are identical [Sharp *et al* 1984b]. The primary structure of the hexaglucosyl glucitol was determined, and the proposed structure confirmed by chemical synthesis [Sharp *et al* 1984c; Ossowski *et al* 1984]. The active hepta-β-glucoside is capable of eliciting phytoalexin accumulation in soybean cotyledons at concentrations of ~10^{-8}M, while closely related structural isomers are only active at a concentration 100-1,000-fold higher. The precise structural requirements for elicitor activity of the active hepta-β-glucoside suggested that plant cells have a specific receptor for this elicitor. Not only can plant cells use carbohydrates as regulatory molecules, but plant cells can also respond differentially to structurally closely related carbohydrates.

The hepta-β-glucoside elicitor was the only active component observed in a mixture of about 300 hepta-β-glucosides generated by partial acid hydrolysis of *Phytophthera megasperma* mycelial walls. The structure of the active heptaglucoside identified the minimum structural requirement for biological activity [Sharp *et al* 1984b]. Addition of a single glucose unit at the non-reducing end of one of the branches of the active hepta-β-glucoside reduces the biological activity of the molecule by about two-thirds. However, it is not known to what extent the β-1,6-linked backbone of the active hepta-β-glucoside can be extended from the reducing end without affecting its biological activity. We and our collaborators have synthesized the active hepta-β-glucoside with various ligands attached to the reducing end. The modified hepta-β-glucoside, even with bulky constituents attached (including aromatic compounds), retains its full elicitor activity, and the structural specificity of the molecule is also unaltered. The reducing-end-modified elicitors are being used to isolate a putative receptor of the elicitor.

Data from several laboratories have demonstrated that fragments of plant cell wall pectic polysaccharides can also elicit plant defense responses. Linear oligosaccharides of α-1,4-linked galactosyluronic acid residues elicit phytoalexin accumulation in soybean, pea and castor bean

[Hahn *et al* 1981; Nothnagel *et al* 1983; Walker-Simmons *et al* 1983; Jin and West 1984]. The oligogalacturonides that result in maximum phytoalexin accumulation range in size between 10 and 13 galactosyluronic acid residues. These oligogalacturonides have been purified to homogeneity. Their phytoalexin-inducing activity is destroyed by treatment with purified *endo*polygalacturonase, providing additional evidence that the oligogalacturonide itself is the elicitor-active species.

Oligogalacturonides have been shown to activate other plant defense responses in addition to eliciting phytoalexin accumulation. Undecagalacturonides elicited lignification in cucumber cells [Robertsen 1986]. The accumulation of protease inhibitors in plant tissues was also elicited by oligogalacturonides [Bishop *et al* 1984]. In the latter case, molecules ranging in size from digalacturonide to hexagalacturonide are active [Bishop *et al* 1984].

Microbial-secreted pectic-degrading enzymes (*endo*polygalacturonase and *endo*pectate lyase) that release biologically active oligogalacturonides have been purified to homogeneity. The oligogalacturonides released by these enzymes have been shown to be similar, if not identical, to the acid-generated oligogalacturonides discussed above. Further, plants have been shown to contain proteins in their cell walls that specifically inhibit the activities of fungal *endo*polygalacturonases [Cervone *et al* , in press]. The *endo*polygalacturonase and inhibitor protein bind to each other, exhibiting a dissociation constant of ~10^{-9}M. Polygalacturonase-inhibiting proteins have been reported to be present in a wide variety of dicotyledonous plants [Hahn *et al* , in press].

In vitro studies with the *endo*polygalacturonases from *Aspergillus niger* and the *endo*polygalacturonase inhibitor protein from *Phaseolus vulgaris* have shown that, over an extended reaction time (24-48 hours), the *endo*polygalacturonase retained residual activity in the presence of even 100-fold excess of the inhibitor. In fact, a 24-hour digestion of polygalacturonic acid by *Aspergillus niger endo*polygalacturonase in the presence of excess inhibitor protein produces oligogalacturonides with an average degree of polymerization of 10-12, precisely the sizes that possess maximum elicitor activity of plant defense responses [Cervone *et al* , in press]. Thus, the plant protein that inhibits *endo*polygalacturonase, by causing the accumulation of elicitor-active oligogalacturonides rather than elicitor-inactive mono- and digalacturonides, the normal products of the *endo*polygalacturonases, appears to subvert the function of fungal *endo*polygalacturonases from that of virulence factors to that of avirulence factors.

Another important discovery was that the hepta-β-glucoside and oligogalacturonide elicitors act synergistically when they are simultaneously applied to plant tissues [Davis *et al* 1986]. When the plant receives simultaneous signals from the pathogen's wall -- the hepta-β-glucoside -- and its own cell wall -- oligogalacturonides -- it requires the presence of less of each of the signals to react defensively. Thus, plants exhibit a high degree of sophistication with regard to their abilities to respond to oligosaccharins.

Our laboratory has two other projects involving oligosaccharins and host-pathogen interactions. A major project in our laboratory, headed by Steven Doares (see extended abstract), involves study of cell wall fragments that kill plant cells. We have demonstrated that pathogen-secreted enzymes release these cell wall fragments, probably from a hemicellulose. It is our hypothesis that these biologically active wall fragments trigger the hypersensitive resistance response. Understanding the trigger of hypersensitive response will lead, we believe, to the elucidation of the molecular basis of race/cultivar specificity in host-pathogen interactions.

The other project involving oligosaccharins and plant defense is a collaborative one with the laboratory of Bernard Fritig (see extended abstract). Our laboratory has partially purified a glucan from the walls of *Phytophthera megasperma* f.sp. *glycinea* which Fritig and his collaborators have used to protect *Nicotiana* species from a variety of viruses [Kopp *et al* , in press]. The hepta-β-glucoside elicitor has no activity in this system. Indeed, the active molecule appears to be considerably larger. Besides the fact that *Nicotiana* can be fully protected by the glucan, the most interesting observation relating to this system is that the mechanism by which the plant defends itself appears to be completely novel, that is, none of the well-characterized defense mechanisms appears to be activated by spraying the plants with the active glucan.

Oligosaccharins not only activate defense responses in plants, they also are able to regulate growth and differentiation. We discovered that a xyloglucan nonasaccharide, at ~10^{-8}M, inhibits auxin-stimulated growth of pea stem segments. A structurally related xyloglucan heptasaccharide had no ability to inhibit the growth. Fry [McDougall *et al* 1986, McDougall *et al, J Exp Bot* and *Plant Physiol,* in press] has confirmed and extended these observations by delineating more exactly the structural requirements of the active nonasaccharide. For example, structurally closely related xyloglucan octa- and decasaccharides are not able to inhibit the auxin-induced growth.

Our laboratory also discovered that acidic (presumably pectic) fragments of plant cell walls are able to inhibit flowering in *Lemna* [Gollin *et al* 1984]. This observation led us to our more recent efforts to study the ability of oligosaccharins to control morphogenesis.

A number of the above studies indicated that oligogalacturonides play an important role in the interaction of plants with their potential pathogens. More recently, our laboratory has shown that oligogalacturonides can also play an extremely important role in plant development. In fact, we have now established that oligogalacturonides of a specific size can regulate plant morphogenesis.

To study the role of oligosaccharins in plant growth and development, we modified and carefully characterized a tobacco thin-cell-layer morphogenesis bioassay and optimized it as a system to test plant cell wall fragments (and other substances) for morphogenesis-regulating activity [Mohnen *et al* , in preparation]. Cell wall pectic fragments were released from suspension-cultured sycamore cells by treatment with the homogeneous *endo*polygalacturonase of *Aspergillus niger*. We have shown that this mixture of pectic fragments has three distinct effects on morphogenesis of tobacco thin-cell-layers. Pectic fragments at a concentration of 1-10 μg/ml inhibit the explants from forming roots when added to a root-inducing medium containing 15 μM IBA and 0.5 μM kinetin. Pectic fragments at a concentration of 1-10 μg/ml alter the position at which roots develop on the explants (from a random distribution over the explant surface to a location at the extreme basal end of the explant) when the fragments are added to a root-inducing medium containing 7 μM IBA and 0.15 μM kinetin. Pectic fragments at 1-10 μg/ml cause marked tissue enlargement at the basal end of the explants when added to a medium containing 1.5 μM IBA and 0.9 μM kinetin. More importantly, the addition of the pectic fragments to this medium induces flowers to form on the explants. The flower-inducing activity of the pectic fragments is stable to boiling and protease digestion, and, after deesterification, is destroyed by treatment with the homogeneous *Aspergillus niger* *endo*polygalacturonase. Pectic fragments isolated from both tobacco and sycamore cell walls induce flowers to form on tobacco explants [Eberhard *et al* , in preparation].

The optimized thin-cell-layer bioassay has enabled us to purify the component of the *endo*polygalacturonase-released sycamore wall pectic fragments that induces flowers to form on the explants. The *endo*polygalacturonase-released sycamore wall fragments are primarily composed of the polysaccharides rhamnogalacturonan I and rhamnogalacturonan II and α-1,4-linked oligogalacturonides. These components were separated by a combination of ion-exchange and gel-permeation chromatographies [Darvill *et al* 1978]. The α-1,4-linked oligogalacturonides purified from the *endo*polygalacturonase-released sycamore cell wall fragment mixture induced flowers on the explants; purified rhamnogalacturonan I and purified rhamnogalacturonan II did not induce flowers. Further, oligogalacturonides generated by partial acid hydrolysis of citrus pectin also induced flowers on the explants. Thus, the flower-inducing component of the pectic fragments was shown to be α-1,4-linked oligogalacturonides.

We have recently established that oligogalacturonides with a degree of polymerization between 10 and 14, at a concentration of $\sim 10^{-7}$M, induce 7 flowers, on average, to form on each explant [Marfà-Riera *et al* , in preparation]. Oligogalacturonides with a degree of polymerization of less than 10, even at higher concentrations, did not induce flowers to form. Indeed, oligogalacturonides with a degree of polymerization of 12-14 are more active than those with a degree of polymerization of 10 and 11. Thus, these data demonstrate that oligogalacturonides of specific sizes are extremely effective inducers of organogenesis in tobacco tissues.

The work of several laboratories has shown that oligogalacturonides are effective oligosaccharins in host-pathogen interactions. Oligogalacturonides of defined size induce various defense responses in plant tissues and can induce flowers to form in tobacco tissues. The tobacco thin-cell-layer bioassay has also provided evidence that oligogalacturonides are effective inhibitors of root formation and that the oligogalacturonides can control the location on the explants where roots form. Thus, it is becoming apparent that oligogalacturonides of specific size are extremely important regulators of plant growth and development; oligogalacturonides may have as many functions as any of the classical phytohormones. Indeed, we believe that oligogalacturonides may be as important and as effective in the control of plant growth and development as auxin, cytokinin, gibberellin, abscisic acid, or ethylene. As with the phytohormones, it remains to be determined how oligogalacturonides control growth and development. Nevertheless, oligogalacturonides are very important regulatory molecules of plants.

Acknowledgments

This work was supported in part by U.S. Department of Energy grants DE-FG09-85ER13426 and DE-FG09-85ER13425 and by DE-FG09-87ER13810 as part of the DOE/NSF/USDA Plant Science Centers program.

References

Bishop PD, Pearce G, Bryant JE, Ryan CA (1984) Isolation and characterization of the proteinase inhibitor-inducing factor from tomato leaves, Identity and activity of poly- and oligogalacturonide fragments. J Biol Chem 259:13172-113177

Cervone F, Hahn MG, De Lorenzo G, Darvill A, Albersheim P (to be published) Host-pathogen interactions XXXIII. A plant protein converts a fungal pathogenesis factor into an elicitor of plant defense responses. Plant Physiol

Darvill AG, McNeill M, Albersheim P (1978) Structure of plant cell walls VIII. A new pectic polysaccharide. Plant Physiol 62:418-422

Davis KR, Darvill AG, Albersheim P (1986) Host-pathogen interactions XXXI. Several biotic and abiotic elicitors act synergistically in the induction of phytoalexin accumulation in soybean. Plant Mol Biol 6:23-32

Eberhard S, Doubrava N, Marfa-Riera V, Mohnen D, Southwick A, Darvill A, Albersheim P (to be published) Pectic cell wall fragments regulate tobacco thin-cell-layers

Gollin DJ, Darvill AG, Albersheim P (1984) Plant cell wall fragments inhibit flowering and promote vegetative growth in *Lemna gibba* G3. Biol Cell 51:275-280

Hahn MG, Bucheli P, Cervone F, Doares SH, O'Neill RA, Darvill A, Albersheim P (to be published) The roles of cell wall constituents in plant-pathogen interactions In: Nester E, Kosuge T (eds) Plant-microbe interactions, vol 3. McGraw-Hill, New York

Hahn MG, Darvill AG, Albersheim P (1981) Host-pathogen interactions XIX. The endogenous elicitor, a fragment of a plant cell wall polysaccharide that elicits phytoalexin accumulation in soybeans. Plant Physiol 68:1161-1169

Jin DF, West CA (1984) Characteristics of galacturonic acid oligomers as elicitors of casbene synthetase activity in castor bean seedlings. Plant Physiol 74:989-992

Kopp M, Rouster J, Fritig B, Darvill AG, Albersheim P (to be published) A fungal glucan protects *Nicotianae* against infection by viruses. Plant Physiol

Marfà-Riera V, Gollin DJ, Darvill A, Albersheim P (in preparation) Galacturonides induce flower formation in tobacco thin-cell layers explants

McDougall GJ, Fry SC (1988) Inhibition of auxin-stimulated growth of pea stem segments by a specific nonasaccharide of xyloglucan. Planta 175:412-416

McDougall GJ, Fry SC (to be published) Anti-auxin activity of xyloglucan oligosaccharides: the role of groups other than the terminal α-L-fucose residue. J Exp Bot

McDougall GJ, Fry SC (to be published) Structure-activity relationships for xyloglucan oligosaccharides with anti-auxin activity. Plant Physiol

Mohnen D, Eberhard S, Marfa-Riera V, Doubrava N, Toubart P, Gollin DJ, Gruber T, Nuri W, Albersheim P, Darvill A (in preparation) Control of root, vegetative shoot and flower morphogenesis in tobacco-thin-cell layers

Nothnagel EA, McNeil M, Albersheim P, Dell A (1983) Host-pathogen interactions XXII. A galacturonic acid oligosaccharide from plant cell walls elicits phytoalexins. Plant Physiol 71:916-926

Ossowski P, Pilotti Å, Garegg PJ, Lindberg B (1984) Synthesis of a glucoheptaose and a glucooctaose that elicit phytoalexin accumulation in soybean. J Biol Chem 259:11337-11340

Robertsen B (1986) Elicitors of the production of lignin-like compounds in cucumber hypocotyls. Physiol Mol Plant Pathol 28:137-148

Sharp JK, Albersheim P, Ossowski P, Pilotti Å, Garegg P, Lindberg B (1984c) Comparison of the structures and elicitor activities of a synthetic and a mycelial-wall-derived hexa (β-D-glucopyranosyl)-D-glucitol. J Biol Chem 259:11341-11345

Sharp JK, McNeil M, Albersheim P (1984b) The primary structures of one elicitor-active and seven elicitor-inactive hexa (β-D-glucopyranosyl)-D-glucitols isolated from the mycelial walls of *Phytophtora megasperma* f. sp. *glycinea*. J Biol Chem 259:11321-11336

Sharp JK, Valent B, Albersheim P (1984a) Purification and partial characterization of a β-glucan fragment that elicits phytoalexin accumulation in soybean. J Biol Chem 259:11312-11320

Walker-Simmons M, Hadwiger L, Ryan CA (1983) Chitosans and pectic polysaccharides both induce the accumulation of the antifungal phytoalexin pisatin in pea pods and antinutrient proteinase inhibitors in tomato leaves. Biochem Biophys Res Comm 110:194-199

PHOSPHORYLATION OF PLANT PLASMA MEMBRANE PROTEINS IN RESPONSE TO OLIGOSACCHARIDES THAT REGULATE PROTEINASE INHIBITOR GENE EXPRESSION

Edward E. Farmer and Clarence A. Ryan.
Institute of Biological Chemistry and Program in Biochemistry and Biophysics, Washington State University, Pullman, WA 99194-6340

INTRODUCTION

Oligo- and polygalacturonides can signal the induction of the expression of defensive genes in tissues of plants from at least 8 families (Bishop et al 1984; West *et al* 1984; Darvill and Albersheim 1984). In dicotyledonous plants polygalacturonides are major constituents of the primary cell wall and oligomeric fragments are released in response to pathogen or insect attacks. The nature of the polygalacturonic acid (PGA) oligomers that are released *in vivo*, and how they interact with target cells to effect gene expression, is poorly understood. Among the chemicals produced in response to oligo- and polygalacturonides are proteins that inhibit the serine class of proteinases (Bishop *et al* 1981; Ryan 1987). The biochemical basis for signalling processes that regulate the synthesis of proteinase inhibitors in response to insect attacks has been a major focus of this laboratory for several years (Bishop *et al* 1981; Ryan 1987).

Little is known about intracellular signal transduction processes leading to the activation of plant defense genes (reviewed in Lamb *et al* 1989), including genes coding for serine proteinase inhibitors. Several lines of evidence have implicated a role for the plasma membrane as a site for the initial reception of plant defense signalling molecules. For example, the fluorescense of chemical probes in the plasma membranes of cultured plant cells has been shown to be altered on addition of crude phytoalexin elicitors (Low and Henstein 1986), and fungal glucans have been shown to bind to a plasma membrane-enriched fraction from soybean (Schmidt and Ebel 1987). Phytoalexin elicitor preparations added to cultured parsley cells inhibit the uptake of both phosphate and amino acids (Strasser and Matern 1986). The plasma membrane of non-plant eukaryotic cells is commonly the site of reception for extracellular signals and hormones. A common event

NATO ASI Series, Vol. H 36
Signal Molecules in Plants and Plant-Microbe Interactions
Edited by B. J. J. Lugtenberg

associated with the reception of a signal (as well as its propagation within the cell) is the covalent modification of the receptor itself or an associated protein. Phosphorylation, mediated by protein kinases, is one of the most frequently observed forms of covalent modification associated, directly or indirectly, with receptor-ligand coupling.

Recently, we have found that a polygalacturonic acid fraction isolated from the cell walls of tomato, herein referred to as PGA, causes an enhancement in the *in vitro* phosphorylation of specific proteins in plasma membranes isolated from leaves of tomato and potato plants (Farmer *et al* 1989). In this chapter we describe in detail the optimal assay conditions for the *in vitro* phosphorylation of plant plasma membrane proteins and report the use of the ATP analogue adenosine γ-^{35}S-(thio)triphosphate to further characterize the PGA-enhanced phosphorylation of a 34 kDa protein in both tomato and potato plasma membranes.

RESULTS AND DISCUSSION

Preparation of plasma membrane from the leaves of tomato and potato.

Plasma membranes were isolated from leaves of young tomato (*L. esculentum*, Castlemart II) and potato (*S. tuberosum*, Russett Burbank) plants by the two-phase method of Yoshida et al 1983. The purified membrane fractions appeared almost colorless indicating a lack of contamination with plastid membranes. Protein content of the membranes was estimated using the BCA assay (Peirce, IL). Membranes (and protein standards) were precipitated with 7.5% TCA, the pellets washed gently with 10% TCA, and then resuspended in the assay reagent. A series of tests were conducted in order to verify the plasma membrane nature of the preparations. Plasma membranes were assayed for Mg^{2+}-dependent ATPase activity (Hodges and Leonard 1974). The specific activity of the ATPase was 5.5 µmol phosphate released/mg/h, but was increased to 42.1 µmol phosphate/mg/h in the presence of 0.02% Triton X-100. The variation in these assays was in the order of 10 %. The ATPase was inhibited 92% by sodium vanadate (50 µM), but not by sodium azide (5 mM, a mitochondrial ATPase inhibitor) or potassium nitrate (100 mM, a vacuolar ATPase inhibitor). Ammonium molybdate (0.5 mM), a cytoplasmic ATPase inhibitor, caused a 16% inhibition. Taken together these results indicate that the plasma membrane is only slightly contaminated with cytoplasmic contents. Examination of the membrane fraction by electron microscopy, with and without phosphotungstic acid staining (not shown), visually confirming that the plasma membrane nature of the preparations. Additionally, the isolated plasma membranes exhibited a characteristically low turbidity at pH 4 (Uemura *et al* 1983). The tomato and potato plasma membrane was screened with monoclonal antibodies (Mabs) raised against crude cellular membranes from *Nicotiana glutinosa* (Norman *et al*

1986). The plasma membrane from both plants cross reacted strongly with Mabs 16.4A6, 16.4B4 and 16.3C1 (data not shown) whereas a control Mab, 17.3A6 directed to an intracellular epitope, did not cross-react. Analysis of the protein components of isolated tomato plasma membranes by SDS gel electrophoresis revealed that the major protein, stained with Coomassie Blue, was at 75 kDa, which corresponds to the major band previously reported for tomato plasma membranes by Grimes and Breidenbach (1987).

Development of an *in vitro* phosphorylation assay.

A highly efficient *in vitro* phosphorylation assay was developed using plasma membrane proteins solubilized with a non-ionic detergent in a pH 6.5 MES buffer containing Mg^{2+}, Mn^{2+} and Ca^{2+} (Farmer *et al* 1989). Phosphorylation was initiated by the addition of γ-^{32}P-labeled ATP. The phosphorylation system was optimized for pH, ionic strength, cation content, and time of incubation in the presence of various detergents and cations. Total ^{32}P incorporation was measured by TCA-precipitation of aliquots of labelled protein onto glass fibre filters and scintillation counting.

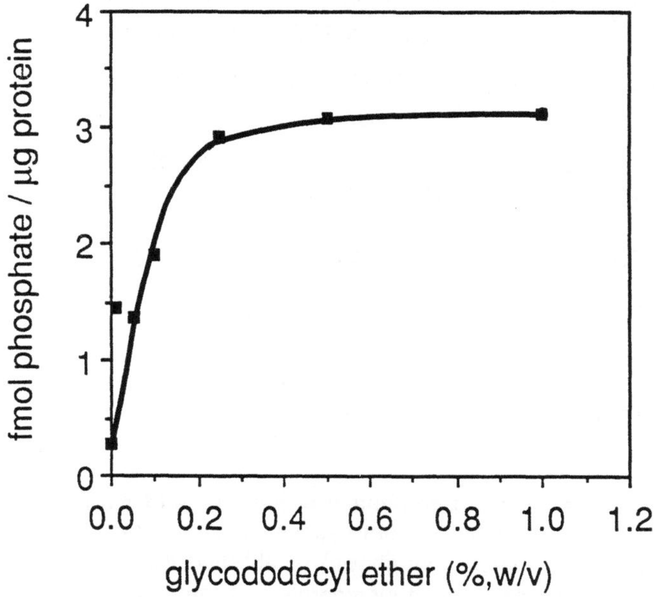

Figure 1. Concentration dependence of the non-ionic detergent octaethylene glycododecyl ether ($C_{12}E_8$) on the incorporation of phosphate from ATP into tomato plasma membrane proteins. Plasma membrane (1µg protein) was incubated for 5 min at 30°C in the presence of γ-^{32}P-ATP (110 nM, 1600 Ci/mmol). Aliquots were transferred to glass fiber filters and immediately washed in ice-cold 10% TCA containing 50 mM sodium pyrophosphate. Filters were dried in ethanol and assayed by scintillation counting.

The incorporation of ^{32}P into membrane proteins was stimulated by nonionic but not ionic detergents. $C_{12}E_8$, (octaethylene glycododecyl ether), Triton X-100 and CHAPS (3-[(3-cholamidopropyl)-dimethylammonio]-1-propanesulfonate), about equally enhanced incorporation of ^{32}P from γ-^{32}P-ATP into protein. However, in the presence of $C_{12}E_8$, more phosphorylated proteins were detected after electrophoresis than with the other nonionic detergents (data not shown). Divalent cations are required for maximal enzyme activity. Three cations were routinely included in the assay, Mg^{2+}, Mn^{2+} and Ca^{2+}. Of these, Mn^{2+} was the most effective and the combination of Mg^{2+} and Mn^{2+} were not additive (Table 1). This is the first report that we are aware of that describes the use of Mn^{2+} in enhancing total phosphate incorporation into plant plasma membrane protein. For some experiments levels of one or more cations may have to be reduced if they are found to inhibit a specific phosphorylation. The effects of pH and duration of the phosphorylation reaction were similar to those described for the plasma membrane ATPase by Schaller and Sussman, 1988. The buffer system chosen for routine analyses had the following composition: 20 mM MES, pH 6.5, containing 10% glycerol (v/v), 5 mM 2-mercaptoethanol, 5 mM $MgCl_2 \cdot 6H_2O$, 0.1 mM $CaCl_2 \cdot 2H_2O$, 1 mM $MnSO_4 \cdot H_2O$ (added after titration of the buffer to pH 6.5 with KOH), 0.4% (w/v) of detergent $C_{12}E_8$ and γ-^{32}P labelled ATP or γ-^{35}S ATP (adenosine γ-^{35}S-(thio)-triphosphate) of various specific activities and concentrations depending on the nature of the experiment.

Table 1 The effect of various cations on the efficiency of phosphate incorporation into tomato plasma membrane protein.

Addition	fmol phosphate/µg protein
Na_2EDTA + EGTA	0.09
Mg^{2+}	1.67
Ca^{2+}	1.87
Mn^{2+}	4.50
Mg^{2+} + Ca^{2+}	1.46
Mg^{2+} + Mn^{2+}	4.31

Tomato plasma membrane (1mg protein) was incubated for 5 min at 30°C in phosphorylation buffer containing divalent cations or chelators (each at 1 mM) listed above. A control experiment (not shown) established that the counter-ions Cl^- and SO_4^{2-} did not significantly affect incorporation at the levels used.

When employing low concentrations of high specific activity ATP, care was taken to ensure that ATP concentration was not limiting. With carrier-free ATP, very low amounts plasma membrane proteins (less than 0.1 µg) could be analysed by gel electrophoresis and autoradiography within the same working day. After phosphorylation, samples were immediately boiled for 2.5 min in SDS PAGE sample buffer, this treatment will often not preserve acyl phosphate linkages which are also destroyed at the pH of Laemmli gels. If acyl phosphate linkages linkages are sought, lower separating pH should be used. The basis of the assay may be of general use for studying other types of covalent modification eg. acetylation, methylation and ADP-ribosylation in response to a variety of signal molecules.

In vitro phosphorylation of plasma membrane proteins in response to tomato leaf PGA and fungal elicitors.

Inclusion of tomato leaf PGA in the *in vitro* phosphorylation assay system with tomato plasma membranes resulted in the increased phosphorylation of two proteins, M_rs 29 kDa and 34 kDa, (Figure 2A). When potato plasma membranes were used, only one protein, M_r 34 kDa, exhibited enhanced phosphorylation (Farmer *et al* 1989). When the ATP analogue adenosine γ-^{35}S-(thio)triphosphate was substituted for γ-^{32}P ATP (Figure 2B) tomato PGA caused the enhanced thiophosphorylation of pp34, but not pp29.

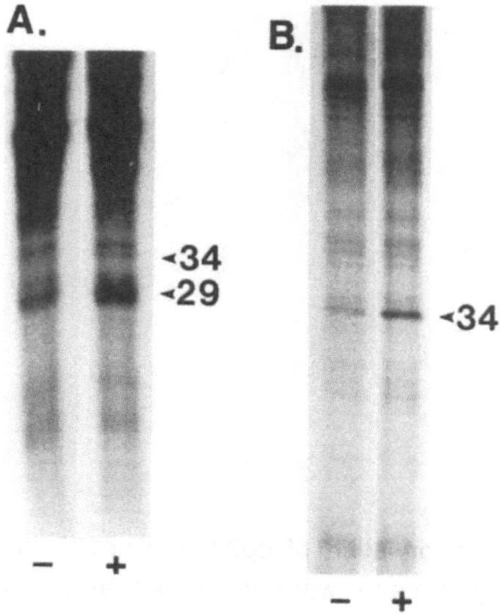

Figure 2. A. The phosphorylation of tomato plasma membrane (1.9 µg protein) in the absence (-) or presence (+) of tomato leaf PGA (0.5 µg). The phosphate donor was γ-^{32}P ATP and phosphorylation was at 30°C for 10 min. B. The phosphorylation of tomato plasma membrane (1µg protein) in the absence (-) or presence (+) of tomato leaf PGA (1µg), in this case the phosphate donor was adenosine γ-^{35}S-(thio)triphosphate and the thiophosphorylation was for 15 min at 30°C. Proteins were resolved on 10% polyacrylamide SDS gels which were dried and autoradiographed. pp34 and pp29 are shown by arrrowhead.

When the electrophoresis time was extended to help visualize larger proteins, a PGA-enhanced thiophosphorylation of a protein of M_r~100 kDa was also seen in tomato plasma membrane proteins (Figure 3). The plasma membrane H^+-ATPase has an M_r of 100 kDa and is known to be phosphorylated at both serine and threonine residues (Schaller and Sussman 1988). The possibility that pp100 is the H^+-ATPase is currently being investigated. Adenosine γ-^{35}S-(thio)triphosphate has been used in animal systems and is known to be an ATP substrate analogue. Thiophosphate is often transferred less efficiently than phosphate to some protein acceptors and in addition thiophosphorylated amino acids are

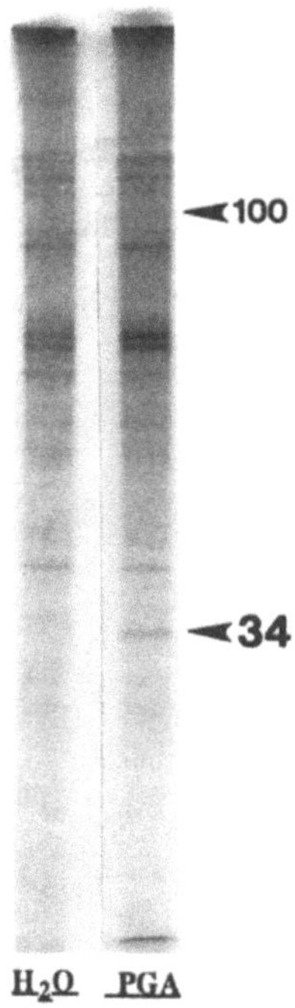

Figure 3. *In vitro* thiophosphorylation of pp34 and pp100 in tomato plasma membranes. Plasma membranes (1µg protein) were incubated at 30°C for 10 min with or without tomato leaf polygalacturonic acid (0.2µg) with adenosine γ-^{35}S (thio)ATP (1 µM, 1350 Ci/mmol). Radiolabelled protein was analysed as in the previous figure.

more stable to phosphatases than phosphoamino acids. One dimensional gels of plasma membrane proteins that have been thiophosphorylated are 'cleaner' than similar gels were γ-^{32}P ATP has been used. Protein kinases are often efficiently autothiophosphorylated

(Cassels and Glasser 1983). pp34 could be a protein kinase subunit ; we are currently trying to isolate pp34 to establish its primary structure and function.

We believe that the *in vitro* phosphorylation of plasma membrane proteins in response to PGA may be related to the signalling process that effects the systemic and localized expression of proteinase inhibitor genes and possibly also phytoalexin genes. Localized defense responses that are triggered by pectic fragments include proteinase inhibitor production, phytoalexin accumulation, lignification etc. and are widespread in the plant kingdom.

Fungal β-glucan (from *Phytophthora Megasperma*) was ineffective in enhancing specific *in vitro* protein phosphorylation in tomato and potato plasma membrane. Chitosan (a glucosamine polymer) virtually abolished plasma membrane protein phosphorylation, perhaps due to charge effects. These preliminary experiments do not necessarily mean that other elicitors do not alter protein phosphorylation in the plasma membrane. It may be that in the presence of degradative enzymes, such as β-glucanase or chitinase that smaller reactive oligomers are produced that would be active. Experiments are in progress to further elucidate the specificity and significance of the phosphorylation of specific plasma membrane proteins in response to oligosaccharide signals.

ACKNOWLEDGEMENT

We thank Dr. T. W.Okita for helpful suggestions concerning the use of detergents. This work was supported by National Science Foundation Grant DCB-8315427 and a grant from EniChem Americas, Inc.

REFERENCES

Bishop PD, Makus DJ, Pearce G, Ryan CA (1981) Proteinase inhibitor-inducing factor activity in tomato leaves resides in oligosaccharides enzymatically released from cell walls. Proc Natl Acad Sci USA 78: 3536-3540.

Bishop PD, Pearce G, Bryant JE, Ryan CA (1984) Isolation and characterization of the proteinase inhibitor inducing factor from tomato leaves. J Biol Chem 259: 13172-13177.

Cassel D, Glaser L (1982) Resistance to phosphatase of thiophosphorylated epidermal growth factor receptor in A431 membranes. Proc Natl Acad Sci USA 79: 2231-2235

Darvill AG, Albersheim P (1984) Phytoalexins and their elicitors- A defense against microbial infection in plants. Ann Rev Plant Physiol 35: 243-275

Farmer EE, Pearce G, Ryan CA (1989) *In vitro* phosphorylation of plant plasma membrane in response to proteinase inhibitor inducing factor. Proc Natl Acad, Sci USA 86: 1539-1542

Grimes HD, Breidenbach RW (1987) Plant plasma membrane proteins: immunological characterization of a major 75 kD protein group. Plant Physiol 87: 1048-1054

Hodges TK, Leonard RT (1974) Purification of plasma membrane bound adenosine triphosphatase from plant roots. Meth Enzymol 32: 392-406

Lamb CJ, Lawton MA, Dron M, Dixon RA (1989) Signals and transduction mechanisms for activation of plant defenses against microbial attack. Cell 56: 215-224

Low PS, Henstein PK (1986) Elicitor stimulation of the defense response in cultured plant cells monitored by fluorescent dyes. Arch Biochem Biophys 249: 472-479

Norman PM, Wingate VPM, Fitter MS, Lamb CJ (1986) Monoclonal antibodies to plant plasma membrane antigens. Planta 167: 452-459

Ryan CA (1987) Oligosaccharide signalling in plants. Ann Rev Cell Biol 3:295-317

Schaller GE, Sussman MR (1988) Phosphorylation of the plasma- membrane H^+-ATPase of oat roots by a calcium-stimulated protein kinase. Plant Physiol 173: 509-518

Schmidt WE, Ebel J (1987) Specific binding of a fungal glucan elicitor to membrane fractions from soybean *Glycine max*. Proc Natl Acad Sci USA 84: 4117-4121

Strasser H, Matern U (1986) Minimal time requirement for lasting elicitor effects in cultured parsley cells. Z Naturforschung 41c: 222-227

Uemura M, Yoshida S (1983) Isolation and identification of plasma membrane from light-grown winter rye seedlings. Plant Physiol 72: 105-114

West CA, Bruce RJ, Jin DF (1984) Pectic fragments of plant cell walls as mediators of stress responses. in *Structure, Function and Biosynthesis of Plant Cell Walls*, eds Dugger WM, and Bartnicki-Garcia S (Waverely, Baltimore) pp. 359-379

Yoshida S, Uemura M, Niki T, Sakai A, Gasta LV (1983) Partition of membrane particles in aqueous two-phase system and its practical use for purification of plasma membranes from plants. Plant Physiol 72: 105-114

BIOSYNTHESIS OF PHYTOALEXINS AND NOD-GENE INDUCING ISOFLAVONES IN SOYBEAN

Hans Grisebach, Lambert Edelmann, Daniela Fischer, Georg Kochs and Roland Welle
Lehrstuhl für Biochemie der Pflanzen
der Universität Freiburg
Schänzlestr. 1
7800 Freiburg, FRG

INTRODUCTION

The accumulation of the pterocarpan derivate glyceollin I is an early response of soybean (Glycine max) roots to infection with an incompatible race of the fungus Phytophthora megasperma f.sp. glycinea (Pmg) and is correlated with the inhibition of fungal growth (Hahn et al 1985). Starting from 4-coumaroyl-CoA and malonyl-CoA the biosynthesis of glyceollin involves nine enzymatic steps, which have all been characterized. For purification of enzymes soybean cell suspension cultures challenged with a Pmg or yeast elicitor were used in most cases.

The series of reactions leading to glyceollin I is shown in Fig. 1. One of the intermediates in this pathway is the isoflavone daidzein. Daidzein and the corresponding trihydroxyisoflavone genistein were identified as inducers of common nod-genes in the Bradyrhizobium japonicum-soybean symbiosis (Kosslak et al 1987). This dual function of isoflavonoids in soybean - as phytoalexins and inducers of nod-genes - increased interest in biosynthesis of these compounds and its regulation. We present here mainly new results on biosynthesis of soybean isoflavonoids and its induction by infection or elicitors. The role of constitutive isoflavonoids in soybean roots is discussed.

BIOSYNTHESIS OF SOYBEAN ISOFLAVONOIDS

Daidzein and glyceollin are 5-deoxyflavonoids which are derived from 2',4',4-trihydroxychalcone (Fig. 1, 1) (Grisebach and Brandner 1961). In vitro synthesis of 6'-deoxychalcone (dCH) from 4-coumaroyl-CoA and malonyl-CoA in presence of NADPH with a crude extract from Glycyrrhiza echinata cells or protoplasts was reported recently (Ayabe et al 1988a, 1988b) and was also found with extracts from cell cultures of Pueraria lobata (Hakamatsuka et al 1988). We found that dCH

NATO ASI Series, Vol. H 36
Signal Molecules in Plants and Plant-Microbe Interactions
Edited by B. J. J. Lugtenberg

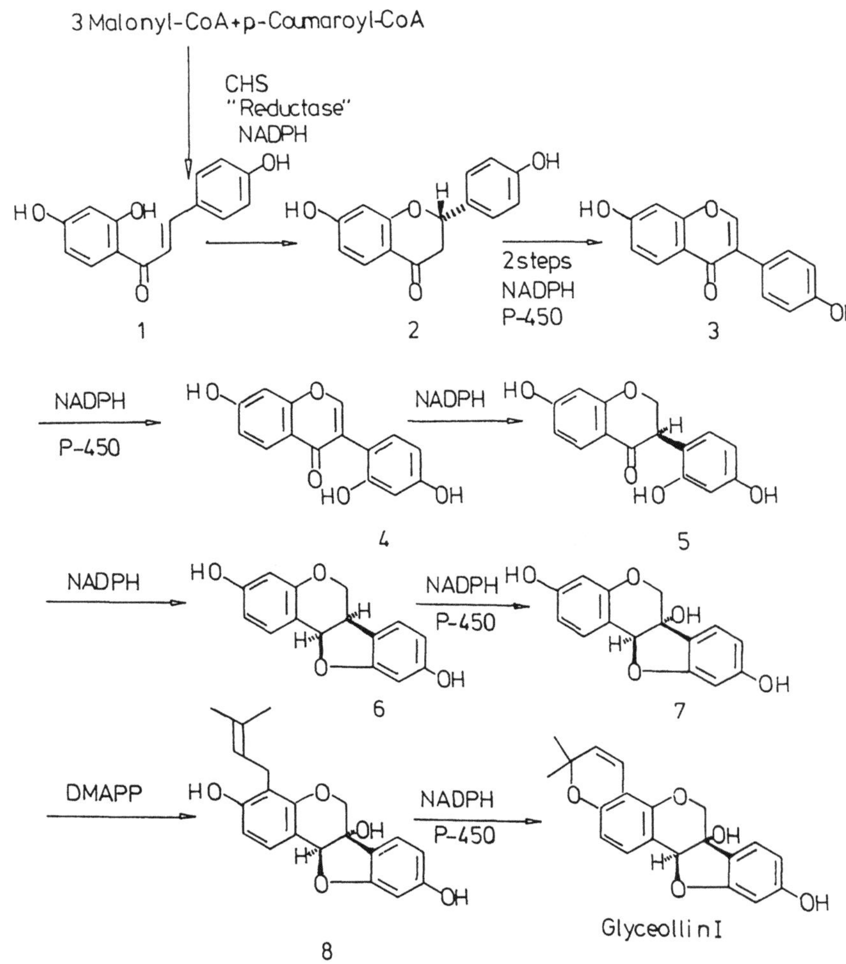

Fig. 1 Biosynthesis of daidzein and glyceollin I in soybean. 1, 2',4',4-trihydroxychalcone; 2 (2S)7,4'-dihydroxyflavanone; 3, daidzein; 4, 2'-hydroxydaidzein; 5, 7,2',4'-trihydroxyisoflavanone; 6, 3,9-dihydroxypterocarpan; 7, 3,6a,9-trihydroxypterocarpan; 8 4-dimethylallyl-3,6a,9-trihydroxypterocarpan

synthesis is not due to a special chalcone synthase (CHS) but requires, in addition to "normal" CHS and NADPH, a 34kDa protein which functions as a reductase (Welle and Grisebach 1988). The reductase was purified to homogeneity from cultured soybean cells challenged with a Pmg-elicitor. Soybean reductase has a high affinity for soybean CHS (K_m 0.75 μM) but also interacts with CHS from parsley and other plants with formation of dCH (Welle and Grisebach 1989). The substrate of the reductase is postulated to be the intermediate enzyme-bound polyketide (Fig. 2).

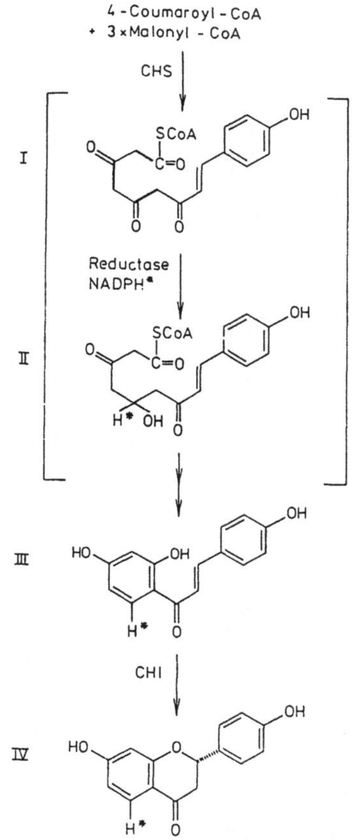

Fig. 2 Postulated reaction mechanism for synthesis of 6'-deoxychalcone by coaction of chalcone synthase and reductase in presence of NADPH. CHI, chalcone isomerase.

A polyclonal anti-soybean reductase antiserum showed no cross-reaction with soybean CHS. Also anti-soybean CHS did not cross-react with reductase. Reductase and CHS activities were determined at various times following Pmg- or yeast elicitor challenge of soybean cells (Welle and Grisebach 1989). Enzyme amounts were analyzed with a quantitative Western blot and mRNA activities were followed by translation of total RNA in presence of L-[^{35}S]-methionine. As shown in Fig. 3A, unchallenged cells contained only very low activities. After Pmg-elicitor challenge, enzyme activities and amounts and activities of mRNA showed a concomitant and transient increase. The ratio of amounts of CHS to reductase protein was between 1.5 and 2. Similar results were obtained with yeast extract (Fig. 3B). The results indicate that the genes for the two enzymes are induced concomitantly.

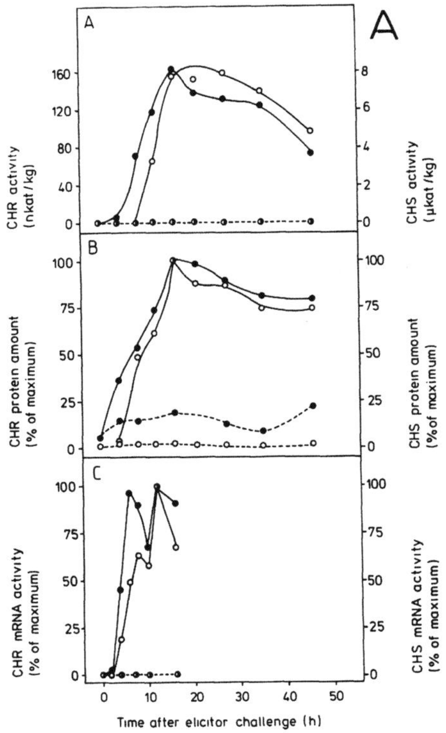

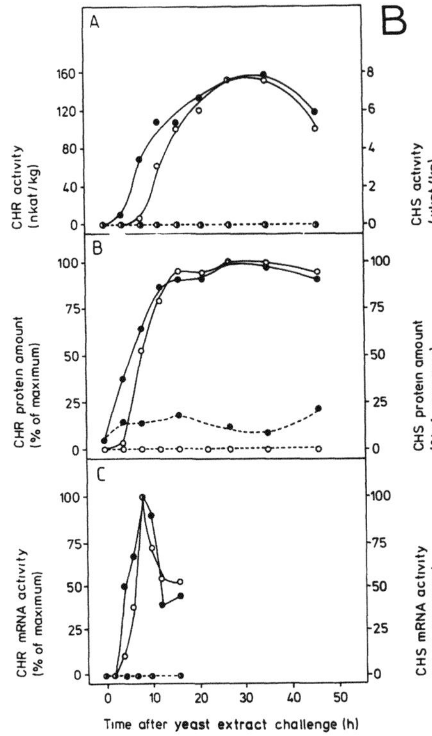

Fig. 3A Changes of enzyme activity (A), protein amount (B) and mRNA activity (C) after Pmg elicitor induction of soybean cells. Reductase (○); CHS (●). Controls with dashed lines.

Fig. 3B Induction with yeast extract.

In contrast to cell cultures, roots of two-day-old soybean seedlings contained significant amounts of reductase protein (Western blots) which did not change after infection with Pmg (Fig. 4). Reductase activity measurements are in progress. It had previously been shown that CHS activity and amount do increase in soybean roots infected with an incompatible race of Pmg (Ebel et al 1986), and these results were confirmed.

6'-Deoxychalcone is cyclized by chalcone isomerase to (2S)7,4'-dihydroxyflavanone, and the latter is converted to daidzein by isoflavone synthase (Kochs and Grisebach 1986). Isoflavone synthase and enzymes catalyzing 3 further reactions leading to glyceollin (Fig. 1) are cytochrome P-450 monooxygenases which are localized in the endoplasmic reticulum (ER).

Daidzein can either be further converted to glyceollin or can be stored in form of its malonylated (?) 7-0-glucoside (see below). The subsequent steps from daidzein to

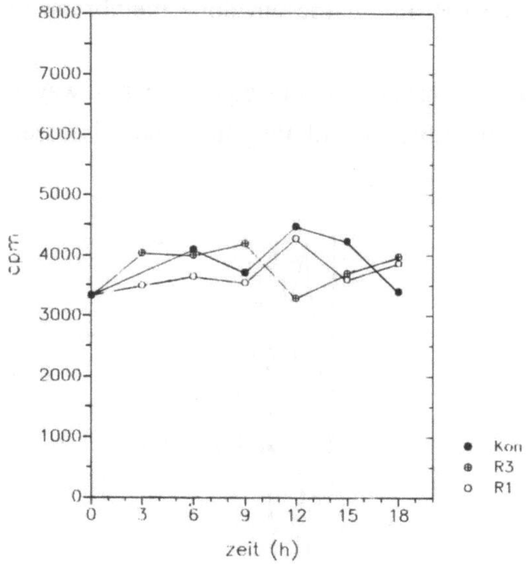

Fig. 4 Changes in amount of reductase in soybean roots after infection with zoospores of Pmg race 1 (o) and race 3 (⊕). Controls (●).

3,9-dihydroxypterocarpan (Fig. 1, 6) are analogous to reactions in synthesis of phytoalexins from chick pea (Cicer arietinum) (Hinderer et al 1987; Bleß and Barz 1988). These involve 2'-hydroxylation of daidzein by a cyt P-450 monooxygenase, NADPH-dependent reduction of 2'-hydroxydaidzein to the corresponding isoflavanone, and NADPH-dependent cyclization to the pterocarpan. The latter two enzymes have been purified from elicitor challenged soybean cells (D. Fischer unpublished). The cyt P-450 monooxygenase responsible for 6a-hydroxylation of dihydroxypterocarpan (Hagmann et al 1984) was detergent solubilized and was purified to homogeneity (Kochs and Grisebach 1989). This enzyme could be separated from cyt P-450 cinnamate 4-hydroxylase. This is the first direct evidence for the existence of several cyt P-450 species in a higher plant.

Final steps in the pathway to glyceollin are prenylation and cyclization of the prenyl residue. The latter enzyme catalyzing the reaction is again a cyt P-450 monooxygenase (Welle and Grisebach 1988). The membrane-bound prenyltransferase was detergent solubilized and has been partially purified. The membrane-bound enzyme peaked at lower density ($1.037/cm^3$) on a Percoll gradient than marker enzymes for the ER ($1.046/cm^3$) and coincided with the maximum of galactolipid synthesis (Fig. 5) (R. Welle unpublished). Since synthesis of galactolipids is associated with the chloroplast envelope (Douce et al 1984), this result indicates association of prenyltransferse with this membrane. It can be concluded that synthesis of glyceollin takes place in the cytoplasm, at the ER and at the envelope membrane of proplastids or chloroplasts.

Specific induction of the enzymes of the glyceollin pathway by infection of soybean seedlings with an incompatible race of Pmg has been discussed previously (Ebel et al 1986).

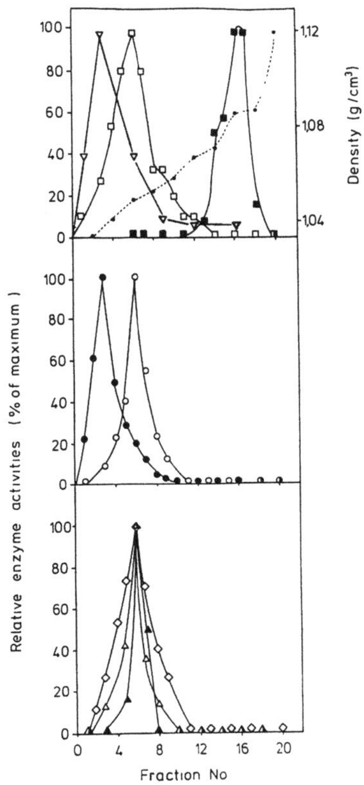

Fig. 5 Separation of microsomal fraction of soybean cells on Percoll gradient. Galactolipid synthesis (▽); NADPH-cytochrome reductase (□); Cytochrome oxidase (■); prenyltransferase (●); cyclase (8 ➞ glyceollin) (○); cinnamate 4-hydroxylase (◇); isoflavone synthase (△); density (- -).

Glyceollin accumulation had also been observed in one type of nodule from soybean which had been infected with a fix⁻strain of Rhizobium japonicum (Werner et al 1985). These nodules were distinguished by an early loss of the peribacteroid membrane. Paniske and Werner (personal communication) have also found glyceollin accumulation in nodules of the combination Glycine soja-B. japonicum 123. Such nodules are fix⁻ Furthermore, these authors found a transient glyceollin accumulation in root hairs of

Glycine max infected with the nod$^+$ fix$^+$ strain of B. japonicum 61-A-101. This finding could explain abortive primary infections which have been observed cytologically in such combinations (Parniske and Werner personal communication).

CONSTITUTIVE ISOFLAVONES IN SOYBEAN

Roots of healthy soybean seedlings contain no glyceollin or only very low amounts of it (Hahn et al 1985). Such roots, however, contain daidzein, mainly in a bound form. HPLC-analysis of 5 day-old seedlings of various soybean cultivars showed the presence of a small amount of free daidzein (2-12 nmol/g. fr.wt.) and larger amounts of bound daidzein (97-150 nmol/g. fr.wt.) in the roots. In the cultivar Harosoy 63 high amounts of bound daidzein were found in the roots (150 nmol/g.fr.wt.) and lower amounts in the hypocotyl (28 nmol/g. fr.wt.) and cotyledons (19 nmol/g. fr.wt.).
It is unknown whether the constitutive pool of bound daidzein contributes to glyceollin accumulation.
A study of soybean root exudate showed among a number of phenolic compounds the presence of daidzein and coumestrol (Arcy-Lameta 1986) which must originate from the constitutive pool of isoflavonoids. This pool could be of special importance for the Bradyrhizobium-soybean interaction.

ACKNOWLEDGEMENT

Our research was supported by Deutsche Forschungsgemeinschaft (SFB 206) and by Fonds der Chemischen Industrie.

REFERENCES

Arcy-Lameta A (1986) Study of soybean and lentil root exudates. II. Identification of some polyphenolic compounds, relation with plantlet physiology. Plant and Soil 92: 113-123
Ayabe S, Udagawa A, Furuya T (1988a) NAD(P)H-Dependent 6'-deoxychalcone synthase activity in Glycyrrhiza echinata cells induced by yeast extract. Arch Biochem Biophys 261: 458-462
Ayabe S, Udagawa A, Furuya T (1988b) Stimulation of chalcone synthase activity by yeast extract in cultured Glycyrrhiza echinata cells and 5-deoxyflavanone formation by isolated protoplasts. Plant Cell Reports 7: 35-38

Bleß W, Barz W (1988) Isolation of pterocarpan synthase, the terminal enzyme of pterocarpan phytoalexin biosynthesis in cell suspension cultures of Cicer arietinum. FEBS Lett 235: 47-50

Douce R, Block M A, Dorne A J, Joyard J (1984) The plastid envelope membranes; their structure, composition, and role in chloroplast biogenesis. In: Roodyn D B (ed) Subcellular biochemistry. Vol 10, Plenum Press New York, p 1-84

Ebel J, Grisebach H, Bonhoff A, Grab D, Hoffmann Ch, Kochs G, Mieth, H, Schmidt W, Stäb M (1986) Phytoalexin synthesis in soybean following infection of roots with Phytophthora megasperma or treatment of cell cultures with fungal elicitor. In: Lugtenberg B (ed) Recognition in microbe-plant symbiotic and pathogenic interactions. NATO ASI series, Vol H4, Springer-Verlag Berlin Heidelberg, p 345-361

Grisebach H, Brandner G (1961) Zur Biogenese der Isoflavone. IV. Mitt.: Über die Biogenese von Biochanin A und Formononetin in der Kichererbse (Cicer arietinum L.) Z Naturforsch 16b: 2-5

Hagmann M L, Heller W, Grisebach H (1984) Induction of phytoalexin synthesis in soybean: Stereospecific 3,9-dihydroxypterocarpan 6a-hydroxylase from elicitor-induced soybean cell cultures. Eur J Biochem 142: 127-931

Hahn M G, Bonhoff A, Grisebach H (1985) Quantitative localization of the phytoalexin glyceollin I in relation to fungal hyphae in soybean roots infected with Phytophthora megasperma f.sp. glycinea. Plant Physiol 77: 591-601

Hakamatsuka T, Noguchi H, Ebizuka Y, Sankawa U (1988) Deoxychalcone synthase from cell suspension cultures of Pueraria lobata. Chem Pharm Bull 36: 4225-4228

Hinderer W, Flentje U, Barz W (1987) Microsomal isoflavone 2'- and 3'-hydroxylases from chickpea (Cicer arietinum L.) cell suspension induced for pterocarpan phytoalexin formation. FEBS Lett 214: 101-106

Kochs G, Grisebach H (1986) Enzymic synthesis of isoflavones. Eur J Biochem 155: 311-318

Kochs G, Grisebach H (1989) Phytoalexin synthesis in soybean. Purification and reconstitution of cytochrome P-450 3,9-dihydroxypterocarpan 6a-hydroxylase and separation from cytochrome P-450 cinnamate 4-hydroxylase. Arch Biochem Biophys, in press

Kosslak R M, Bookland R, Barkei J, Paaren, H E, Appelbaum E R (1987) Induction of Bradyrhizobium japonicum common nod genes by isoflavones isolated from Glycine max. Proc Natl Acad Sci USA 84: 7428-7432

Tiemann K, Hinderer W, Barz W (1987) Isolation of NADPH: isoflavone oxidoreductase, a new enzyme of pterocarpan phytoalexin biosynthesis in cell suspension cultures of Cicer arietinum. FEBS Lett 213: 324-328

Welle R, Grisebach H (1988a) Isolation of a novel NADPH-dependent reductase which coacts with chalcone synthase in the biosynthesis of 6'-deoxychalcone. FEBS Lett 236: 221-225

Welle R, Grisebach H (1988b) Induction of phytoalexin synthesis in soybean: enzymatic cyclization of prenylated pterocarpans to glyceollin isomers. Arch Biochem Biophys 263: 191-198

Welle R, Grisebach H (1989) Phytoalexin synthesis in soybean cells: Elicitor induction of reductase involved in biosynthesis of 6'-deoxychalcone. Arch Biochem Biophys 272: 97-102

Werner D, Mellor R B, Hahn M G, Grisebach H (1985) Soybean root response to symbiotic infection. Glyceollin I accumulation in an ineffective type of soybean nodules with an early loss of the peribacteroid membrane. Z. Naturforsch 40c: 179-181

THE POTENTIAL USE OF THE GLUCURONIDASE REPORTER GENE AND ANTISENSE TECHNOLOGIES TO UNRAVEL BIOLOGICAL FUNCTIONS OF FLAVONOIDS

J.N.M. Mol, R.E. Koes, R. van Blokland, F. Quattrocchio, A.J. van Tunen, A.R. van der Krol and A.R. Stuitje
Department of Genetics,
Section Biosynthesis Secondary Metabolites,
Vrije Universiteit,
De Boelelaan 1087
1081 HV Amsterdam
The Netherlands

Flavonoids are higher plant secondary metabolites which have a key function in floral pigmentation (Harborne and Mabry 1982), defence against phytopathogens (Lamb et al 1989) and induction of nodulation (Firmin et al 1986; Zaat et al 1987). Furthermore, they have been implicated in UV-protection (Schmelzer et al 1988) and regulation of auxin transport (Jacobs and Rubery 1988). The multitude of functions suggest complex regulation and one might wonder if more functions are to be discovered. We propose that use of the ß-glucuronidase (GUS) reporter gene in combination with antisense technologies may lead to a better understanding of the 'signalling' function(s) of flavonoid compounds.

Flavonoids are synthesized via the phenylpropanoid pathway, narigenin chalcone being the first specific intermediate. It is formed via condensation of 3 molecules of malonyl coA with 1 unit of 4 coumaroyl CoA by the enzyme chalcone synthase (CHS). Subsequent isomerization yields narigenin flavanone, which is the central precursor for the synthesis of different classes of flavonoids, (isoflavonoids, flavonols, flavanones, anthocyanins etc.). CHS and CHI genes have been studied in several plant species and corresponding cDNA and chromosomal clones have been isolated (Mol et al 1988). In the inbred *Petunia hybrida* line V30, 11 CHS genes/gene fragments have been identified, isolated and characterized (Koes et al 1989a and b). Two CHS genes (CHS-A and CHS-J) have been shown to be active in pigmented floral tissues and under UV stress conditions in seedlings. Two additional CHS genes (CHS-B and CHS-G) are only expressed at low levels in UV-light. In the same *P. hybrida* line only 2 CHI genes are present, which are differentially expressed (van Tunen et al 1988). CHI-A is expressed in the corolla and tube of the flower and UV-treated seedlings. In mature anthers a longer CHI-A transcript is

NATO ASI Series, Vol. H 36
Signal Molecules in Plants and Plant-Microbe Interactions
Edited by B. J. J. Lugtenberg

found, which results from transcription from an upstream promoter (van Tunen et al 1989). CHI-B is expressed in immature anthers only.

To get further insight into the cell-type specificity of the different expressed CHS and CHI genes we fused their promoters to the bacterial ß-glucuronidase (GUS) gene and started to study the expression of the resulting chimaeric genes in *Petunia* at the macroscopic and microscopic level.

Table I: Activity in various tissues of chimaeric GUS genes driven by different CHS and CHI gene promoters

		CHS-A	CHS-J	CHS-G	CHI-A
	corolla	++	++	-	++
flower	tube	++	++	-	++
	anthers	+	+	-	-
flower stem		+	+	ND	ND
ovary		++	++	-	+
seed pot		+++	+++	-	+
leaf		-	-	-	+
root		-	-	-	-

The data summarize GUS-activities in tissues of multiple (7-20) independent transformants for each construct. The number of + gives an indication for the amount of GUS activity, - is no activity. ND is not determined.

As expected, fusions containing the CHS-A or CHS-J promoter are expressed in pigmented tissues like the flower corolla and tube and the stem of the flower, but not in leaves or roots (Table 1). The CHI-A GUS fusion exhibits a similar expression pattern and in addition is expressed in leaves. This is not unexpected since the endogenous CHI-A gene is also expressed in leaves, albeit at a low level. Histochemical analysis shows that expression in corolla is localised in both inner and outer epidermis as well as the mesophyl cells in between. In the flower tube however, expression is localised in the inner epidermis mainly (but not strictly). Early during flower development strong expression is found also in mesophyl cells in addition to the epidermal cell layers.

These results are somewhat surprising since anthocyanins are known to be accumulated in epidermal cells only. We believe that the strong expression in mesophyl cells is related to biosynthesis of flavonols, although we cannot exclude that these cells produce (precursors of) anthocyanins which are subsequently transported into the epidermis.

In addition to the pigmented tissues considerable GUS activity is found in the reproductive organs (ovary, anthers, seed pots). Expression of CHS-A and CHS-J is localised in the tapetum cell layer and associated parenchymal cells of anthers, which coincides with the site of anthocyanin accumulation. As expected the CHI-A fusion is not expressed in anther tissue.

Expression in ovaries is localised in the ovules mainly and to a lesser extent in the tissues immediately surrounding them (see Fig. 1). After pollination high expression of CHS-A, CHS-J and CHI-A fusions is found in developing seeds, predominantly in the seed coat and somewhat less in an internal cell layer immediately surrounding the embryo. These results indicate that flavonoid-specific enzymes and flavonoid products themselves are formed in these organs/tissues, which may serve yet unknown functions. Initial analyses indicate that ovaries indeed contain considerable amounts of flavonols, but no anthocyanins.

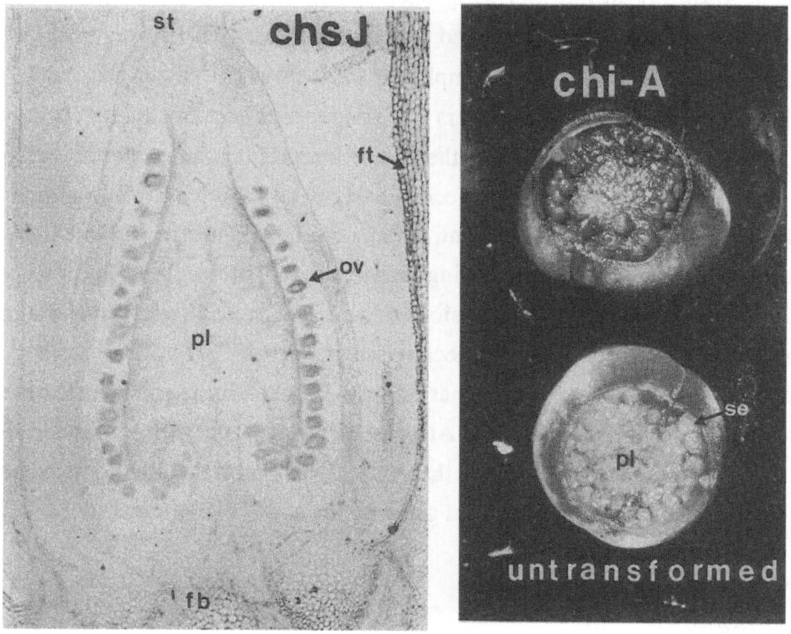

Fig. 1. Histochemical localization of flavonoid gene expression in ovaries and seed pots.
Left panel: Activity of CHS-J-GUS in ovaries. The dark colour of ovules and the outer layers of the placenta are the result of GUS activity.
Right panel: Activity of CHI-A-GUS in seed pots. The dark colour of CHI-A transformed seeds (compare to untransformed controls) represents GUS activity.
Abbreviations: st = style, ov = ovule, ft = flower tube, fb = flower base, pl = placenta, se = seed.

One way to study the function(s) of flavonoids in developmental processes is to modulate or prevent their synthesis. Such studies are however hampered by the unavailability of CHS mutants. As an alternative flavonoid synthesis can be blocked using antisense RNA strategies. Using antisense genes it has been demonstrated that target gene expression can effectively be inhibited in a number of different organisms varying from bacteria and slime molds to mammalian cells and plants (for review see Green et al 1986; Inouye 1988; van der Krol et al

1988a and 1988b). It is generally accepted that antisense genes function by synthesis of antisense RNA which forms a duplex with the target RNA. The target RNA is then either inaccessible for nuclear processing, rapidly degraded by RNases or blocked for translation (Van der Krol et al 1988a and b). To evaluate the applicability of the antisense technology in manipulation of valuable plant traits *Petunia* and its flavonoid genes are a useful model systen since the effects of antisense flavonoid genes can be readily scored visually.

Introduction in *Petunia* of a full length antisense CHS cDNA, driven by the CaMV 35S promoter, resulted in flower tissue in which CHS mRNA and protein levels are reduced to a variable extent, resulting in the following phenotypes: (i) completely coloured flowers (no visible effect) or (ii) flowers with reduced pigmentation, or (iii) flowers with pigmentation patterns (white rings or sectors) or (iv) completely white flowers (Van der Krol et al 1988). The different pigmentation patterns in flowers of independent transformants are ascribed to a qualitative position effect, revealing that the site of integration can confer ectopic patterns of expression within one tissue. Some transformants show a highly variable degree of inhibition between different flowers on the same plant, which could be related to variable light conditions and internal hormone concentrations during flower development. In addition to an effect on CHS mRNA and protein level, the expression of the antisense CHS gene causes a reduction of CHI enzymic activity by a yet unknown mechanism, but not of CHI mRNA levels.

Taken together the data clearly show that flavonoid biosynthesis can be effectively blocked by the introduction of antisense genes. At present we are analyzing whether synthesis of flavonoids in other organs (ovary, seeds) than the flower corolla is also reduced and whether this affects characteristics like fertility, seed germination and UV-light sensitivity.

REFERENCES

Firmin JL, Wilson KE Rossen L, Johnston AWB (1986) Flavonoid activation of nodulation genes in Rhizobium reversed by other compounds in plants. Nature 324: 90-92

Green PJ, Pines O and Inouye M (2986 The role of antisense in gene regulation - a review. Ann.Rev.Biochem. 55: 569-597.

Harborne JB and Mabry TJ (eds) (1982) The Flavonoids. Alan R Liss, New York

Inouye M (1988) Antisense RNA: its function and applications in gene regulation - a review. Gene 72: 25-34.

Jacobs M, Rubery PH (1988) Naturally occurring auxin transport regulators. Science 241: 346-349

Koes RE, Spelt CE, Mol JNM (1989a) The chalcone synthase multigene family of Petunia hybrids (V30): differential, light-regulated expression during flower development and UV light induction. Plant Mol Biol 12: 213-225

Koes RE, Spelt CE, van den Elzen PE, Mol JNM (1989b) Cloning and Molecular characterization of the chalcone synthase multigene family of Petunia hybrida. Gene in press

Lamb CJ, Lawton MA, Dron M, Dixon RA (1989) Signals and transduction mechanisms for activation of plant defences against microbial attack. Cell 56: 215-224

Mol JNM, Stuitje AR, Gerats AGM, Koes RE (1988) Cloned genes of plant phenyl propanoid metabolism. Plant Mol Biol Rep 6: 274-279

Schmelzer E, Jahnen W, Hahlbrock K (1988) In situ localization of light-induced chalcone synthase mRNA, chalcone synthase and flavonoid end products in epidermal cells of parsley leaves. Proc Natl Acad Sci USA 85: 2989-2993

Van der Krol AR, Lenting PE, Veenstra K, van der Meer IM, Koes RE, Gerats AGM, Mol JNM, Stuitje AR (1988) An anti-sense chalcone synthase gene in transgenic plants inhibits flower pigmentation. Nature 333: 866-869

Van Tunen AJ, Koes RE, Spelt CE, van der Krol AR, Stuitje AR, Mol JNM (1988) Cloning of the two chalcone flavanone isomerase genes from *Petunia hybrida*: coordinate light regulated and differential expression of flavonoid genes.

Van Tunen AJ, Hartman SA, Mur LA, Mol JNM. (1989) Regulation of chalcone flavanone isomerase (CHI) gene expression in *Petunia hybrida*: the use of alternative in corolla, anthers and pollen Plant Mol Biol 12 (in press).

Van der Krol AR, Mol JNM, Stuitje AR (1988a) Modulation of Eukaryotic gene expression by complementary RNA or DNA sequences. BioTechniques 6: 958-976

Van der Krol AR, Mol JNM and Stuitje AR (1988b) Antisense genes in plants; an overview. Gene 72: 45-50

Zaat SAJ, Wijffelman CA, Spaink HP, van Brussel AAN, Okker RJH, Lugtenberg BJJ (1987) Induction of the nodA promoter of Rhizobium leguminosarum Sym plasmid pRL1JI by plant flavanones and flavones. J Bacteriol 169: 198-204

SIGNALLING ELEMENTS IN HIGHER PLANTS: IDENTIFICATION AND MOLECULAR ANALYSIS OF AN AUXIN-BINDING PROTEIN, GTP-BINDING REGULATORY PROTEINS AND CALCIUM SENSITIVE PROTEINS

Klaus Palme, Thomas Diefenthal, Thomas Hesse, Klaus Nitschke, Narciso Campos, Joachim Feldwisch, Christine Garbers, Friederike Hesse, Sybil Schwonke, and Jeff Schell

Max-Planck-Institut für Züchtungsforschung
Abteilung Genetische Grundlagen der Pflanzenzüchtung
Carl-von-Linné-Weg 10
D-5000 Köln 30
F.R.G.

INTRODUCTION

In eucaryotic cells external signals detected by receptors are translated into a limited repertoire of intracellular second messengers. Occupancy of these receptors initiates the production of active messengers, including the well studied cyclic adenosine monophosphate (cAMP) as well as the recently discovered messenger molecules that are derived from phosphoinositides such as arachidonic acid, inositol-1,4,5-triphosphate, and 1,2-diacylglycerol (for review see: Berridge 1986; Newton and Brown 1986; Boss and Morre 1989). These messengers are capable of regulating a vast array of physiological and biochemical processes either by direct interaction with distinct proteins or indirectly by activating enzymes which trigger conformational changes in the final target proteins. However, the number of second messengers in eucaryotic cells appears to be surprisingly small, indicating that most probably only a limited number of internal signal pathways are needed, albeit remarkably universally in all eucaryotes analyzed up to now, to transduce these signals to their final biological destination. From analysis of animal growth factors it has become clear that specific agonists are not necessarily limited to controlling only a single physiological activity, but rather form part of a complex cellular signalling language, in which the individual agonists and second messengers appear to be the

NATO ASI Series, Vol. H 36
Signal Molecules in Plants and Plant-Microbe Interactions
Edited by B. J. J. Lugtenberg

equivalent of characters of a code or an alphabet. Thus, the information content may not reside only in the individual agonists or growth factors, but in the pattern or set of regulatory molecules to which a cell is exposed. Therefore it is conceivable that combinations selected from a number of signalling molecules can increase the amount of information which can be transmitted, and furthermore, the redundancy of individual or sets of signalling molecules may promote error-free transmission of information in the presence of noise (Shannon and Weaver 1949; Ashby 1956).

In comparison to animal cells, plants have many unique developmental processes including the absence of a separate germ line and the ability to reprogram in the postembryonic plant the morphological patterns for the various organs (Steeves and Sussex 1988). Moreover, the totipotency of many plant cells indicates that irreversible determination may not play a major role in plant development. This leads to the interesting question whether the intracellular recognition and signalling systems controlling these processes share any evolutionary relationship with other eucaryotes such as yeast and higher animals. Thus it will be of fundamental interest to determine if plants have evolved communication systems which are either unique or share, at least with respect to domains recognizing these signals, strong evolutionary relationships with those of all other eucaryotes.

Perhaps the most difficult problem remaining in plant molecular biology is the identification and molecular analysis of mechanisms responsible for the integration of the signals that are expected to be transmitted from specific cell surface receptors to the interior of the plant cell in response to the limited number of plant hormones. How does a plant cell read sense from the auxins, cytokinins, abscisic acid, gibberellins and ethylene? The solution to this question will be of major importance in the coming decade. Although plant physiologists have been inspired by animal systems to define the action of plant specific signalling molecules ultimately in the context of hypothetical receptor proteins, unfortunately we know only very little about the mechanism of action of plant hormones at both the molecular and cellular level. In order to understand the biochemical and molecular events of transduction of such signals, we have started to isolate and characterize various genes which may be involved in such processes. As a basis for study of auxin functions and molecular mechanisms of auxin action we have isolated genes encoding auxin-binding proteins, GTP-binding regulatory proteins, and several calcium sensitive elements including calmodulin and calcium regulated protein kinases.

STRUCTURAL ANALYSIS OF GENES ENCODING AUXIN-BINDING PROTEINS

The most widely studied group of plant hormones are the auxins which elicit a wide variety of responses, including cell enlargement, root differentiation, xylem formation, and control of apical dominance (for review see Davies 1987). Rapid responses to auxins have been observed in organelles such as the cell wall, the plasma membrane, the Golgi apparatus, the tonoplast and the endoplasmic reticulum (Rubery 1981; Theologis 1986; Key et al 1986; Barbier-Brygoo et al 1989). Although many studies have indicated the presence of auxin-binding sites not only in membranes from maize coleoptiles, but also in membrane preparations from various other tissues and species (Hertel et al 1972; Venis 1985; Libbenga et al 1986), it is not clear up to now whether these hormone binding sites represent functional receptors. This is mostly due to the failure to identify primary functional responses which are coupled to such hormone binding sites. As auxins rapidly affect the transmembrane potential of many plant cells (Marre et al 1974; Nelles 1977; Bates and Goldsmith 1983; Ephritikine et al 1987), Barbier-Brygoo et al (1989) used the measurement of auxin induced transmembrane potential difference (Em) of evacuolated tobacco protoplasts as a functional assay for this hormone. Interestingly this electrical response could be suppressed by antibodies raised against both a yeast plasmalemma ATPase and an auxin-binding protein containing fraction from maize microsomes (Barbier-Brygoo et al 1989). This may indicate that a functionally active protein related to the site I specific auxin-binding protein from maize (Hertel et al 1972) is present in tobacco protoplasts as well.

In order to identify and characterize auxin-binding proteins from maize we have purified the major auxin-binding protein from maize coleoptiles to homogeneity (Hesse et al 1989). The protein has an apparent molecular weight of 22 kDa and binds 1-naphthylacetic acid with a K_D of 2.40 x 10^{-7} M. The stoichiometry of the 1-naphthylacetic acid binding was determined from Scatchard plots of its binding curve. The isolated protein bound 895 pmol 1-naphthylacetic acid per nmol protein at pH 5.5, indicating the presence of one hormone binding site per protein. The binding of 1-naphthylacetic acid was competitively inhibited by other auxins such as 3-indoleacetic acid. Hydrophobic compounds and nonfunctional auxin analogs such as benzoic acid and 2-naphthylacetic acid were not able to displace bound 1-naphthylacetic acid efficiently. Hormone binding analysis of this homogeneous auxin-binding protein, as determined by various chromatographic and protein sequence analysis techniques, revealed pK_D values which were in agreement with those obtained for the site I specific auxin-binding protein (Ray et al 1977; Dohrmann et al 1978; Hesse et al 1989; Palme et al 1989). Additional antigenically related proteins, present in very low amounts and differing in charge and size from the major auxin-binding protein, could be detected using antisera raised against the auxin-binding protein 1 (Hesse et al 1989). Extensive protein sequence analysis of the auxin-binding protein 1 allowed the construction of several synthetic oligonucleotide probes which were used to isolate a cDNA coding for this protein. The cDNA corresponds to a mRNA with a 3'-poly(A)+ sequence and a single, long open reading frame of 603 bases. The open reading frame, starting 34 residues from the 5'-end of the cDNA, predicts a 21,990 Da protein of 201 amino acids. Comparison of the deduced amino acid sequence with the partial amino acid sequences of the purified auxin-binding protein, revealed a perfect match over 53 amino acid residues. The primary amino acid sequence includes a 38 amino acids long N-terminal hydrophobic leader sequence which could represent a signal for translocation of this protein to the lumen of the endoplasmic reticulum.

An additional signal is located at the C-terminal end consisting of the amino acids Lys-Asp-Glu-Leu (KDEL) known to be responsible for preventing secretion of proteins from the lumen of the endoplasmic reticulum in eucaryotic cells (Munro and Pelham 1987; Pelham 1988; Pelham et al 1988). The signal K/HDEL is well known in organisms like yeast, Xenopus, and mouse. However, this is the first detection of such a sequence motif in plants. We expect that this sequence performs a similar function in plants as it does in other eucaryotes. This is of particular interest as the auxin-binding protein isolated is supposed to be associated with membrane fractions comigrating with marker enzymes for the endoplasmic reticulum on density gradient centrifugation (Ray 1977). As deduced from the amino acid sequence of this protein and Chou-Fasman plots of its hydrophobicity properties we were not able to identify hydrophobic domains which might be responsible for the stable insertion of this protein in membranes. Furthermore, the only hydrophobic sequence located to the N-terminus, representing an eucaryotic leader sequence, is removed in vivo and in vitro during translocation in the endoplasmic reticulum. Consistent with the proposed location of this auxin-binding protein to the endoplasmic reticulum is its high mannose type glucan composition. The high mannose type N-glycanes identified for auxin-binding protein 1 are consistent with the observation of the KDEL signal found at the C-terminal end, which prevents the entry to the Golgi-membranes. Structural analysis of this protein indicates that this particular protein corresponds to the site I specific auxin-binding protein, first identified by Hertel et al (1972), which was located in the endoplasmic reticulum and displayed auxin-binding characteristics similar to the protein isolated.

Other genes encoding related auxin-binding proteins have been isolated. For example, the axr^4 gene in comparison to the axr^1 gene has an amino acid sequence similarity of 75% and its amino-terminal signal sequence is 41 amino acids long and different with respect to both charge and amino acid composition. Only the amino acids around the cleavage site for the signal peptidase are strictly conserved. Detailed **in situ** expression analysis of these genes using gene specific probes will provide us in future with information on cell type specific alterations of expression during development of a maize plant.

STRUCTURAL ANALYSIS OF GENES ENCODING RAS-RELATED GTP-BINDING REGULATORY PROTEINS

The super-family of GTP-binding proteins is still growing, and consists of a collection of proteins which transduce hormonal and sensory signals either across the plasma membrane or inside the cell. As the distribution of the GTP-binding proteins has been extended from the plasma membrane to the cell interior it has become clear that effects of important regulatory stimuli on the control of many physiological processes can be achieved by using these switch-like proteins. Functionally, these GTP-binding proteins bind and hydrolyse GTP, which serve to control the switching between the two different protein conformations. GTP-binding proteins are generally turned on in the GTP-bound form and turned off by hydrolysis, but not dissociation of GTP. In the GTP-binding proteins the switch serves to propagate and amplify regulatory signals.

Despite our extensive knowledge of the structure and function of GTP-binding proteins in many eucaryotes, their presence and functional relevance in plant cells have been indicated by only a few, indirect studies (Hasunuma and Funadera 1987; Millner 1987). The ras genes especially are likely to play a fundamental role in basic cellular functions based on their high degree of conservation throughout evolution (Chardin 1988). Independent of their phylogenetic origin, they encode membrane associated proteins that bind and hydrolyze GTP. They have been implicated in the control of important signal transduction pathways involved in cellular proliferation, differentiation and secretory processes.

Our strategy for isolation of ras related genes from plants involved the use of synthetic probes to isolate cDNA clones, and sequencing of the region hybridising to these probes to verify the presence of the conserved sequence motifs in an open reading frame. Oligonucleotides were chosen from sequence comparisons of all recently isolated members of the ras super-gene family by multiple alignments of the amino acids of the GTP-binding domain. Comparison of amino acid sequences representing the GTP-binding domain of all presently known members of the ras super-gene family points to three consensus elements, GxxxxGKSsxl, DTAGQE, and lxgNKxDL. As indicated by the three dimensional structure of the c-H-ras-p21 oncogene (De Vos et al 1987), the first two domains are involved in binding the phosphate moiety of the GTP-molecule, whereas elements located further downstream of the sequence are involved in determining guanine nucleotide specificity. More extensive amino acid sequence conservation is detectable only in subfamilies. To isolate plant related members of the ras super-gene family, we used mixtures of oligonucleotides to screen lambda gt10 cDNA libraries under optimum stringency conditions. We were able to isolate a series of cDNAs revealing 25-35% similarity to the H-ras genes and 40-78% to the YPT gene family. The percentage of amino acids strictly conserved in the YPTm (Zea mays) family isolated from Zea mays is mainly restricted to the GTP-binding domain where four regions corresponding to residues 10-21, 58-69, 118-125, and 146-150 are almost identical. In addition, amino acids 70-80 are highly conserved within the YPT subfamily, including some of the rab proteins, clearly distinguishing this group from the ral and rho genes (Palme et al 1989b). As in all other ras-proteins two cysteine residues are found near the carboxyl-terminus of the maize proteins. Stable association with membranes seems to be the result of a postranslational modification of one of these cysteines in which palmitic acid is bound via a thioester linkage. It is interesting to note that the C-termini of all YPT-related proteins currently isolated from plants consist of hydroxylated amino acids (Ser, Thr). These amino acids most probably determine substrate specificity of plant fatty acid acylases. This indicates that the plant YPT genes represent a new and separate branch of the ras super-gene family.

STRUCTURAL ANALYSIS OF GENES ENCODING CALCIUM SENSITIVE PROTEINS

Calcium has long been recognized as an important component in regulating cell growth. Calcium is necessary for sustaining cell viability and progression through the cell cycle which is particularly sensitive to variations in the intracellular calcium concentration (for review see: Cheung 1983; Hepler and Wayne 1985; Poovaiah and Reddy 1987).

Recent studies have shown that a variety of developmental processes in plants can be influenced by changes in the intracellular calcium concentrations (Hepler and Wayne 1985). Extracellular signals such as red light, gravity and hormones such as auxins apparently modify transiently the calcium concentrations by cleavage of L-α-phosphatidylinositol-4,5-diphosphate and production of inositol 1,4,5-triphosphate (De Silva et al 1985; Boss and Massel 1986; Ettlinger and Lehle 1987; Roux and Serlin 1987).

Many of the effects of calcium ions in both plant and animal cells are transduced by a large family of calcium-binding regulatory proteins (Carafoli 1987). The expression of members of this gene family is mostly restricted to a small number of tissues or cells in a given organism. Calmodulin is the most well known member of the calcium-binding regulatory proteins (Roberts et al 1986). Calmodulin is a highly conserved, heat stable, acidic protein with a molecular mass of 18 kDa. This protein has been highly conserved during evolution and represents the major calcium receptor of the cells. The single 148 amino acid long polypeptide chain contains four calcium-binding domains, which undergo conformational changes after binding of the calcium allowing calmodulin to interact with specific enzymes peptides and hydrophobic agents like trifluorophenazin. Specifically in plants, enzymes such as NAD kinase, ATPases, and protein kinases are known to be regulated by calcium and calmodulin (Roberts et al 1986). Although the calmodulin protein has been isolated from various plants its gene structure and developmental expression is not yet known.

To understand and analyze its developmental regulation we have isolated a cDNA for calmodulin from Nicotiana tabacum. The open reading frame of the cDNA isolated encodes a protein of 148 amino acids which shows a high degree of homology to other eucaryotic proteins such as calmodulin from bovine brain (91.4%). Protein sequence similarity is less well conserved with respect to the calmodulins from lower eucaryotes (amino acid sequence similarity to calmodulin from Saccharomyces cerevisiae is 61%). Northern blot analysis demonstrated that the Nicotiana tabacum cDNA hybridized to several mRNAs, 0.7, 1.9, and 4.9 kb in length. Differential expression was observed in both Nicotiana tabacum and maize tissues with a specifically strong expression in maize root tissue. Using the isolated cDNA, the tissue and cell type specific expression of this gene with respect to agonists such as auxins can now be analyzed in detail.

CONCLUSION

Throughout the life cycle a plant has to respond to environmental, developmental and metabolic signals. Although signals such as hormones, light, and gravity are known to influence these diverse processes, it is not known how the plant cell detects and converts these signals into the given biological response. However, detailed biochemical, molecular, and genetic studies have established the role of growth factors, hormones, and ions in the control of various signal transduction pathways in animal cells. With both the concepts and molecular techniques that have been developed with animal hormone systems it is now possible to study the possible relationships between the plant signalling systems and those of other eucaryotes. Therefore it will be of fundamental importance to determine if plants have evolved hormone coupled recognition-communication systems which share evolutionary relationships with animals. This will include the isolation and molecular analysis of plant hormone-binding proteins, which may represent candidates for putative receptors, and the identification of second messengers and proteins which bring about the medley of changes necessary for plant growth and development.

Here we have summarized some of our studies to analyze such signal transduction elements on the molecular level, including the characterization of GTP-binding regulatory proteins, calcium sensitive elements, and auxin-binding proteins. The striking homology between protein domains involved in the recognition of signals such as GTP or ions indicates strongly that the overall principles of plant signal transduction may be similar to those known from other eucaryotic systems. The detailed analysis of proteins involved in signal transduction processes will lead us to understand the various developmental strategies of plants and how they differ specifically in their coupling to the hormone responses. As a basis for the analysis of the mechanism of auxin action we have begun to identify systematically proteins and genes encoding auxin-binding proteins.

As auxin action is supposed to be mediated by specific receptors (Rubery 1981) the analysis of such genes may help to identify possible candidates for such receptors. The structure of the site I specific auxin-binding protein has been determined, indicating that it is located to the lumen of the endoplasmic reticulum (Hesse et al 1989; Palme et al 1989; Tillmann et al 1989). There are indications that this protein may be a receptor protein (Walton and Ray 1981), but conclusive interpretation of these results has to await a critical analysis using molecular strategies including the construction of transgenic plants. We can expect that the precise functional analysis of genes encoding various essential proteins involved in the control of signal transduction pathways will facilitate the fundamental conceptual framework for plant signalling in the next few years.

REFERENCES

Ashby WR (1956) An Introduction to Cybernetics. Wiley, New York.

Barbier-Brygoo H, Ephritikine G, Klämbt D, Ghishlain M and Guern J (1989) Functional evidence for an auxin receptor at the plasmalemma of tobacco mesophyll protoplasts. Proc Natl Acad Sci USA 86: 891-895

Bates GW and Goldsmith MHM (1983) Rapid response of the plasma membrane potential in oat coleoptiles to auxin and other weak acids. Planta 159. 231-237

Berridge MJ (1987) Inositol triphosphate and diacylglycerol: Two interacting second messengers. Ann Rev Biochem 56: 159-193

Boss WF, Massel MO (1985) Polyphosphoinositides are present in plant tissue cells. Biochem Biophys Res Commun 132: 1018--1023

Boss WF and Morre DJ (1989) Second messengers in plant growth and development. Alan R Liss Inc, New York

Carafoli E (1987) Intracellular calcium homeostasis. Ann Rev Biochem. 56: 395-433

Chardin P (1988) The ras superfamily protein. Biochimie 70: 865-868

Cheung WY (1979) Calmodulin plays a pivotalrole in cellular regulation. Science 207: 19-27

Davies PJ (1987) Plant hormones and their role in plant growth and development. Martinus Nijhoff Publishers, Kluwer Academic Publishers Group, Dordrecht

De Silva DLR, Cox RC, Hetherington AM and Mansfield TA (1985) Suggested involvement of calcium and calmodulin in the responses of stomata to abscisic acid. New Phytol 101: 555-563

De Vos AM, Tong L, Milburn MV, Matias PD, Jancarik J, Noguchi S, Nishimura S, Miura K, Ohtsuka E, Kim KH (1988) Three--dimensional structure of an oncogene protein: catalytic domain of human c-Ha-ras p21. Science 239: 888-893

Dohrmann U, Hertel R and Kowalik H (1978) Properties of auxin binding sites in different subcellular fractions from maize coleoptiles. Planta (Berl) 140: 97-106

Ephritikine G, Barbier-Brygoo H, Muller JF and Guern J (1987) Auxin effect on the transmembrane potential difference of wild type and mutant tobacco protoplasts exhibititng a differential sensitivity to auxin. Plant Physiol 83: 801-804

Ettlinger C and Lehle L (1987) Auxin induces rapid changes in phosphatidylinositol metabolites. Nature (London) 331: 176-178

Hasunuma K and Funadera K (1987) GTP-binding proteins in green plant Lemna paucicostata. Biochem Biophys Res Comm 143: 908-912

Hepler PK and Wayne RO (1985) Calcium and plant development. Ann Rev Plant Physiol 36: 397-439

Hertel R, Thomson KSt and Russo VEA (1972) In vitro auxin binding to particulate cell fractions from corn coleopti les. Planta 107: 325-340

Hesse T, Feldwisch J, Balshüsemann D, Bauw G, Vandekerckhove J, Puype M, Löbler M, Klämbt D, Schell J and Palme K (1989) Molecular cloning and structural analysis of a gene from Zea mays (L.) coding for a putative receptor for the plant hormone auxin. EMBO J, in press

Key JL, Kroner P, Walker J, Hong JC, Ulrich TH, Ainley WM, Gantt JS and Nagao RT (1986) Auxin-regulated gene expression. Philos Trans R Soc Lond B Biol Sci 314: 427-440

Libbenga KR, Maan AC, Van der Linde PCG and Mennes AM (1986)

Auxin Receptors. In Hormones, receptors and cellular interactions in plants (Chadwick CM and Garrod DR, eds) pp 1-67 Cambridge University Press, Cambridge.
Marre E, Lado P, Ferroni A and Ballarin-Denti A (1974) Transmembrane potential increase induced by benzyladenine and fusicoccin. Correlation with proton extrusion and cell enlargement. Plant Sci Lett 2: 257-265
Millner PA (1987) Are guanine nucleotide-binding proteins involved in regulation of thylakoid protein kinase activity? FEBS Lett 226: 155-160
Munro S and Pelham HRB (1987) A C-terminal signal prevents secretion of luminal ER proteins. Cell 48: 899-907
Nelles A (1977) Short term effects of plant hormones on membrane permeability of dwarf coleoptile cells (Zea mays L.) in comparison with growth responses. Planta 137: 293-298
Newton RP and Brown EG (1986) The biochemistry and physiology of cyclic AMP in higher plants. In Hormones, receptors and cellular interactions in higher plants (Chadwick CM, Garrod DR, eds) pp 115-153. Cambridge University Press
Palme K, Diefenthal T, Hesse T, Feldwisch J and Schell J (1989) Genes involved in plant differentiation: identification of gene families from Zea mays encoding auxin--binding proteins and the ras-related YPT-proteins. In Plant Gene Transfer (Beachy R and Lamb C, eds) Alan R Liss Inc, New York, in press
Palme K, Diefenthal T, Sander C, Vingron M and Schell J (1989b) Identification of guanine-nucleotide binding proteins in plants: structural analysis and evolutionary comparison of the ras-related YPT-gene family from Zea mays. In The guanine-nucleotide binding proteins: common structural and evolutionary principles (Bosch L, Kraal B, Parmeggiani A, eds) Plenum Press, New York, in press
Pelham HRB (1988) Evidence that luminal ER proteins are sorted from secreted proteins in a post-ER compartment. EMBO J. 7: 913-918
Pelham HRB, Hardwick KG, Lewis MJ (1988) Sorting of soluble ER proteins in yeast. EMBO J 7: 1757-1762
Poovaiah BW and Reddy ASN (1987) Calcium messenger system in plants. CRC Critical Reviews in Plant Sciences 6: 47-103
Ray PM (1977) Auxin-binding sites of maize coleoptiles are localized on membranes of the endoplasmic reticulum. Plant Physiol 59: 594-599
Ray PM and Dohrmann U (1977) Characterization of naphthaleneacetic acid binding to receptor sites on cellular membranes of maize coleoptile tissue. Plant Physiol 59: 357-364
Roberts DM, Lukas TJ and Watterson DM (1986) Structure, function, and mechanism of action of calmodulin. CRC Critical Reviews in Plant Sciences 4: 311-339
Roux SJ and Serlin BS (1987) Cellular mechanisms controlling light-stimulated gravitropism: role of calcium. CRC Critical Reviews in Plant Sciences 5: 205-237
Rubery PH (1981) Auxin receptors. Ann Rev Plant Physiol 32: 569-596
Shannon CE and Weaver W (1949) The Mathematical Theory of Communication. University of Illinois Press, Urbana.
Steeves TA and Sussex I (1988) Patterns of Plant Development.

2nd Edition. Cambridge University Press, New York
Theologis A (1986) Rapid gene regulation by auxin. Ann Rev Plant Physiol 37: 407-438
Tillmann U, Viola G, Kayser B, Hesse T, Palme K, Löbler M and Klämbt D (1989) cDNA clones of the auxin-binding protein from corn coleoptiles (Zea mays L.): isolation and characterization by immunological methods. EMBO J, in press
Venis MA (1985) Hormone binding sites in plants. Longman Ltd, New York, London
Walton DJ and Ray PM (1981) Evidence for receptor functions of auxin binding sites in Maize. Red light inhibition of mesocotyl elongation and auxin binding. Plant Physiol 68: 1334-1338

Release of Phytoalexin Elicitor-Active Oligogalacturonides by Microbial Pectic Enzymes

F. Cervone [1], G. De Lorenzo [1], G. Salvi [1], C. Bergmann, M.G. Hahn, Y. Ito, A. Darvill and P. Albersheim

Complex Carbohydrate Research Center, The University of Georgia, 220 Riverbend Road, Athens, Georgia 30602, USA

INTRODUCTION

Linear α-1,4-linked oligogalacturonides of chain lengths between 10 and 13 have been shown to elicit phytoalexin accumulation and lignification in plants, but shorter oligogalacturonides do not have this effect (Hahn *et al* 1981; Nothnagel *et al* 1983; Jin & West 1984; Davis *et al* 1986; Robertsen 1986; De Lorenzo *et al* 1987b; Cervone *et al* 1989). Phytoalexin elicitor-active oligogalacturonides are thought to be released from the plant cell wall by the action of two groups of microbial pectic-degrading enzymes, the endopolygalacturonases and the endopectate lyases. These enzymes release elicitor-active oligogalacturonides from polygalacturonic acid *in vitro* but also quickly depolymerize the elicitor-active chains, converting them into shorter, elicitor-inactive molecules (Bruce & West 1982; Nothnagel *et al* 1983; Cervone *et al* 1987). We will discuss two mechanisms by which the action of fungal endopolygalacturonases and bacterial endopectate lyases may be regulated such that the formation of elicitor-active oligogalacturonides is favored.

REGULATION OF FUNGAL ENDOPOLYGALACTURONASE ACTIVITY BY A PLANT-DERIVED POLYGALACTURONASE-INHIBITING PROTEIN

An endopolygalacturonase-inhibiting protein (PGIP) has been purified and shown to be localized in the apoplast of *Phaseolus vulgaris* L. Functionally identical inhibitor proteins have been extracted from each of a wide variety of dicots that have been examined (Hahn *et al* 1989). The action of the *P. vulgaris* PGIP on endopolygalacturonases purified from *Aspergillus niger* (Cervone *et al* 1987) and *Fusarium moniliforme* (De Lorenzo *et al* 1987a) was

[1] Permanent Address: Dipartimento di Biologia Vegetale, Università di Roma, "La Sapienza," Roma, Italy

NATO ASI Series, Vol. H 36
Signal Molecules in Plants and Plant-Microbe Interactions
Edited by B. J. J. Lugtenberg

investigated *in vitro*. The effect of PGIP on endopolygalacturonases was to decrease the rate of depolymerization of polygalacturonic acid by the endopolygalacturonase (Fig. 1A). An additional effect of PGIP on endopolygalacturonase was to prolong the lifetime of oligogalacturonides that were able to elicit phytoalexin synthesis in soybean cotyledons. In the absence of PGIP, a one-minute exposure of polygalacturonic acid to endopolygalacturonase resulted in the production of elicitor-active oligogalacturonides. However, the enzyme depolymerized essentially all of the polygalacturonic acid substrate to elicitor-inactive oligogalacturonides within 15 minutes. When the digestion of polygalacturonic acid was carried out with the same amount of enzyme in the presence of excess PGIP, the rate of production of elicitor-active oligogalacturonides was dramatically altered. The amount of elicitor-active oligogalacturonides steadily increased for 24 h. It was only after about 48 h that the enzyme converted the polygalacturonic acid into short, elicitor-inactive oligomers (Fig. 2A).

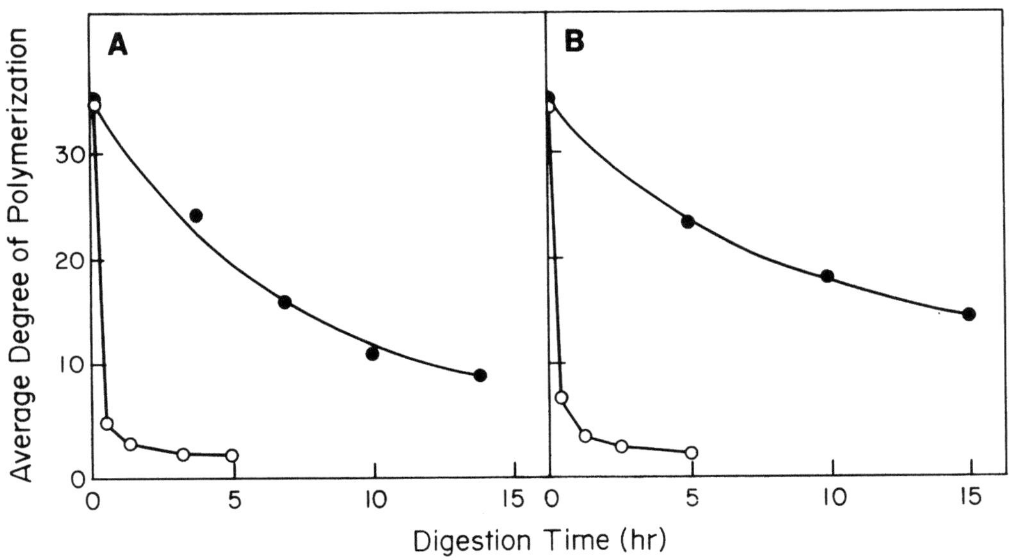

Figure 1. Time course of digestion of polygalacturonic acid by *Aspergillus niger* endopolygalacturonase (**A**) or *Erwinia carotovora* endopectate lyase (**B**). Digestions with endopolygalacturonase (**A**) were carried out either in the presence (●) or absence (○) of *Phaseolus vulgaris* PGIP as described (Cervone *et al.*, 1989). Digestions with endopectate lyase (**B**) were carried out either at pH 5.75 (●) or pH 8.5 (○).

The effect of PGIP isolated from different cultivars of *Phaseolus vulgaris* on endopolygalacturonases purified from three races of *Colletotrichum lindemuthianum* was also investigated. No difference in the ability of PGIPs from four cultivars of *P. vulgaris* to inhibit the endopolygalacturonases from three races of *C. lindemuthianum* was detected. Also, no appreciable differences in the rate or pattern of degradation of polygalacturonic acid were found using the various endopolygalacturonase-PGIP combinations, confirming that the interaction between the two proteins is unlikely to be a determining factor in race-cultivar specificity (Anderson & Albersheim 1972).

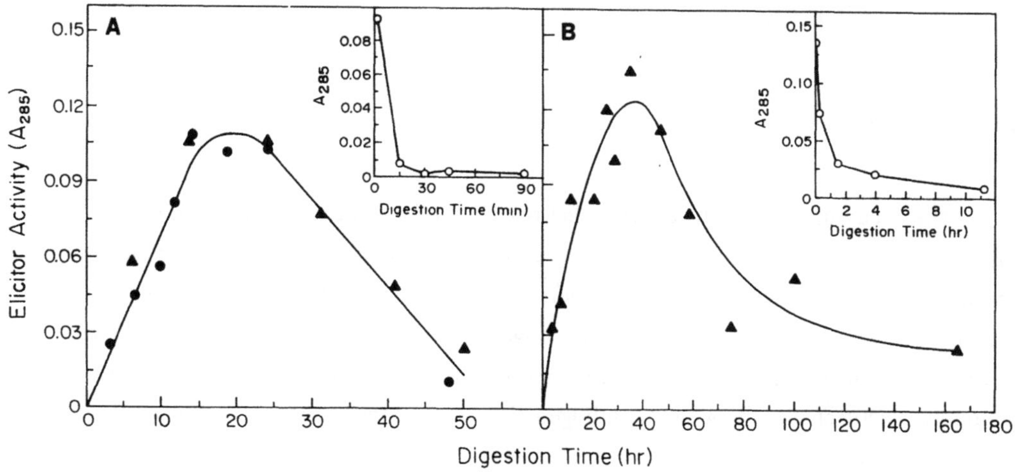

Figure 2. Elicitor activity of oligogalacturonides produced by treatment of polygalacturonic acid with either *Aspergillus niger* endopolygalacturonase (**A**) or *Erwinia carotovora* endopectate lyase (**B**). Digestions with endopolygalacturonase (**A**) were carried out either in the absence (inset) or the presence of a 10-fold molar excess of *Phaseolus vulgaris* PGIP as described (Cervone *et al* 1989). Digestions with endopectate lyase (**B**) were carried out either at pH 8.5 (inset) or pH 5.75. Elicitor activity of the digestion products was determined using the soybean cotyledon bioassay (Hahn *et al* 1981).

REGULATION BY PLANT CELL WALL pH OF THE ACTIVITY OF A BACTERIAL ENDOPECTATE LYASE

Endopectate lyase was purified from *Erwinia carotovora* (Davis *et al* 1984), and its ability to convert polygalacturonic acid into elicitor-active oligogalacturonides was studied at various pH values. The enzyme depolymerized polygalacturonic acid *in vitro* at the maximum rate when the pH was between 8 and 9; the lyase had very little activity between pH 5 and 6. Digestion products obtained from polygalacturonic acid treated with endopectate lyase at pH 5.75, the pH of soybean cotyledon exudates, were tested for their phytoalexin-eliciting activity in the soybean cotyledon bioassay. The half-life of the oligogalacturonides with a DP between 10 and 20 was longer than 24 h when the digestion of polygalacturonic acid was carried out at pH 5.75 (Fig. 1B). The elicitor activity of the digestion products produced at pH 5.75 increased during the first 40 h of incubation and decreased thereafter as degradation to shorter elicitor-inactive oligomers progressed. On the other hand, digestion of polygalacturonic acid with the same amount of the lyase at pH 8.5 resulted, in less than 20 minutes, in depolymerization to elicitor-inactive oligogalacturonides, that is, oligogalacturonides containing less than five galactosyluronic acid residues (Fig. 2B).

CONCLUSIONS

The accumulation of phytoalexin elicitor-active oligogalacturonides by fungal endopolygalacturonases is greatly affected by the presence of PGIP. Endopolygalacturonase rapidly degrades polygalacturonides into oligomers too short to possess elicitor activity. In the presence of excess PGIP, the elicitor-active oligogalacturonides are maintained in the reaction mixture for many hours rather than a few minutes. In dicots, PGIP may function by converting endopolygalacturonase, an essential pathogenicity factor of the pathogen, into a trigger of plant defense responses.

The release of phytoalexin elicitor-active oligogalacturonides by bacterial endopectate lyase is greatly affected by the pH of plant cell walls. It is known that plant cell walls have a physiological pH of between 5 and 6. Endopectate lyases have optimum pH between 8 and 9. The lyases would not be expected to be very active at the physiological pH of the plant cell wall. However, the residual activity of endopectate lyases between pH 5 and 6 is sufficient to generate elicitor-active oligogalacturonides.

ACKNOWLEDGMENTS

This work was supported by grants from the National Science Foundation (#DMB8518488) and the Department of Energy (#DE FG09-85ER13425), and a NATO-CNR fellowship to G. De Lorenzo. The Complex Carbohydrate Research Center is supported in part by Department of Energy grant DE FG09-87ER13810 as a part of the USDA/DOE/NSF Plant Science Centers Program.

REFERENCES

Anderson AJ, Albersheim P (1972) Host-pathogen interactions. V. Comparison of the abilities of proteins isolated from three varieties of *Phaseolus vulgaris* to inhibit the endopolygalacturonases secreted by three races of *Colletotrichum lindemuthianum*. *Physiol Plant Pathol* 2:339-346

Bruce RJ, West CA (1982) Elicitation of casbene synthetase activity in castor bean. The role of pectic fragments of the plant cell wall in elicitation by a fungal endopolygalacturonase. *Plant Physiol* 69:1181-1188

Cervone F, De Lorenzo G, Degrà L, Salvi G (1987) Elicitation of necrosis in *Vigna unguiculata* Walp. by homogeneous *Aspergillus niger* endo-polygalacturonase and by α-D-galacturonate oligomers. *Plant Physiol* 85:626-630

Cervone F, Hahn MG, De Lorenzo G, Darvill A, Albersheim P (1989) Host-pathogen interactions. XXXIII. A plant protein converts a fungal pathogenesis factor into an elicitor of plant defense responses. *Plant Physiol* 90:542-548

Davis KR, Lyon GD, Darvill AG, Albersheim P (1984) Host-pathogen interactions. XXV. Endopolygalacturonic acid lyase from *Erwinia carotovora* elicits phytoalexin accumulation by releasing plant cell wall fragments. *Plant Physiol* 74:52-60

Davis KR, Darvill AG, Albersheim P, Dell A (1986) Host-pathogen interactions. XXIX. Oligogalacturonides released from sodium polypectate by endopolygalacturonic acid lyase are elicitors of phytoalexins in soybean. *Plant Physiol* 80:568-577

De Lorenzo G, Salvi G, Degrà L, D'Ovidio R, Cervone F (1987a) Induction of extracellular polygalacturonase and its mRNA in the phytopathogenic fungus *Fusarium moniliforme*. *J Gen Microbiol* 133:3365-3373

De Lorenzo G, Ranucci A, Bellincampi D, Salvi G, Cervone F (1987b) Elicitation of phenylalanine ammonia-lyase in *Daucus carota* by oligogalacturonides released from sodium polypectate by homogenous polygalacturonase. *Plant Sci* 51:147-150

Hahn MG, Darvill AG, Albersheim P (1981) Host-pathogen interactions. XIX. The endogenous elicitor, a fragment of a plant cell wall polysaccharide that elicits phytoalexin accumulation in soybeans. *Plant Physiol* 68:1161-1169

Hahn MG, Bucheli P, Cervone F, Doares SH, O'Neill RA, Darvill A, Albersheim P (to be published) The roles of cell wall constituents in plant-pathogen interactions. In: Nester E, Kosuge T (eds) Plant-Microbe Interactions, Volume 3. McGraw Hill, New York

Jin DF, West CA (1984) Characteristics of galacturonic acid oligomers as elicitors of casbene synthetase activity in castor bean seedlings. *Plant Physiol* 74:989-992

Nothnagel EA, McNeil M, Albersheim P, Dell A (1983) Host-pathogen interactions. XXII. A galacturonic acid oligosaccharide from plant cell walls elicits phytoalexins. *Plant Physiol* 71:916-926

Robertsen B (1986) Elicitors of the production of lignin-like compounds in cucumber hypocotyls. *Physiol Mol Plant Pathol* 28:137-148

ELICITATION OF PHYTOALEXINS BY SYNTHETIC OLIGOGLUCOSIDES, SYNTHETIC OLIGOGALACTURONIDES, AND THEIR DERIVATIVES

M.G. Hahn [1,2], J.-J. Cheong [1,3], W. Birberg [4], P. Fügedi [4], Å. Piloti [4], P. Garegg [4], N. Hong [5], Y. Nakahara [5], and T. Ogawa [5]

[1] Complex Carbohydrate Research Center & Departments of Botany[2] and Biochemistry[3], University of Georgia, 220 Riverbend Road, Athens, GA 30602, USA.

[4] Arrhenius Laboratory, Department of Organic Chemistry, University of Stockholm, S-106 91 Stockholm, Sweden

[5] RIKEN, Institute of Physical and Chemical Research, Wako, Saitama, 351-01, Japan.

INTRODUCTION

Plants respond to invasive microorganisms in a variety of ways (Bell 1981). Among these, the synthesis and accumulation of toxic molecules called phytoalexins is a response observed in a number of plant-pathogen interactions (Dixon 1986; Ebel 1986). Phytoalexins accumulate rapidly at the site of infection, and reach levels inhibitory to microbial growth quickly enough to play a role in defense against disease (Hahn *et al* 1985). Several studies have demonstrated that phytoalexin accumulation is limited, in incompatible (plant resistant) responses, to the immediate vicinity of the infection (Mansfield *et al* 1974; Moesta *et al* 1982; Mayama & Tani 1982; Hahn *et al* 1985). These observations are important since phytoalexins are toxic to both plant and microbial cells (Smith 1982; Weinstein & Albersheim 1983).

Phytoalexin synthesis and accumulation is observed not only after microbial infection, but also upon treatment of plant tissues with elicitors. The term "elicitor," originally used to refer to molecules and other stimuli that induce phytoalexin synthesis in plant cells (Keen 1975), is now commonly used for molecules that stimulate any plant defense mechanism. A number of different cell wall constituents originating either from the host plant or from the invading microbe can induce phytoalexin accumulation and other plant defense responses (Hahn *et al* 1989). Crude elicitor fractions have been prepared from many different sources (West 1981; Darvill & Albersheim 1984). Elicitors that have been purified to apparent homogeneity include fragments of fungal (Sharp *et al* 1984c; Sharp *et al* 1984b) and plant (Nothnagel *et al* 1983) cell walls, and enzymes, secreted by fungi (Bruce & West 1982; Cervone *et al* 1987; Farmer & Helgeson 1987) and bacteria (Davis *et al* 1984), that release elicitors from plant cell walls.

NATO ASI Series, Vol. H 36
Signal Molecules in Plants and Plant-Microbe Interactions
Edited by B. J. J. Lugtenberg

The smallest elicitor-active fragment from fungal cell walls that has been purified to homogeneity and structurally characterized is a branched hexa-β-glucosyl glucitol (Sharp *et al* 1984c; Sharp *et al* 1984b). This hexaglucosyl glucitol was purified from a complex mixture of oligosaccharides released by partial acid hydrolysis from the mycelial walls of the soybean pathogen *Phytophthora megasperma* f. sp. *glycinea*. The elicitor activity of the hexa-β-glucosyl glucitol is identical to the activity of the corresponding unreduced hepta-β-glucoside (Sharp *et al* 1984a). The structure of the elicitor-active hepta-β-glucoside was confirmed by its chemical synthesis (Sharp *et al* 1984a; Ossowski *et al* 1984; Fügedi *et al* 1987). The hepta-β-glucoside elicitor is the most active elicitor of phytoalexin accumulation yet isolated and is active at concentrations between 10^{-8} and 10^{-9} molar.

Elicitor-active fragments of plant origin were first identified in the material released by partial acid hydrolysis of plant cell walls (Hahn *et al* 1981). It was later shown that the same, or very similar, oligomers could be released by treatment of plant cell walls with purified microbial pectic-degrading enzymes (Bruce & West 1982; Davis *et al* 1984; Davis *et al* 1986b). The elicitor-active plant cell wall fragments have been purified to apparent homogeneity and are linear α-(1→4)-linked oligomers of D-galacturonic acid (Nothnagel *et al* 1983; Jin & West 1984; Davis *et al* 1986b). Oligogalacturonides that elicit maximum phytoalexin accumulation range in length between 10 and 13 galactosyluronic acid residues and are active at micromolar concentrations (Hahn *et al* 1981; Nothnagel *et al* 1983; Jin & West 1984; Walker-Simmons *et al* 1983).

Plant cell wall-derived oligogalacturonides and the fungal cell wall-derived hepta-β-glucoside, in addition to being separately active as elicitors, act synergistically to induce phytoalexin accumulation when applied in combination to plant tissues (Davis *et al* 1986a; Davis & Hahlbrock 1987). That is, half maximum elicitor activity is obtained with a mixture of the two elicitors at concentrations that are too low for either elicitor alone to stimulate phytoalexin accumulation. The synergistic action of plant and fungal wall-derived elicitors may play a role in the plant's recognition of potential pathogens. The molecular basis for the synergism of these two elicitors remains to be determined.

The observation that only one of some 300 structurally similar hepta-β-glucosides was found to be active as an elicitor of phytoalexin accumulation in plants suggested that a specific receptor for this molecule exists in plants (Sharp *et al* 1984a). Our current efforts are directed at identifying and characterizing such a receptor. Initially, we are obtaining additional information about the structural features of the hepta-β-glucoside elicitor that are required for its activity by comparing the activities of a variety of structurally related synthetic oligoglucosides. We are also determining to what extent the structure of the hepta-β-glucoside elicitor can be modified, without altering its activity, in order to prepare radiolabeled elicitor for receptor binding studies.

RESULTS & DISCUSSION

We have synthesized several oligoglucosides, ranging in size from hexamer to decamer, that are structurally related to the hepta-β-glucoside elicitor (Fig. 1).

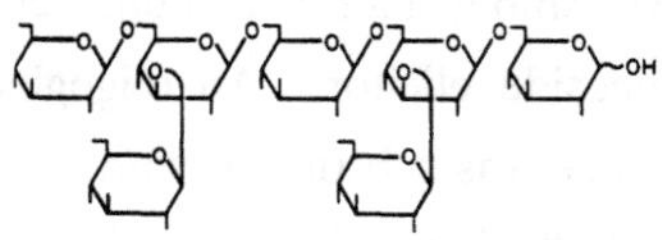

1

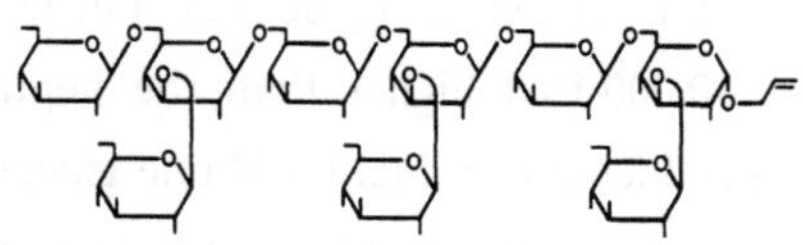

2

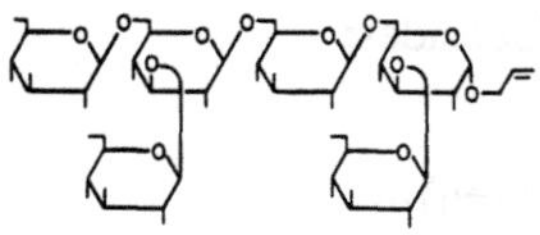

3

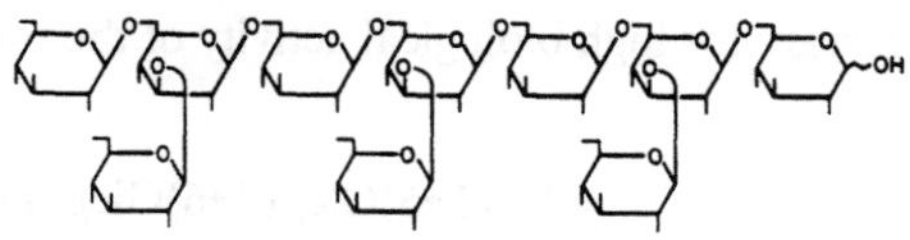

4

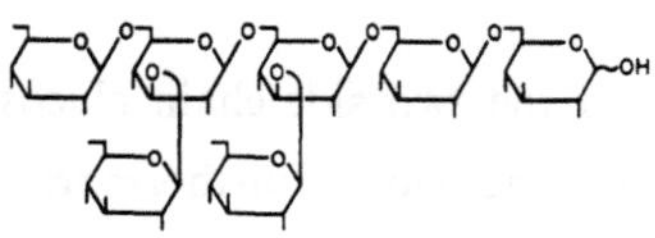

5

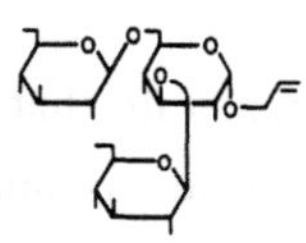

6

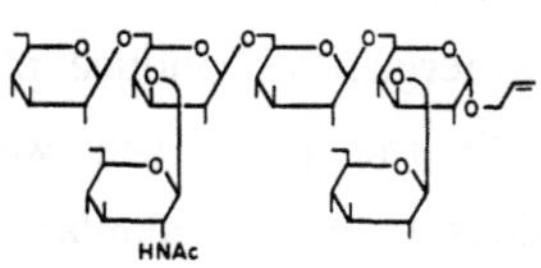

7

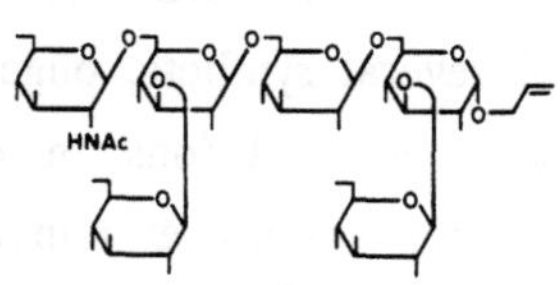

8

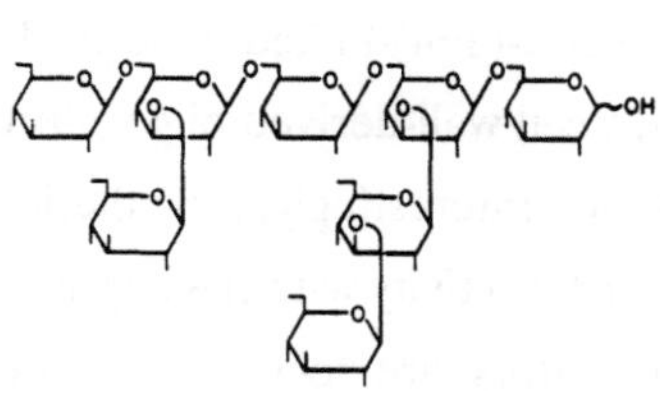

9

Figure 1. Structures of synthetic oligoglucosides. Hydroxyl groups not involved in linkages have been omitted for clarity. Syntheses of several of these oligoglucosides have been reported: hepta-β-glucoside **1** (Ossowski *et al* 1984; Fügedi *et al* 1987); deca-β-glucoside **4** (Birberg *et al* 1989); hepta-β-glucoside **5** (Fügedi *et al* 1988). Syntheses of the α-allyl-oligo-β-glucosides **2**, **3**, **7**, and **8** are unpublished work of N.H. and T.O.

The ability of each of the synthetic oligoglucosides to induce phytoalexin accumulation in soybean cotyledon tissue has been determined. The results demonstrate that the synthetic hepta-β-glucoside elicitor (structure **1**) and oligoglucosides **2**, **3**, **4**, and **9** exhibit half maximum activity of approximately 10^{-8} molar. A related heptaglucoside (structure **5**) and an oligoglucoside, in which the terminal side-chain glucosyl residue closest to the non-reducing end was substituted with a β-*N*-acetylglucosamine residue (structure **7**), exhibit half maximum activity at concentrations approximately 1000-fold higher than the hepta-β-glucoside elicitor. A structurally related α-allyl-tri-β-glucoside (structure **6**) exhibits half maximum activity at a concentration 10,000-fold higher than the hepta-β-glucoside elicitor. An oligoglucoside in which the non-reducing terminal backbone glucosyl residue was substituted with a β-*N*-acetyl-glucosamine residue (structure **8**), was the least active of all of the oligoglucosides tested (half maximum activity > 10^{-4} molar. From these results, we deduce that the following structural motif is necessary for high biological activity of the oligo-β-glucosides:

```
Glcβ(1→6)Glcβ(1→6)Glcβ(1→6)Glcβ(1→6)-R
             |                  |
  Glcβ(1→3)--┘       Glcβ(1→3)--┘
```

The bioassay results demonstrate that the positioning of the two side-chain glucosyl on the backbone of the oligoglucoside is important for elicitor activity. Furthermore, the non-reducing terminal backbone glucosyl residue and the glucosyl side-chain closest to the non-reducing end of the oligoglucoside are essential for maximum activity.

The ability of several synthetic oligoglucosides to synergistically induce phytoalexin accumulation in soybean cotyledons in combination with the plant cell wall-derived oligogalacturonide elicitor was also examined. Oligoglucosides **1** - **4** show similar synergistic activity with the oligogalacturonide elicitor prepared from soybean cell walls (Hahn *et al* 1981). In addition, a synthetic linear α-(1→4)-linked dodecagalacturonide (Nakahara & Ogawa 1989) was as effective as the plant cell wall-derived oligogalacturonide in acting synergistically with the synthetic hepta-β-glucoside. Interestingly, a synthetic decagalacturonide (Nakahara & Ogawa 1987) did not show synergistic activity with the hepta-β-glucoside elicitor.

Current efforts in our laboratory are focussed on preparing a radiolabeled hepta-β-glucoside elicitor derivative possessing both high specific radioactivity and the same biological activity as the underivatized elicitor. To that end, several derivatives of the hepta-β-glucoside elicitor (structure **1**) with bulky substituents attached to the reducing end of the oligoglucoside have been prepared. The structures of these derivatives are shown in Figure 2. Each of the hepta-β-glucoside derivatives retains full elicitor activity when compared with the underivatized hepta-β-glucoside. A tyramine conjugate of maltoheptaose was not active as an elicitor, even when assayed at a concentration of 10^{-4} molar. These data, together with the activity data on longer oligoglucosides reported above, suggests that the nature of the substituent attached to the reducing end has no apparent effect on the elicitor activity of the oligo-β-glucosides. We are presently optimizing the reaction conditions for the radioiodination of the tyramine conjugate of the hepta-β-glucoside elicitor (structure **12**). The radioiodinated derivative will be used for the receptor-ligand binding studies necessary to identify and purify the receptor for the hepta-β-glucoside elicitor.

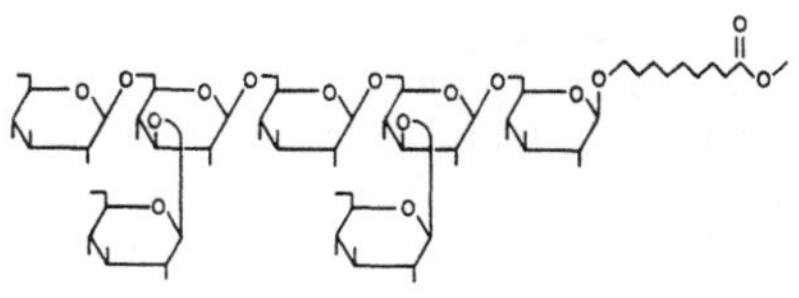

10

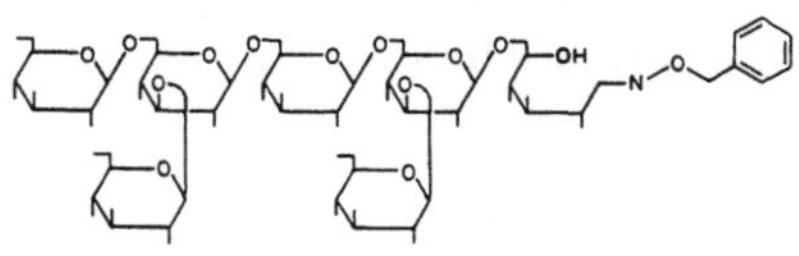

11

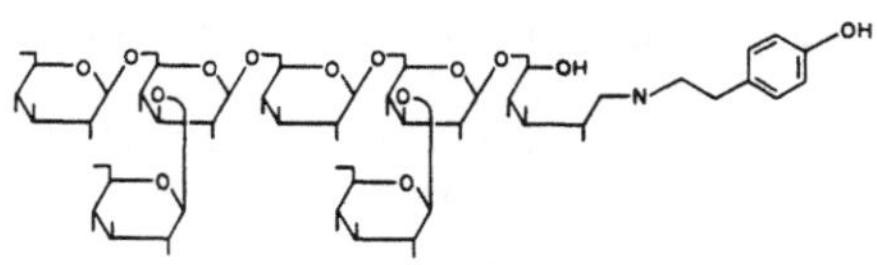

12

Figure 2. Structures of three derivatives of the hepta-β-glucoside elicitor (compound **1**). Hydroxyl groups not involved in linkages have been omitted for clarity. Synthesis of compound **10** has been reported elsewhere (Birberg *et al* 1989). Compound **11** was prepared by reacting hepta-β-glucoside **1** with *O*-benzylhydroxylamine to form the oxime, which was then reduced with sodium cyanoborohydride (Mester & El Khadem 1980; Chaves das Neves *et al* 1986). Compound **12** was prepared by reductive amination as described (Wang *et al* 1984).

ACKNOWLEDGEMENTS

The authors wish to thank Drs. Peter Albersheim and Alan Darvill for their encouragement and assistance in initiating this work. This research was supported by the National Science Foundation (grant number DMB 87-04022 to M.G.H.). The Complex Carbohydrate Research Center is supported in part by Department of Energy grant DE-FG09-87ER13810 as part of the USDA/DOE/NSF Plant Science Centers Program.

REFERENCES

Bell AA (1981) Biochemical mechanisms of disease resistance. *Ann Rev Plant Physiol* 32:21-81

Birberg W, Fügedi P, Garegg PJ, Pilotti Å (1989) Syntheses of a heptasaccharide β-linked to an 8-methoxycarbonyl-oct-1-yl linking arm and of a decasaccharide with structures corresponding to the phytoelicitor active glucan of *Phytophthora megasperma* f.sp. *glycinea*. *J Carbohydr Chem* 8:47-57

Bruce RJ, West CA (1982) Elicitation of casbene synthetase activity in castor bean. The role of pectic fragments of the plant cell wall in elicitation by a fungal endopolygalacturonase. *Plant Physiol* 69:1181-1188

Cervone F, De Lorenzo G, Degrà L, Salvi G (1987) Elicitation of necrosis in *Vigna unguiculata* Walp. by homogeneous *Aspergillus niger* endo-polygalacturonase and by α-D-galacturonate oligomers. *Plant Physiol* 85:626-630

Chaves das Neves HJ, Riscado AMV, Frank H (1986) Derivatives for the analysis of monosaccharides by capillary G.L.C.: Trimethylsilylated deoxy(methoxyamino)alditols. *Carbohydr Res* 152:1-6

Darvill AG, Albersheim P (1984) Phytoalexins and their elicitors - A defense against microbial infection in plants. *Ann Rev Plant Physiol* 35:243-275

Davis KR, Lyon GD, Darvill AG, Albersheim P (1984) Host-pathogen interactions. XXV. Endopolygalacturonic acid lyase from *Erwinia carotovora* elicits phytoalexin accumulation by releasing plant cell wall fragments. *Plant Physiol* 74:52-60

Davis KR, Darvill AG, Albersheim P (1986a) Host-pathogen interactions. XXXI. Several biotic and abiotic elicitors act synergistically in the induction of phytoalexin accumulation in soybean. *Plant Mol Biol* 6:23-32

Davis KR, Darvill AG, Albersheim P, Dell A (1986b) Host-pathogen interactions. XXX. Characterization of elicitors of phytoalexin accumulation in soybean released from soybean cell walls by endopolygalacturonic acid lyase. *Z Naturforsch* 41c:39-48

Davis KR, Hahlbrock K (1987) Induction of defense responses in cultured parsley cells by plant cell wall fragments. *Plant Physiol* 85:1286-1290

Dixon RA (1986) The phytoalexin response: elicitation, signalling and control of host gene expression. *Biol Rev* 61:239-291

Ebel J (1986) Phytoalexin synthesis: The biochemical analysis of the induction process. *Ann Rev Phytopathol* 24:235-264

Farmer EE, Helgeson JP (1987) An extracellular protein from *Phytophthora paprsitica* var. *nicotianae* is associated with stress metabolite accumulation in tobacco callus. *Plant Physiol* 85:733-740

Fügedi P, Birberg W, Garegg PJ, Pilotti Å (1987) Syntheses of a branched heptasaccharide having phytoalexin-elicitor activity. *Carbohydr Res* 164:297-312

Fügedi P, Garegg PJ, Kvarnström I, Svansson L (1988) Synthesis of a heptasaccharide, structurally related to the phytoelicitor active glucan of *Phytophthora megasperma* f. sp. *glycinea*. *J Carbohydr Chem* 7:389-397

Hahn MG, Darvill AG, Albersheim P (1981) Host-pathogen interactions. XIX. The endogenous elicitor, a fragment of a plant cell wall polysaccharide that elicits phytoalexin accumulation in soybeans. *Plant Physiol* 68:1161-1169

Hahn MG, Bonhoff A, Grisebach H (1985) Quantitative localization of the phytoalexin glyceollin I in relation to fungal hyphae in soybean roots infected with *Phytophthora megasperma* f. sp. *glycinea*. *Plant Physiol* 77:591-601

Hahn MG, Bucheli P, Cervone F, Doares SH, O'Neill RA, Darvill A, Albersheim P (1989) The roles of cell wall constituents in plant-pathogen interactions. In: Nester E, Kosuge T (eds) Plant-Microbe Interactions, Volume 3. McGraw Hill, New York, in press

Jin DF, West CA (1984) Characteristics of galacturonic acid oligomers as elicitors of casbene synthetase activity in castor bean seedlings. *Plant Physiol* 74:989-992

Keen NT (1975) Specific elicitors of plant phytoalexin production: Determinants of race specificity in pathogens? *Science* 187:74-75

Mansfield JW, Hargreaves JA, Boyles FC (1974) Phytoalexin production by live cells in broad bean leaves infected with *Botrytis cinerea*. *Nature* 252:316-317

Mayama S, Tani T (1982) Microspectrophotometric analysis of the location of avenalumin accumulation in oat leaves in response to fungal infection. *Physiol Plant Pathol* 21:141-149

Mester L, El Khadem HS (1980) Hydrazine derivatives and related compounds. In: Pigman W, Horton D, Wander JD (eds) The Carbohydrates, Vol. IB, Chemistry and Biochemistry. Academic Press, New York, p 929-988

Moesta P, Seydel U, Lindner B, Grisebach H (1982) Detection of glyceollin on the cellular level in infected soybean by laser microprobe mass analysis. *Z Naturforsch* 37c:748-751

Nakahara Y, Ogawa T (1987) Synthetic studies on plant cell-wall glycans. 3. Stereocontrolled, total synthesis of α-D-GalA-[(1→4)-α-D-GalA]$_8$-(1→4)-β-D-GalA-1→OPr, a synthetic model for phytoalexin elictor-active oligogalacturonic acids. *Carbohydr Res* 167:c1-c7

Nakahara Y, Ogawa T (1989) Total synthesis of galactododecaosiduronic acid, an endogenous phytoalexin elicitor isolated from soybean cell wall. *Tetrahedr Lett* 30:87-90

Nothnagel EA, McNeil M, Albersheim P, Dell A (1983) Host-pathogen interactions. XXII. A galacturonic acid oligosaccharide from plant cell walls elicits phytoalexins. *Plant Physiol* 71:916-926

Ossowski P, Pilotti Å, Garegg PJ, Lindberg B (1984) Synthesis of a glucoheptaose and a glucooctaose that elicit phytoalexin accumulation in soybean. *J Biol Chem* 259:1137-1140

Sharp JK, Albersheim P, Ossowski P, Pilotti Å, Garegg PJ, Lindberg B (1984a) Comparison of the structures and elicitor activities of a synthetic and a mycelial-wall-derived hexa(β-D-glucopyranosyl)-D-glucitol. *J Biol Chem* 259:11341-11345

Sharp JK, McNeil M, Albersheim P (1984b) The primary structures of one elicitor-active and seven elicitor-inactive hexa(β-D-glucopyranosyl)-D-glucitols isolated from the mycelial walls of *Phytophthora megasperma* f. sp. *glycinea*. *J Biol Chem* 259:11321-11336

Sharp JK, Valent B, Albersheim P (1984c) Purification and partial characterisation of a β-glucan fragment that elicits phytoalexin accumulation in soybean. *J Biol Chem* 259:11312-11320

Smith DA (1982) Toxicity of phytoalexins. In: Bailey JA, Mansfield JW (eds) Phytoalexins. Blackie, Glasgow and London, p 218-252

Walker-Simmons M, Hadwiger L, Ryan CA (1983) Chitosans and pectic polysaccharides both induce the accumulation of the antifungal phytoalexin pisatin in pea pods and antinutrient proteinase inhibitors in tomato leaves. *Biochem Biophys Res Commun* 110:194-199

Wang WT, LeDonne NC Jr, Ackerman B, Sweeley CC (1984) Structural characterisation of oligosaccharides by high-performance liquid chromatography, fast-atom bombardment-mass spectrometry, and exoglygosidase digestion. *Anal Biochem* 141:366-381

Weinstein LA, Albersheim P (1983) Host-pathogen interactions. XXIII. The mechanism of the antibacterial action of glycinol, a pterocarpan phytoalexin synthesized by soybeans. *Plant Physiol* 72:557-563

West CA (1981) Fungal elicitors of the phytoalexin response in higher plants. *Naturwissenschaften* 68:447-457

ARABIDOPSIS THALIANA AS A MODEL SYSTEM FOR STUDYING PLANT-PATHOGEN INTERACTIONS

Keith R. Davis[1], Eric Schott, Xinnian Dong, and Frederick M. Ausubel
Department of Molecular Biology, Massachusetts General Hospital
Boston, Massachusetts 02114 USA

INTRODUCTION

Although plants are continuously exposed to a variety of potentially pathogenic microorganisms, successful infections are rare. Plants utilize a diverse array of defense mechanisms to prevent microbial infections. Some of these defense strategies involve constitutively expressed physical and chemical barriers that provide a first line of defense against potential pathogens (Mansfield 1983). Other defense mechanisms are specifically induced upon attempted infection. These induced defense responses include the synthesis of polyphenolic lignins and hydroxyproline-rich glycoproteins which are incorporated into plant cell walls, causing the walls to be more resistant to microbial invasion; the synthesis of the hydrolases β-1,3-glucanase and chitinase, which may inhibit fungi by degrading their cell walls; and the synthesis and accumulation of antimicrobial compounds called phytoalexins (Bell 1981, Collinge and Slusarenko 1987, Darvill and Albersheim 1984, Hahlbrock and Scheel 1987).

Some of the most detailed studies on the regulation of defense-related genes have involved the characterization of the induction of genes encoding enzymes involved in phytoalexin biosynthesis in bean (Lamb et al 1986) and parsley (Hahlbrock and Scheel 1987, Kombrink et al 1986). The results of these studies indicate that there are groups of coordinately regulated genes that differ in the timing of their induction. Moreover, several of these genes: phenylalanine ammonia lyase

[1]Present address; Biotechnology Center, Rightmire Hall, Ohio State University, Columbus, OH 43210-1002 USA

NATO ASI Series, Vol. H 36
Signal Molecules in Plants and Plant-Microbe Interactions
Edited by B. J. J. Lugtenberg

(PAL), 4-coumarate:CoA ligase (4CL), and chalcone synthase (CHS) have been shown to be transcriptionally activated (Lamb et al 1986, Kombrink et al 1986). Other defense-related genes have also been shown to be associated with increased steady-state levels of their mRNAs (Collinge and Slusarenko 1987), suggesting that transcriptional activation may play a major role in the expression of many defense-related genes.

Although significant progress has been made in identifying specific genes that are activated during attempted infection and/or treatment with compounds derived from microorganisms (elicitors; see Darvill and Albersheim 1984), very little is known about the mechanisms involved in the activation of these defense-related genes. To provide further insight into the regulation of defense-related genes, we have initiated studies of the regulation of defense responses in *Arabidopsis thaliana*. Our major goal is to use a combination of genetic and biochemical/molecular biological approaches to elucidate the mechanisms whereby potential pathogens are perceived, and how the recognition signals are transduced to the genome, causing the transcriptional activation of specific defense-related genes.

A. thaliana, a small crucifer that has received considerable attention as a model plant system, offers many experimental advantages (Meyerowitz 1987). *A. thaliana* is a self-fertilizing diploid with a small genome size and is therefore well suited for genetic studies. There are many mapped mutations that can be utilized for mapping new mutations. Moreover, several groups are developing RFLP and physical maps of the *A. thaliana* genome and correlating these maps with the genetic map already available. As the genetic and physical maps become more detailed, it will be possible to identify and clone specific genes which have only been identified by mutation. Another important advantage of *A. thaliana* is that it can be genetically engineered using standard *Agrobacterium* transformation protocols. This makes it possible to use molecular genetic approaches to study gene regulation, and to conduct genetic complementation experiments--a very powerful tool for the functional characterization of cloned genes.

At the time these studies were initiated, a disadvantage of A. thaliana for studying plant-microbe interactions was that there were no known pathogens of A. thaliana. Also, no data were available concerning the types of defense responses elaborated by A. thaliana in response to microbial attack. Our initial efforts towards overcoming this lack of information concerning disease resistance in A. thaliana have focused on the following major areas: 1) the establishment of a tissue culture system in which putative defense responses can be induced by elicitors; 2) the cloning of putative defense-related genes from A. thaliana; and 3) the identification of bacterial strains which are either pathogenic or induce a hypersensitive response (HR) when infiltrated into leaves of A. thaliana. Here, we summarize our progress in developing A. thaliana as a model system for studying plant-pathogen interactions and describe genetic and molecular genetic approaches that will be complementary to the molecular biological and biochemical approaches used by other workers.

ELICITOR-INDUCED RESPONSES OF CULTURED CELLS OF ARABIDOPSIS

Our initial studies have focused on the response of suspension-cultured cells, since cultured cells have proven to be valuable model systems for characterizing the induction of defense-related genes in several plant species (Ebel et al 1984, Kombrink et al 1986, Lamb et al 1986). The responses of cultured Arabidopsis cells (ecotype Fi-3) treated with two different microbial elicitors, either crude cell wall components from Phytophthora megasperma pv glycinea (Pmg elicitor) or the pectin-degrading enzyme endopolygalacturonic acid lyase (PGA lyase) from Erwinia carotovora pv carotovora, were very similar to the elicitor-induced responses of cultured parsley cells (Davis and Hahlbrock 1987). Both Pmg elicitor and PGA lyase induced PAL enzyme activity in a dose-dependent fashion, as well as cell browning and decreased growth rates. At doses sufficient to cause maximum induction, Pmg elicitor and PGA lyase induced 30-fold and 20-fold increases, respectively, in PAL activity compared to control cultures.

More detailed studies using PGA lyase as an elicitor demonstrated that in addition to PAL, several other enzymes involved in phenylpropanoid biosynthesis were also induced. Cell cultures treated with PGA lyase exhibited the coordinate, transient induction of PAL and 4CL activities which reached peak levels approximately 6 to 8 h after elicitor treatment. Caffeic acid O-methyl transferase (CMT) and peroxidase (PO) activities were also induced by PGA lyase. These two enzymes exhibited similar induction kinetics; peak levels were observed approximately 24 h after elicitor treatment and were maintained for at least 48 h. The induction of PAL and 4CL enzyme activities was preceded by the transient induction of their corresponding mRNAs. Both PAL and 4CL mRNA levels peaked at 3 h after elicitor treatment. In addition, increased levels of β-1,3-glucanase mRNA were also observed in cultures treated with PGA lyase. The induction of the β-1,3-glucanase mRNA was distinct from that observed for PAL and 4CL mRNAs; maximum levels were observed 8 to 16 hours after elicitor treatment.

CHARACTERIZATION OF DEFENSE-RELATED GENES FROM *A. THALIANA*

Results from the tissue culture experiments indicated that cDNAs of PAL and β-1,3-glucanase isolated from bean and tobacco, respectively, detected mRNAs of the expected sizes in elicitor-treated *Arabidopsis* cells. Therefore, we used these cDNAs to identify clones in a library constructed from *A. thaliana* (ecotype Landsberg erecta) genomic DNA.

Three independently isolated clones that hybridized to the bean PAL cDNA were shown to contain the same genomic sequences, as determined by restriction fragment analysis. Sequence analysis of one of these clones demonstrated that the protein coding sequences of this *A. thaliana* PAL gene are approximately 80% similar to the bean PAL gene. Genomic DNA blot analysis using fragments from protein coding regions of the *A. thaliana* PAL gene as a probe indicated that there may be two or three additional PAL genes present in *A. thaliana*.

Five independently isolated clones that hybridized to the tobacco β-1,3-glucanase cDNA were analyzed in detail and were found to contain three different genes that appear to be clustered within a 12 kb region. Sequence analyses demonstrated that the protein-coding regions of the three A. thaliana β-1,3-glucanase genes are 65 to 70% similar to the tobacco gene. Genomic DNA blot analysis indicated that there are only three β-1,3-glucanase genes present in A. thaliana.

IDENTIFICATION OF BACTERIAL PATHOGENS OF A. THALIANA

Our approach for identifying bacterial pathogens of A. thaliana has been to screen known pathogens of other crucifers for virulence on different ecotypes of A. thaliana. The first strains tested were Xanthomonas campestris pv campestris. All five strains of Xcc infiltrated into leaves of A. thaliana were virulent on all of the ecotypes of A. thaliana tested. Infiltrated leaves developed spreading necrotic lesions with chlorotic margins in which bacterial populations increased 3 to 5 logs within 48 h. Several days after infiltration, the infected leaf would become completely chlorotic and die. Systemic infection was never observed, suggesting that Xcc inoculated by infiltration techniques cannot move through the vasculature of A. thaliana.

Since we could not identify any ecotype of A. thaliana that was resistant to any of the Xcc strains tested, a prerequisite for characterizing an active defense response, we tested another species of crucifer pathogen, Pseudomonas syringae pv maculicola for virulence on A. thaliana. We have found that the various Psm strains tested cause one of three phenotypes; a null response in which there are no visible effects and the bacteria do not seem to grow within the leaf tissue; a hypersensitive reaction (HR) in which the bacterial population increases 1 to 1.5 logs; and an apparent susceptible reaction in which water-soaked lesions with chlorotic margins develop concomitantly with an increase in bacterial numbers of 3 to 4 logs. All of the ecotypes of A. thaliana tested exhibited the same phenotype when infiltrated with a particular strain of Psm. Thus, we have not identified any plant-bacterial strain combinations that resemble the race-specific interactions that have been characterized in other plant-pathogen interactions.

GENETIC APPROACHES FOR STUDYING THE REGULATION OF DEFENSE-RELATED GENES

The major advantage that _A. thaliana_ offers as an experimental system is that it is well suited for genetic studies. Many of the plants in which defense mechanisms have been studied in detail are not amenable to genetic approaches. A major goal of our research is to use genetic approaches to identify genes that are involved in the regulation of defense-related genes. One approach currently being pursued is to mutagenize seeds from transgenic plants that contain defense-gene promoters fused to β-glucuronidase (GUS; see Jefferson et al 1987). The progeny of the mutagenized plants will then be screened for aberrant GUS expression; either constitutive expression or loss of inducibility by elicitors or bacteria. These experiments should yield mutants containing promoter mutations, as well as mutations in _trans_-acting loci that are involved in various steps of the regulatory pathway. Mutations in _trans_-acting loci can be identified by examining the expression of the endogenous gene for aberrant expression, or alternatively, the mutations can be mapped and compared to the mapped position of the promoter-GUS gene fusion. The putative _trans_-acting genes will be the major focus of further research.

A more direct genetic approach is to screen mutagenized plants for individuals that have become susceptible to a phytopathogenic bacterium to which the wild-type plants are normally resistant. To date, several such mutants have been identified and the progeny of these mutants are currently being tested to determine if the susceptible phenotype is reproducible. This screen will supply another set of mutations that can be compared to those obtained by screening for aberrant expression of gene fusions. A major question that these experiments will address is whether or not mutations which affect disease resistance have any effects on the regulation of the previously described defense-related genes.

CONCLUSIONS AND PROSPECTS

Our studies using cultured cells of _Arabidopsis_ demonstrated that the induction of putative defense responses by microbial elicitors is very similar to that observed in other plant systems. Elicitor-treated Arabidopsis cells exhibited increased levels of several enzymes involved in phenylpropanoid metabolism, namely PAL, 4CL, CMT, and PO, as well as increased levels of β-1,3-glucanase mRNA. The timing and magnitude of the induction of these gene products were quite similar to those reported for elicitor-treated parsley cells (Davis and Hahlbrock 1987).

We have also made significant progress in developing a collection of phytopathogenic bacteria with which to study disease resistance in A. thaliana. Analyses of the different isolates of the bacterial crucifer pathogen Psm infiltrated into leaves of A. thaliana have identified strains that cause a HR or cause disease. Thus, we can directly compare the activation of various defense-related genes during a resistance reaction and pathogenesis. Preliminary results indicate that PAL, 4CL, and one of the β-1,3-glucanase mRNAs are induced to several fold higher levels in leaves infiltrated with Psm strains that cause a HR when compared to pathogenic strains.

Overall, our results suggest that A. thaliana is an excellent model system for studying disease resistance in plants. The observation that elicitor-treated cultured cells and plants infiltrated with various phytopathogenic bacteria respond in a manner similar to that observed in other plants suggests that new insights provided by studies of A. thaliana will be directly applicable to agronomically important plants. Thus, the utility of A. thaliana as an experimental system may allow the identification of specific genes critically involved in conferring disease resistance, a prerequisite for using molecular biological techniques to improve disease resistance in crop plants.

ACKNOWLEDGEMENTS

The authors thank Drs. Carol Cramer, Klaus Hahlbrock, and Fred Meins for providing cDNAs, Daniel Voytas for the genomic library, and Paul Williams, Brian Staskawicz, and Michael Daniels for bacterial strains.

REFERENCES

Bell AA (1981) Biochemical mechanisms of disease resistance. Ann Rev Plant Physiol 32:21-81

Collinge DB, Slusarenko AJ (1987) Plant gene expression in response to pathogens. Plant Mol Biol 9:389-410

Darvill AG, Albersheim P (1984) Phytoalexins and their elicitors - a defense against microbial infection in plants. Ann Rev Plant Physiol 35:243-275

Davis KR, Hahlbrock K (1987) Induction of defense responses in cultured parsley cells by plant cell wall fragments. Plant Physiol 85:1286-1290

Ebel J, Schmidt WE, Loyal R (1984) Phytoalexin synthesis in soybean cells: elicitor induction of phenylalanine ammonia-lyase and chalcone synthase mRNAs and correlation with phytoalexin accumulation. Arch Biochem Biophys 232:240-248

Hahlbrock K, Scheel D (1987) Biochemical responses of plants to pathogens. In: Chet I (ed) Innovative approaches to plant disease control, John Wiley & Sons, New York, pp. 229-254

Jefferson RA, Kavanagh TA, Bevan MW (1987) GUS fusions: Beta-glucuronidase as a sensitive and versatile gene fusion marker in higher plants. EMBO J 6:3901-3907

Kombrink E, Bollmann J, Hauffe KD, Knogge W, Scheel, Schmelzer E, Somssich I, Hahlbrock K (1986) Biochemical responses of non-host plant cells to fungi and fungal elicitors. In: Bailey JA (ed) Biology and Molecular Biology of Plant-Pathogen Interactions, NATO ASI Series, Vol. H1, Springer-Verlag, Berlin, pp. 253-262

Lamb, CJ, Corbin DR, Lawton MA, Sauer N, Wingate VPM (1986) Recognition and response in plants:pathogen interactions. In: Lugtenberg B (ed) Recognition in microbe-plant symbiotic interactions. Springer-Verlag, Berlin, pp 333-344

Mansfield, J (1983) Antimicrobial compounds. In: Callow JA (ed) Biochemical plant pathology. John Wiley & Sons Ltd, New York, pp. 237-265

Meyerowitz EM (1987) Arabidopsis thaliana. Ann Rev Genet 21:93-111

Ca^{2+} IS AN IMPORTANT BUT NOT THE ONLY SIGNAL IN CALLOSE SYNTHESIS INDUCED BY CHITOSAN, SAPONINS AND POLYENE ANTIBIOTICS

H. Kauss[1], T. Waldmann[1], W. Jeblick[1], G. Euler[1], R. Ranjeva[2], A. Domard[3]

[1]Universität Kaiserslautern, FB Biologie, Postfach 3049, D-6750 Kaiserslautern, Bundesrepublik Deutschland
[2]Centre de Physiologie Végétale, Université Paul Sabatier, 118 route de Narbonne, F-31062 Toulouse Cédex, France
[3]Université Claude Bernard, Laboratoire des Matériaux Plastiques et des Biomatériaux, 43 Boulevard du 11 Novembre 1918, F-69622 Villeurbanne Cédex, France

CALLOSE INDUCTION AS A MODEL FOR RAPID DEFENSE REACTIONS

The 1,3-ß-glucan callose is rarely a constituent in the normal cell wall of higher plants. Since callose can be easily recognized cytochemically by several more or less specific techniques, many examples are known where 1,3-ß-glucans are deposited on the cell wall as a result of various types of stress, often in a very localized manner (for lit. see Kauss 1987, 1990). A striking example for the function of callose is the sealing, presumably within minutes, of plasmadesmata or sieve pores after mechanical injury. Implication of callose in defense reactions against pathogens is best illustrated by the formation of "papillae". These cell wall appositions are deposited from the plasma membrane basipetal to the sites of fungal penetration. When e.g. *Phytophthora infestans* infects tubers of resistant potato cultivars, the papillae form earlier then in tissue from susceptible cultivars, and are correlated with a decrease in fungal penetration (Hächler and Hohl 1984). Basipetal to an appressorium of *Erysiphe graminis* on epidermal cells of barley coleoptiles from a resistant line, the formation of papillae begins about 1 h <u>before</u> fungal penetration pegs become microscopically visible (Gold et al 1986). As a result, penetration is greatly suppressed. Although successful papillae in this (Aist et al 1988) and other systems likely contain several chemical constituents, they are always rich in callose. Formation of callose can, therefore, be regarded as a model to study the biochemical details involved in signal perception and transmission, leading to rapid defense reactions.

THE 1,3-ß-GLUCAN SYNTHASE AS A TARGET FOR Ca^{2+}

The plasma membrane-located 1,3-ß-glucan synthase (called GS II by plant cytochemists) is strictly dependent on Ca^{2+} *in vitro* (Kauss 1987, Fink et al 1987). In the presence

NATO ASI Series, Vol. H 36
Signal Molecules in Plants and Plant-Microbe Interactions
Edited by B. J. J. Lugtenberg

of 0.2 mM spermine and/or 4 mM Mg^{2+}, halfmaximal activity is attained at about 0.8 μM Ca^{2+}. This type of enzyme activation is readily reversible and not dependent on nucleotides and, therefore, appears to be allosteric. As a working hypothesis it has been suggested that the enzyme remains latent in undisturbed cells as long as the cytoplasmic concentration of free Ca^{2+} remains in the 0.1 μM range but becomes activated by Ca^{2+} if the cells are damaged or rendered permeable to Ca^{2+}. As a consequence, callose may directly be deposited from the surface of the plasma membrane to the cell wall. This requires a vectorial arrangement of the enzyme. There is some indirect evidence available that the binding sites for the substrate UDP-glucose and the effector Ca^{2+} are indeed located at the cytoplasmic side of the plasma membrane (Fink et al 1987, Kauss 1987, 1990). In plasmolyzed cells undergoing callose synthesis, polymer accumulation at the plasma membrane is clearly evident by fluorescence microscopy. The 1,3-ß-glucan synthase is present with high activity in homogenates of numerous plants but has not yet been purified. A final and direct proof for the vectorial arrangement of the enzyme awaits, therefore, enzyme purification.

INITIAL INTERACTION OF THE CHEMICALLY UNRELATED CALLOSE ELICITORS WITH THE PLASMA MEMBRANE

Early experiments on the induction of callose synthesis focused on physical stress such as a bending of root hairs or compressing epidermal cells (for literature see Eschrich 1956, Kauss 1987). For biochemical studies on the mechanism of callose induction, however, a system is needed where callose synthesis in all cells of a population could be simultaneously induced. We have, therefore, searched for chemical substances able to elicit callose production in suspension cells.

In the course of several years, a number of callose elicitors have been empirically found, examples are listed in Table 1. The substances may be taken as models for diffusible or, in the case of chitosan, surface-exposed recognition signals of pathogen origin which can trigger defense reactions in plant cells. It should be noted that the potency of these callose elicitors varies considerably and that the effect of the amphipaths is not due to a general detergent action as, e.g. Triton-X-100, CHAPS and Octylglucosid are almost inactive. The substance groups in Table 1 were introduced to indicate possible modes of first interaction of the elicitors with the plasma membrane. Such an artificial system bears, however, the danger that action mechanisms suggested in the literature on the basis of hemolytic or fungistatic properties might be incorrectly implicated also in callose induction. The saponins and polyene antibiotics, for instance, are widely considered to change membrane functions by their interaction with sterols but it becomes more and more clear that this cannot be their only action mechanism. Polyenes (Bolard 1986) and digitonin (Gögelein und Hüby 1984) also change several properties of

Table 1
Some examples of callose elicitors.

Chemical Nature	Examples
Polycations	Chitosan, Poly-L-Orn, Poly-L-Lys
Steroid saponins	Digitonin, Tomatin (N-containing)
Triterpene saponins	α-Hederin, α-Escin
Polyene antibiotics	Amphotericin B, Filipin
Acylated cyclic peptides	Polymyxin B, Echinocandin B
Amphipaths	Acylcarnitine, Na-Dodecylsulphate

sterol-free artificial membranes indicating that these substances can also interact with the lipophilic part of phospholipids. This property is shared with acylated cyclic polypeptides (e.g. Polymyxin B; Storm et al 1977) and acylcarnitine. The latter compound inhibits, for instance, the phospholipid-sensitive protein kinase C, presumably by competition with the activator (Wise et al 1982).

For induction of callose synthesis, all the substances indicated in Table 1 are active over a narrow concentration range, which must be experimentally determined based on cell type, cell density and composition of the suspension medium particularly with regard to divalent ion concentration. It appears that certain membrane functions need only to be slightly perturbed to trigger the metabolic effects described below. Supramaximal doses lead to an early and severe damage of many cells resulting in a greatly decreased callose production by the population.

As discussed below, early and presumably initial effects of the callose elicitors appear to be changes in ion transport rates. One could envisage, therefore, that the low doses of amphipathic callose elicitors act by disturbing the lipid environment of membrane proteins that are involved in mediating or regulating ion transport. A similar mode of action might be concluded for the polycationic callose elicitor chitosan from the molecular properties governing its potency. We have recently compared chitosan preparations differing in degree of polymerization and N-acetylation in regard to their activity for the induction of callose synthesis in suspension-cultured cells and protoplasts. Fig. 1A shows the dose-response curves for three different chitosan preparations. At the 50 μM concentration of Ca^{2+} used in the experiment reported in Fig. 1, the curves rapidly decline above a certain chitosan concentration. At higher Ca^{2+}-concentrations (2 mM) a broader apparent optimum is found (Kauss et al 1989). In any case, for the

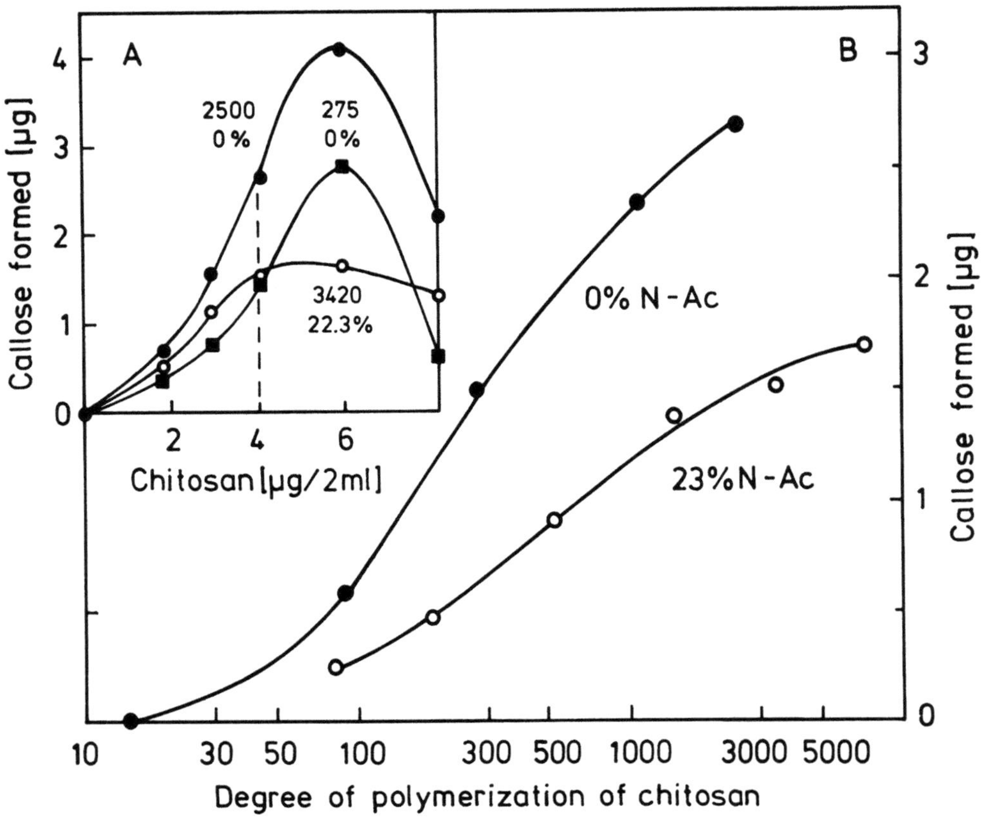

Fig. 1 Callose formation in protoplasts as influenced by the average degree of polymerization (DP) and N-acetylation (% N-Ac) of chitosan. A: Dose-response curves for chitosan preparations of DP 2500/0 % N-Ac (●), DP 275/0 % N-Ac (■) and DP 3420/22.3 % N-Ac (o). B: Callose induction by various fully N-deacetylated (●) and about 23 % N-acetylated (o) chitosan preparations. Protoplasts from *Catharanthus roseus* were used and each batch ($4x10^5$ in 2 ml of 25 mM Bistris/Mes buffer, pH 5.7, 0.4 M sucrose, 50 μM $CaCl_2$) was supplied with 4 μg of the chitosans, standardized to the same polyglucosamine content. The callose formed within 3 h was determined fluorometrically and was expressed in μg pachyman-equivalents. Chitosan of DP 5000, as an example, has an average molecular weight of 10^6 Da and average chain lenght of 2.5 μm. For experimental details see Kauss et al 1989.

comparison of different chitosan preparations, one must carefully ascertain that a concentration at the increasing flank of the dose-response curves is choosen. If this requirement is fulfilled, it becomes evident (Fig. 1B) that chitosan oligomers are inactive, and the ability to induce callose formation increases with the chain length of chitosan up to about 2.5 μm (DP 5000). Binding over broad areas of the surface appears to be required leading to the physiological effects in a cooperative manner. At a comparable DP, the fragments exhibiting about 23 % N-acetylation are less effective than fully deacetylated fragments (Fig. 1B). Similar results were found with cells, although far higher chitosan concentrations were required, presumably due to binding of some of the chitosan to cell wall polymers and also to a certain exclusion effect of the cell wall for the largest part of the chitosan molecule population (Kauss et al 1989). Taken together, these results suggest that if the row of amino groups in the polyglucosamine molecule is interrupted statistically at every fourth to fifth residue, that is after 2-3 nm, the binding is obviously disfavored. The points to which chitosan binds on the plasma membrane must not only be very abundant but must also not be separated by more than a few nm, suggesting that the polycation binds by electrostatic or ternary interactions with the negatively charged head groups of phospholipids. It is easy to imagine that such a binding of rigid polymers might change overall membrane fluidity and thereby influence the phospholipid density at the borders of membrane transport proteins either functioning themselves in ion transport, or regulating the function of ion channels. This possible action mechanism bears obvious parallels to the opening of specific ion channels due to physical stretching of membranes, as explored recently by patch-clamp experiments with animal (Christensen 1987) and plant (Falke et al 1988) membranes.

ALL THE CALLOSE ELICITORS INDUCE Ca^{2+} UPTAKE AND CHANGES IN TRANSPORT RATES OF K^+ AND H^+

Due to the short times required for callose induction, the cells can be suspended in buffers containing the essential ions at low concentrations. Specific ion concentrations can, therefore, be determined, with ion-selective electrodes. All the elicitors listed in Table 1 have been shown to cause a net Ca^{2+} uptake, K^+ leakage and external alkalinization, amphotericin B is shown as one example in Fig. 2.

It is difficult to evaluate quantitatively the net uptake of Ca^{2+} by cells induced by chitosan as this polycation also displaces Ca^{2+} from the cell wall (Young and Kauss 1983). Ca^{2+} uptake can be better studied employing noncharged elicitors. The digitonin-elicited net Ca^{2+} uptake in soybean cells correlated in time with callose formation and was preceded by the leakage of ions, mainly K^+, from the cells (Waldmann et al 1988). The Ca^{2+} taken up by these cells would correspond to an internal increase of about 120 μM, calculated on total cell water and assuming that all Ca^{2+} remains free. Such a high cytoplasmic Ca^{2+} concentration is not compatible with cellular metabolism. The only compartment that might accommodate > 100 μM Ca^{2+} is the vacuole. The movement of Ca^{2+} through the cytoplasm to the vacuole might result in a transient increase in cytoplasmic [Ca^{2+}]. Determination of the concentration of free Ca^{2+} in the cytoplasm is, however, only possible yet in certain single plant cells but not in populations of cells or protoplasts under conditions where they can be induced to form callose.

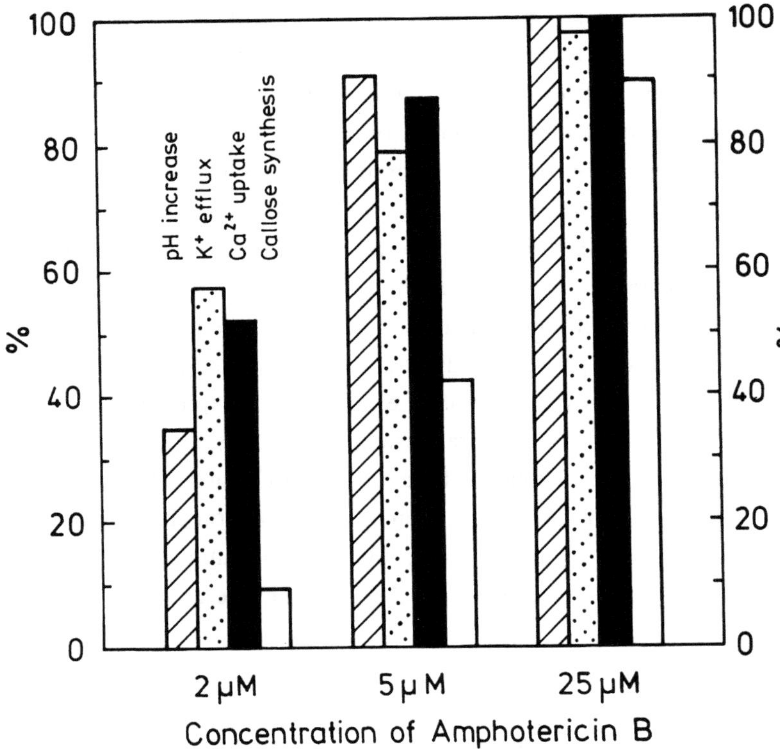

Fig. 2 Induction of callose formation and changes in ion transport rates by amphotericin B in suspension-cultured cells of *Catharanthus roseus*. The cells (300 mg) were suspended in 5 ml of 1 mM Bistris/Mes buffer, pH 6.0, containing 4 % (v/v) growth medium and 3 % (w/v) sucrose which results in a $[K^+]$ of 1.5 mM and a $[Ca^{2+}]$ of 47 μM. Amphotericin B was added at the indicated concentrations in dimethylsulfoxide (final conc. of the solvent 0.2 %, v/v). The concentration of the ions in the suspension fluid of a control and the samples was determined after 2 h with ion-selective electrodes. The increase in pH (alkalinization), increase in $[K^+]$, decrease in $[Ca^{2+}]$ and the callose production are calculated taking the effects caused by 50 μM amphotericin B as 100 %.

Callose induction by digitonin and amphotericin B (Fig. 2) is also associated with an external alkalinization, exhibiting a time course similar to K^+ efflux (data not shown). As far as can be determined for the weakly buffered suspension used in the experiments shown in Fig. 2, the K^+ efflux and the degree of external alkalinization also correlated in the dose-response studies with amphotericin B. The phenomenon of a K^+ efflux and an associated external alkalinization has also been observed in experiments on the induction of other defense reactions, such as the induction of the hypersensitive response by bacteria (Atkinson et al 1986) and the phytoalexin induction by glucan-type elicitors in soybean (Osswald et al 1985). The effect appears to be associated with a decrease in intracellular pH (Ojalvo et al 1987) and with changes in the size of the plasma membrane potential (Pavlovkin et al 1986, Pelissier et al 1986, Mayer and Ziegler 1988).

Interpretation of the data on K^+ leakage and pH-changes is still controversial. A single antiporter may mediate the coupled movement of K^+ and H^+ (Atkinson et al 1986), or the effects may be indirectly coupled and involve the H^+-ATPase.

The net Ca^{2+} uptake induced by digitonin or chitosan can be inhibited by the Ca^{2+} channel blockers nifedipine (Waldmann et al 1988) and flunarizine (data not shown), indicating that the Ca^{2+} uptake may occur through Ca^{2+}-specific channels. It was suggested above that elicitor-induced Ca^{2+} uptake may result directly from the binding of the callose elicitors to general membrane components. Alternatively, more complex mechanisms of signal transduction such as G-proteins may be involved. The considerable time gap of up to 10 min between Ca^{2+} uptake and K^+ efflux in soybean cells (Waldmann et al 1988) makes an electrical coupling by changes in the membrane potential less likely, although the methods used to determine ion fluxes may not be sensitive enough to detect small initial effects on the K^+ and Ca^{2+} transport rates. This would be necessary to decide whether threshold phenomena exist.

CA^{2+} INFLUX IS ESSENTIAL FOR CALLOSE SYNTHESIS

Early evidence for a role of Ca^{2+} as a signal in callose synthesis came from experiments in which the [Ca^{2+}] in the suspension buffer of chitosan-induced suspension cells was greatly lowered by the addition of EGTA (Köhle et al 1985). Under these conditions callose synthesis stopped, indicating that external Ca^{2+} was essential. The addition of EGTA in the presence of the elicitor, however, also appeared to increase ion leakage, suggesting that external Ca^{2+} not only provides a pool for Ca^{2+} functioning as second messenger but additionly may function by stabilizing membranes. The problem was partly overcome by decreasing the external [Ca^{2+}] with ion exchange beads, which have no direct access to the membrane surface and, therefore, caused less additional ion leakage, although callose formation was greatly diminished (Köhle et al 1985). That Ca^{2+} uptake is required for callose synthesis was also demonstrated with the Ca^{2+}-channel blockers nifedipine (Waldmann et al 1988) and flunarizine (data not shown). These substances inhibited the net Ca^{2+} uptake induced by digitonin and chitosan and in parallel also the callose synthesis.

CA^{2+} IS NOT SUFFICIENT AS A SIGNAL IN CALLOSE SYNTHESIS

The above experiments indicate that Ca^{2+} influx is a prerequisite for callose synthesis. Surprisingly, however, the Ca^{2+} ionophores A23187 and ionomycin induced only little callose formation (Waldmann et al 1988). These resuls were recently confirmed. At a comparable net Ca^{2+} uptake, A23187 caused only about 10 % of the callose synthesis effected by digitonin (data not shown). That Ca^{2+} cannot be the only signal in callose induction becomes evident also from dose-response studies with amphotericin B (Fig. 2). Half-maximal changes in K^+ and Ca^{2+} transport rates were obtained at 2 μM amphotericin B. The comparatively lower effect on pH might have resulted from the use of slightly buffered cell suspensions. Rapid K^+ leakage and Ca^{2+} uptake in yeast cells were also induced by 1 μM nystatin, another polyene antibiotic (Eilam and Grossowicz 1982). In contrast to the effects on ion flux, callose induction at 2 μM amphotericin B was low and about 50 μM was required to fully induce callose synthesis. These results suggest that in addition to Ca^{2+} uptake, induction of callose synthesis requires triggering of another process which needs higher amphotericin B concentrations.

The question thus arises which signals may, in addition to Ca^{2+}, be required to activate the 1,3-ß-glucan synthase. Delmer (1987) has speculated that *in vivo* the callose synthase complex might gain activity on a depolarisation of the plasma membrane. This appears unlikely since callose synthesis in suspension cells is increased by the presumably hyperpolarizing toxin fusicoccin and decreased by several presumably depolarizing protonophores (results not shown). One other possibility for additional stimulation might be that the callose elicitors not only cause changes in the transport rates of Ca^{2+} and other ions but also have a direct stimulatory effect on the 1,3-ß-glucan synthase. This enzyme requires phospholipids for activity and has been shown to be activated *in vitro* to a certain degree by low concentrations of certain free polyunsaturated fatty acids, lysophosphatidylcholine, platelet activating factor, acylcarnitine and Echinocandin B (Kauss und Jeblick 1986). The major problem with the assay of the 1,3-ß-glucan synthase is that in homogenates the enzyme is mostly present in the form of sealed right-side-out vesicles. Detergents like digitonin have, therefore, to be added at about 130 μM concentration to allow an access of the substrate UDPglucose and of effectors (Mg^{2+}, spermine, cellobiose) to the cytoplasmic side. It is difficult, therefore, to decide to which extent the detergent might directly activate the synthase *in vitro* and contribute *in vivo* to callose synthesis in induced cells. Similarly, chitosan and other polycations also cause some direct stimulation of the 1,3-ß-glucan synthase in digitonin-permeabilized microsomes (Kauss und Jeblick 1985). Whether this is relevant to *in vivo* callose synthesis is also not known yet. If such a direct activation of the 1,3-ß-glucan synthase by the substances listed in Table 1 contributes to callose synthesis *in vivo*, similar substances must arise endogenously when callose synthesis is induced by mechanical stress (Kauss 1987).

Still other events in addition to Ca^{2+} uptake might, however, be involved in the regulation of callose synthesis. It appears possible that an increase of cytoplasmic [Ca^{2+}] could occur only if additionally the vacuolar Ca^{2+} accumulation is disturbed. Such an effect could be caused by the inositolphosphate system (Ranjeva et al 1988) but whether this second message system is triggered by the callose elicitors is also not known.

INDUCTION OF SECONDARY METABOLITES BY CALLOSE ELICITORS

Under conditions where chitosan (Köhle et al 1984) and digitonin (Kauss 1987) induced a rapid callose synthesis in soybean suspension cells, we also observed a slower formation of the phytoalexin glyceolline. This has been confirmed by Bonhoff and Grisebach (1988) with digitonin and tomatin, using soybean roots as an experimental system. More details on the induction of secondary metabolites by chitosan have been recently reported for the synthesis of coumarin derivatives in parsley cell suspensions (Conrath et al 1989). In conditioned growth medium, induction of callose and coumarin synthesis both increased with the degree of polymerization of chitosan, and the cells became more sensitive to chitosan with addition of 1 mM reduced glutathione. These results suggest that both processes have initial steps in common, although fully different pathways and regulatory features are involved. The induced Ca^{2+} uptake might directly represent such a common event. Alternatively or in addition, endogenous phytoalexin elicitors (oligosaccharides, free polyunsaturated fatty acids) might be formed from the cell wall or cell membrane due to necrosis, which occurs to a certain degree when the cells are induced to form callose by the substances listed in Table 1.

ACKNOWLEDGEMENTS

Financial support by the Deutsche Forschungsgemeinschaft, the Fonds der Chemischen Industrie and by a cooperative NATO research grant is acknowledged. We also thank T.J. Buckhout for editing the English manuscript.

REFERENCES

Aist JR, Gold RE, Bayles CJ, Morrison GH, Chandra S, Israel HW (1988) Evidence that molecular components of papillae may be involved in ml-o resistance to barley powdery mildew. Physiol Mol Plant Pathol 33:17-32

Atkinson MM, Baker CJ, Collmer A (1986) Transient activation of plasmalemma K^+ efflux and H^+ influx in tobacco by a pectate lyase isozyme from *Erwinia chrysanthemi*. Plant Physiol 82:142-146

Bolard J (1986) How do the polyene macrolide antibiotics affect the cellular membrane properties? Biochim Biophys Acta 864:257-304

Bonhoff A, Grisebach H (1988) Elicitor-induced accumulation of glyceollin and callose in soybean roots and localized resistance against *Phytophthora megasperma* f. sp. *glycinea*. Plant Sci 54:203-209

Christensen O (1987) Mediation of cell volume regulation by Ca^{2+} influx through stretch-activated channels. Nature 330:66-68

Conrath U, Domard A, Kauss H (1989) Chitosan-elicited synthesis of callose and of coumarin derivatives by parsley cell suspensions. Plant Cell Reports, in press

Delmer DP (1987) Cellulose biosynthesis. Annu Rev Plant Physiol 38:259-290

Eilam Y, Grossowicz N (1982) Nystatin effects on cellular calcium in *Saccharomyces cerevisiae*. Biochim et Biophys Acta 692:238-423

Eschrich W (1956) Kallose. Protoplasma 47:487-530

Falke LC, Edwards KL, Pickard BG, Misler S (1988) A stretch-activated anion channel in tobacco protoplasts. FEBS Lett 237:141-144
Fink J, Jeblick W, Blaschek W, Kauss H (1987) Calcium ions and polyamines activate the plasma membrane-located 1,3-ß-glucan synthase. Planta 171:130-135
Gögelein H, Hüby A (1984) Interaction of saponin and digitonin with black lipid membranes and lipid monolayers. Biochim Biophys Acta 773:32-38
Gold RE, Aist JR, Hazen BE, Stolzenburg MC, Marshall MR, Israel HW (1986) Effects of calcium nitrate and chlortetracycline on papilla formation, ml-o resistance and susceptibility of barley to powdery mildew. Physiol Plant Pathol 29:115-129
Hächler H, Hohl HR (1984) Temporal and spatial distribution patterns of collar and papillae wall appositions in resistant and susceptible tuber tissue of *Solanum tuberosum* infected by *Phytophthora infestans*. Physiol Plant Pathol 24:107-118
Kauss H (1987) Some aspects of calcium-dependent regulation in plant metabolism. Annu Rev Plant Physiol 38:47-72
Kauss H (1990) Role of the plasma membrane in host/pathogen interactions. In: Larsson Ch, Møller IM (eds) The Plant Plasma Membrane - Structure, Function and Molecular Biology. Springer-Verlag Berlin Heidelberg, in press
Kauss H, Jeblick W (1985) Activation by polyamines, polycations, and ruthenium red of the Ca^{2+}-dependent glucan synthase from soybean cells. FEBS Lett 185:226-230
Kauss H, Jeblick W (1986) Influence of free fatty acids, lysophosphatidylcholine, platelet-activating factor, acylcarnitine, and Echinocandin B on 1,3-ß-D-glucan synthase and callose synthesis. Plant Physiol 80:7-13
Kauss H, Jeblick W, Domard A (1989) The degrees of polymerization and N-acetylation of chitosan determine its ability to elicit callose formation in suspension cells and protoplasts of *Catharanthus roseus*. Planta, in press
Köhle H, Young DH, Kauss H (1984) Physiological changes in suspension-cultured soybean cells elicited by treatment with chitosan. Plant Sci Lett 33:221-230
Köhle H, Jeblick W, Poten F, Blaschek W, Kauss H (1985) Chitosan-elicited callose synthesis in soybean cells as a Ca^{2+}-dependent process. Plant Physiol 77:544-551
Mayer MG, Ziegler E (1988) An elicitor from *Phytophthora megasperma* f. sp. *glycinea* influences the membrane potential of soybean cotyledonary cells. Physiol and Mol Plant Pathol 33:397-407
Ojalvo I, Rokem JS, Navon G, Goldberg I (1987) ^{31}P NMR study of elicitor treated *Phaseolus vulgaris* cell suspension cultures. Plant Physiol 85:716-719
Osswald WF, Zieboll S, Elstner EF (1985) Comparison of pH changes and elicitor induced production of glyceollin isomers in soybean cotyledons. Z Naturforsch 40c:477-481
Pavlovkin J, Novacky A, Ullrich-Eberius CI (1986) Membrane potential changes during bacteria-induced hypersensitive reaction. Physiol Mol Plant Pathol 28:125-135
Pelissier B, Thibaud JB, Grignon C, Esquerré-Tugayé MT (1986) Cell surfaces in plant-microorganism interactions. VII. Elicitor preparations from two fungal pathogens depolarize plant membranes. Plant Sci 46:103-109
Ranjeva R, Carrasco A, Boudet AM (1988) Inositol trisphosphate stimulates the release of calcium from intact vacuoles isolated from *Acer* cells. FEBS Lett 230:137-141
Ride JP, Drysdale RB (1972) A rapid method for the chemical estimation of filamentous fungi in plant tissue. Physiol. Plant Pathol 2:7-15
Storm DR, Rosenthal KS, Swanson PE (1977) Polymyxin B and related peptide antibiotics. Annu Rev Biochem 46:723-763
Waldmann T, Jeblick W, Kauss H (1988) Induced net Ca^{2+} uptake and callose biosynthesis in suspension-cultured plant cells. Planta 173:88-95
Wise BC, Glass DB, Jen Chou CH, Raynor RL, Katoh N, Schatzman RC, Turner RS, Kibler RF, Kuo JF (1982) Phospholipid-sensitive Ca^{2+}-dependent protein kinase from heart. J Biol Chem 257:8489-8495
Young DH, Kauss H (1983) Release of calcium from suspension-cultured *Glycine max* cells by chitosan, other polycations, and polyamines in relation to effects on membrane permeability. Plant Physiol. 73:698-702

MOLECULAR BIOLOGY OF CHEMOTAXIS IN *AGROBACTERIUM*

C.H. Shaw, G.J. Loake, A.P. Brown & C.S. Garrett
Department of Biological Sciences, University of Durham, Science Laboratories, South Road, Durham DH1 3LE, England

INTRODUCTION

Our interest in chemotaxis stems from the concept that unless bacteria are present in huge numbers, and evenly distributed throughout the soil, they must arrive from somewhere to interact with a plant. *Agrobacterium tumefaciens* thus possesses a multifunctional system capable of triggering either chemotaxis or gene-induction depending upon the ligand concentration, as part of an intricate mechanism for sensing and guiding *A. tumefaciens* towards wounded cells.

MOTILITY AND CHEMOTAXIS IN *AGROBACTERIUM TUMEFACIENS*

A. tumefaciens is vigorously motile, with speeds up to 60 μm.sec^{-1}, over >500 μm runs (Loake *et al* 1988). The motility pattern is reminiscent of *Rhizobium* (Götz & Schmitt 1987; Loake *et al* 1988) distinct from *E. coli* and suggests these organisms use unidirectional clockwise flagellar rotation.

A. tumefaciens is chemotactic to a range of sugars and amino acids found in plant extracts/exudates (Loake *et al* 1988). The most powerful and sensitive response is to sucrose (peak chemotaxis: 10^{-6}M; threshold: $<10^{-7}$M) 1000X more sensitive than the equivalent attraction in *E. coli* (Adler 1973).

In the Enterobacteriaceae, chemotactic responses are effected by inner membrane proteins known as methyl-accepting chemotaxis proteins (MCPs; Springer *et al* 1979). In tactic responses, methyl groups are transferred from S-adenosyl methionine to sites in the MCP C-terminal domains (Kehry & Dahlquist 1982; Kehyr *et al* 1983; Terwilliger *et al* 1986. The methylation level is determined by the activities of the *che*B methylesterase (Springer & Koshland 1977; Stock & Koshland 1978).

A similar system operates in *A. tumefaciens* (Loake *et al* 1989). Methionine auxotrophs, though motile, without methionine supplementation are non-chemotactic. The degree of restoration of chemotaxis by methionine analogues is related to their affinity for S-adenosyl-methionine synthetase. Methyl groups are transferred from methionine to

NATO ASI Series, Vol. H 36
Signal Molecules in Plants and Plant-Microbe Interactions
Edited by B. J. J. Lugtenberg

MCP-like proteins and olignucleotide probes from the conserved domain of *E.coli* MCPs hybridise to specific fragments in *Agrobacterium* DNA.

We have created a battery of Tn5 mutants deficient in motor function (*mot*) flagellum production (*fla*) or chemotaxis (*che*; Loake & Shaw 1989). Using pulsed-field gel electrophoresis, we are mapping the insertions onto a chromosomal restriction map of *A. tumefaciens*. From 6 of the mutants the Tn5 flanking sequences have been used to clone 3 of the wild type genes. Early indications are of extensive sequence similarity between *Agrobacterium* and *Rhizobium fla, mot* and *che* genes. One mutant (*che-2*) is intriguing. It is motile, tumbles excessively, thus non-chemotactic, and exhibits reduced MCP methylation (Loake *et al* 1989). In *E. coli* such mutants show increased MCP methylation and have *cheZ* or *cheB* lesions.

Ti-PLASMID-DETERMINED CHEMOTAXIS

Wounded plant tissue exudes a mixture of aromatic and aliphatic chemicals (Stachel *et al* 1985). Octopine and nopaline Ti-plasmids confer upon *A. tumefaciens* chemotaxis to *vir*-inducing phenolics (Ashby *et al* 1987; 1988; Shaw *et al* 1988). Plant wound phenolics fall into three groups: strong *vir*-inducers (eg acetosyringone, syringic acid or sinapinic acid) which require the Ti-plasmid for chemotaxis; weak or non-inducers (eg catechol and vanillyl alcohol) which are attractants for cured strains; and non *vir*-inducing, non attracting compounds (eg vanillic acid). Thus, *vir*-inducers are chemoattractants only with the Ti-plasmid present, these responses being much more sensitive than those evoked by non-inducers. For strong *vir*-inducers, the response threshold is $>10^{-8}$M, with the peak at 10^{-7}M, some 100X lower than the maximal *vir*-inducing concentration (Stachel *et al* 1985). Furthermore, for optimal chemotaxis/*vir*-induction, 4' hydroxyl (Bolton *et al* 1986) 3' and 5' O-methyl, and 1' polar side chains are essential (Ashby *et al* 1988). There is a separate less sensitive chromosomally-encoded phenolic receptor (Ashby *et al* 1987; 1988; Parke *et al* 1987).

The Ti-plasmid loci required for chemotaxis towards acetosyringone are *vir*A and *G* (Shaw *et al* 1988) indicating a multifunctional role for *virA/G*; triggering chemotaxis or *vir*-induction depending on the ligand concentration. The constitutive level of *virA/G* expression (Rogowsky *et al* 1987) is sufficient to effect chemotaxis, as under peak chemotaxis conditions, *vir*-induction is unlikely to occur. Indeed, *vir*-induction at high acetosyringone concentrations may suppress chemotaxis or motility (Shaw *et al* 1988).

virA is sufficient to confer acetosyringone chemotaxis upon *E. coli*, implying an *E. coli* function complements for *virG*. This response requires at least *motA & B*, *cheA*, *B*, *Y & Z* (Shaw *et al* 1989) indicating that the chemotaxis systems of *E. coli* and *A. tumefaciens* are functionally conserved.

Due to the similarity of *virA/G* to other component chemoreceptor-regulators, eg. *envZ/ompR* and *ntrB/C* (Winans *et al* 1986; Melchers *et al* 1986; Leroux *et al* 1987) it is thought that *virA* is a chemoreceptor, and tranduces a signal to the *virG* transcriptional regulator. *virA* resembles MCPs structurally (Leroux *et al* 1987) and has conservation with *cheA* (Stock *et al* 1988). *virG* is similar to *cheB* and *Y* (Winans *et al* 1986; Melchers *et al* 1986). Possibly *virA* acts as a binding protein, or MCP during chemotaxis and *virG* transmits a signal to the flagellar apparatus. Phosphorylation cascades are involved in *NtrB* regulation of gene expression (Ninfa & Magasanik 1986; Ninfa *et al* 1987) and *cheA* mediated chemotaxis (Hess *et al* 1988; Oosawa *et al* 1988). Thus, *virA/G* may be part of a phosphorylation cascade, with a branch point allowing chemotaxis or *vir*-induction.

CHEMOTAXIS IN *RHIZOBIUM*

Nod-inducing flavones are good chemoattractants for *Rhizobium* (Caetano-Anolles *et al* 1988; Aguilar *et al* 1988; Armitage *et al* 1988). Both *nod*-induction and chemotaxis in response to flavones require *nodD* (Caetano-Anolles *et al* 1988). Different *Rhizobium* strains recognise distinct flavones as attractants, implying chemotaxis may feature in host-range determination. The commonality of chemotaxis and gene regulation in response to exudates suggests such responses may be a common factor in plant/microbe interactions.

AGROBACTERIUM IN THE RHIZOPHERE

Attraction of Agrobacteria towards sugars and amino acids in plant exudates may help explain its prevalence in the rhizosphere. Accumulation of *A. tumefaciens* to plant roots occurs (Schroth *et al* 1971). Mutants have been isolated which are deficient in chemotaxis towards root cap cells or exudates (Hawes *et al* 1988). Soluble factors from homogenates of tobacco, *Kalanchoë* and wheat roots and shoots evoke strong chemotactic responses from *A. tumefaciens* which are enhanced by Ti-plasmid presence (Ashby *et al* 1988). This supports the involvement of Ti-plasmid specified chemotaxis in the early interactions between *Agrobacterium* and plants (Shaw *et al* 1986). It also suggests that attraction towards the plant is not the failure in monocot interactions.

Attraction to *vir*-inducers is probably only important on the micrometre scale, in the rhizosphere or on the rhizoplane. Migration up the concentration gradient, towards the wounded plant cells (and higher inducer concentrations) will effect *vir-induction*. Thus, *A. tumefaciens* has an extremely sensitive mechanism to attract and guide it towards plant wounds. As *acetosyringone* release is an active process (M. Van Montagu & A. Darvill, personal communications) not requiring cell wall degradation (Ashby *et al* 1988) it is likely that this has evolved to exclude signals from leaf litter, which is a potent source of phenolic lignin degradation products. Therefore, *A. tumefaciens* would be drawn towards susceptible and not dying cells in a wounded site.

REFERENCES

Adler J (1973) A method for measuring chemotaxis and use of the method to determine optimum conditions for chemotaxis by *Escherichia coli*. *J. Gen Micro* 74:77-91

Aguilar JMM, Ashby AM, Richards AJM, Loake GJ, Watson MD, Shaw CH (1988) Chemotaxis of *Rhizobium leguminosarum* biovar *phaseoli* towards flavonoid inducers of the symbiotic nodulation genes. *J Gen Micro* 134:2741-2746

Armitage JP, Gallagher A, Johnston AWB (1988) Comparison of the chemotactic behaviour of *Rhizobium leguminosarum* with and without the nodulation plasmid. *Molec Micro* 2:743-749

Ashby AM, Watson MD, Shaw CH (1987) A Ti-plasmid determined function is responsible for chemotaxis towards the plant wound product aceto-syringone. *FEMS Micro Letts* 41:189-192

Ashby AM, Watson MD, Loake GJ, Shaw CH (1988) Ti-plasmid specified chemotaxis of *Agrobacterium tumefaciens* $C58C^1$ to *vir*-inducing phenolics and soluble factors from monocotyledonous and dicotyledonous plants. *J. Bacteriol* 170:4181-4187

Bolton GW, Nester EW, Bordon MP (1986) Plant phenolic compounds induce expression of the *Agrobacterium tumefaciens* loci needed for virulence. *Science* 232:983-985

Caetano-Anolles G, Crist-Estes DK, Bauer D (1988) Chemotaxis of *Rhizobium meliloti* to the plant flavone luteolin requires functional nodulation genes. *J Bacteriol* 170:3164-3169

Gotz R, Schmitt R (1987) *Rhizobium meliloti* swims by unidirectional intermittent rotation of right-handed flagellar helices. *J Bacteriol* 169:3146-3150

Hawes MC, Smith LY, Howarth AJ (1988) *Agrobacaterium tumefaciens* mutants deficient in chemotaxis to root exudates. *Molec Plant-Microbe Interact* 1:182-186

Hess JF, Oosawa K, Kaplan N, Simon MI (1988) Phosphorylation of three proteins in the signalling pathway of bacterial chemotaxis. *Cell* 53: 79-87

Kehry MR, Dahlquist FW (1982) The methyl-accepting chemotaxis proteins of *Escherichia coli*: identification of the multiple methylation sites on methyl-accepting chemotaxis protein I. *J Biol Chem* 257:10378-10386

Kehry MR, Engstrom P, Dahlquist FW, Hazelbauer GL (1983) Multiple covalent modifications of Trg a sensory transducer of *Escherichia coli*. *J Biol Chem* 258:5050-5055

Leroux B, Yanofsky MF, Winans SC, Ward JE, Ziegler SF, Nester EW (1987) Characterisation of the *vir*A locus of *Agrobacterium tumefaciens*: a transcriptional regulator and host range determinant. *EMBO Journal* 6:849-856

Loake GJ, Ashby AM, Shaw CH (1988) Attraction of *Agrobacterium tumefaciens* C58C^1 towards sugars involves a highly sensitive chemotaxis system. *J Gen Micro* 134:1427-1432

Loake GJ, King TJ, Shaw CH (1989) Involvement of a methionine-dependent MCP system in chemotactic responses of *Agrobacterium tumefaciens*. Manuscript in preparation

Loake GJ, Shaw CH (1989) Isolation and characterisation of chemotaxis and motility mutants of *Agrobacterium tumefaciens*. Manuscript in preparation

Melchers LS, Thompson DV, Idler KB, Schilperoort RA, Hooykaas PJJ (1986) Nucleotide sequence of the virulence gene *vir*G of the *Agrobacterium tumefaciens* octopine Ti-plasmid: significant homology between *vir*G and the regulatory genes *omp*R *pho*B and *dye* of *Escherichia coli*. *Nucl Acids Res* 14:9933-9942

Ninfa AJ, Magasanik B (1986) Covalent modification of the *gln*G product NRI by the *gln*L product NRII regulates the transcription of the *gln*ALG operon in *Escherichia coli*. *Proc Natl Acad Sci USA* 83:5909-5913

Ninfa AJ, Reitzer LJ, Magasanik B (1987) Initiation of transcription at the bacterial *gln*Ap2 promoter by purified *E.coli* components is facilitated by enhancers. *Cell* 50:1039-1046

Oosawa K, Hess JF, Simon MI *che*A mutants defective in chemotaxis show modified protein phosphorylation. *Cell* 53:89-96

Parke D, Ornston LN, Nester EW (1987) Chemotaxis in plant phenolic inducers of virulence genes is constitutively expressed in the absence of the Ti-plasmid in *Agrobacterium tumefaciens*. *J Bacteriol* 169:5336-5338

Rogowsky PM, Close TJ, Chimera JA, Shaw JJ, Kado CI (1987) Regulation of the *vir* genes of *Agrobacterium tumefaciens* plasmid pTiC58. *J Bacteriol* 169:5101-5112

Schroth MN, Weinhold AR, McCain AH, Hildebrand DC, Ross N (1971) Biology, control of *Agrobacterium tumefaciens*. *Hilgardia* 40:536-552

Shaw CH, Ashby AM, Watson MD (1986) Plant tumour induction. *Nature* 324:415

Shaw CH, Ashby AM, Brown A, Royal C, Loake GJ, Shaw CH (1988) *VirA & G* are the Ti-plasmid functions required for chemotaxis of *Agrobacterium tumefaciens* towards acetosyringone. *Molec Micro* 2:413-418

Shaw CH, King TJ, Duffell A, Loake GJ (1989) Acetosyringone chemotaxis in *Ecoli* requires *virA cheA B Y & Z motA & B* but not *virG*. Manuscript in preparation

Springer MS, Goy MF, Adler J (1979) Protein methylation in behavioural control mechanisms and in signal transduction. *Nature* 280:279-284

Springer WR, Koshland DE (1977) Identification of a protein methyltransferase as the *che*R gene product in the bacterial sensing system. *Proc Natl Acad Sci USA* 74:533-537

Stachel SE, Messens E, Van Montagu M, Zambryski P (1985) Identification of the signal molecules produced by wounded plant cells that activate T-DNA transfer in *Agrobacterium tumefaciens*. *Nature* 318:624-629

Stock A, Chen T, Welsh D, Stock J (1988) *Che*A protein a central regulator of bacterial chemotaxis belongs to a family of proteins that control gene expression in response to changing environmental conditions. *Proc Natl Acad Sci USA* 85:1403-1407

Stock JB, Koshland DE (1978) A protein methylesterase involved in bacterial chemosensing. *Proc Natl Acad Sci USA* 75:3659-3663

Terwilliger TC, Wang JY, Koshland DE (1986) Surface structure recognised for covalent modification of the aspartate receptor in chemotaxis. *Proc Natl Acad Sci USA* 83:6707-6710

Winans SC, Ebert PR, Stachel SE, Gordon MP, Nester EW (1986) A gene essential for *Agrobacterium* virulence is homologous to a family of positive regulatory loci. *Proc Natl Acad Sci USA* 83:8278-8282

CONDITIONAL REQUIREMENT FOR CHEMOTAXIS IN INDUCTION OF ROOTY TUMORS BY AGROBACTERIUM TUMEFACIENS ON ROOTS OF PISUM SATIVUM

Martha C. Hawes
Departments of Plant Pathology/Molecular and Cellular Biology
104 Forbes Building
University of Arizona
Tucson, AZ 85721
U.S.A

The rhizosphere is a zone of intense microbial activity where the successful establishment of symbiotic or pathogenic organisms can have a dramatic impact on plant health. Little is known about production and regulation of plant factors that influence the rhizosphere environment. Components of exudates from roots can influence the gene expression and growth of soilborne microorganisms (for ex Stachel et al 1985, Peters et al 1986, reviewed in Curl & Truelove 1986), and can act as chemoattractants (for ex Ashby et al 1988, Caetano-Anolles et al 1988, Hawes et al 1988). The primary sources of such exudates in healthy plants of some species are secretory cells from the root cap (Oades 1978). Sloughed root cap (SRC) cells, which can be isolated nondestructively (Hawes and Wheeler 1982), and can be cultured (Hawes and Pueppke 1986), have been shown to survive for some time in the rhizosphere independently of the root (Vermeer and McCully 1982). Not unexpectedly, carbon-rich SRC cells can support the growth of bacteria in culture (Hawes and Pueppke 1989), and have been shown to act as nuclei for microbial colonies in soil. However, a series of studies in my laboratory suggest that the impact of these cell populations may go beyond the nonspecific stimulation of microbial growth. In studies with several fungal and bacterial pathogens, I have found that the cells from different plant species and genotypes exhibit dramatic selectivity in binding, chemotaxis, and susceptibility to infection (Goldberg et al 1989, Hawes 1983, 1989, Hawes & Pueppke 1985,1986,1989, Hawes et al 1989, Hawes & Wheeler 1982). I am using Agrobacterium tumefaciens as a model system to test the hypothesis that the expression of such properties in SRC cells may act to regulate microbial populations in the rhizosphere.

NATO ASI Series, Vol. H 36
Signal Molecules in Plants and Plant-Microbe Interactions
Edited by B. J. J. Lugtenberg

In preliminary experiments, we studied the relationship between A. tumefaciens binding to SRC cells and root disease or rhizosphere colonization. Binding ability of SRC cells from different plant species is strongly correlated with susceptibility to crown gall (Hawes & Pueppke 1985,1987,1989a). Thus, cells from plants susceptible to tumorigenesis bind large numbers of A. tumefaciens cells, whereas plants whose cells do not bind do not develop tumors. In reciprocal experiments, mutants or strains which do not bind to root cap cells of a particular plant do not develop tumors on the roots of that plant. A dramatic reduction in rhizosphere colonization ability of chromosomal virulence (chv) mutants of A. tumefaciens, which are deficient in binding to SRC and other plant cells (Douglas et al 1982, Hawes & Pueppke 1989a) is consistent with the hypothesis that inability to form stable interactions with roots or SRC cells reduces ability to colonize (Hawes & Pueppke 1989). However, the pleiotropic effect of the chv mutation makes it impossible to draw conclusions about the relative contributions of binding and chemotaxis to reduced colonization. I therefore set out to isolate mutants with more specific defects in recognition.

Mutants deficient in chemotaxis to root exudates were selected and used to study the ecological importance of the phenomenon in crown gall pathogenesis (Hawes et al 1988, Hawes & Smith, unpublished). Populations of SRC cells provided a developmentally and experimentally defined source of root exudates, and excised root tips were used as a source of chemoattractants from wounded tissue. Mutants selected for deficiency in chemotaxis to SRC cells or to excised root tips include nonmotile, slow-migrating, and fully motile but chemotaxis-deficient strains (Hawes et al 1988). One "wound-specific" mutant is attracted to excised root tips but not to SRC cells. Genes controlling the phenotypes in all mutants are chromosomally encoded, and the Ti plasmid does not appear to influence mutant behaviour under the experimental conditions in these studies. The mutants all exhibit wild type binding abilities.

Virulence of the mutants was compared with that of the parent by two methods: an in vitro growth pouch assay, and a soil tumorigenesis assay. In the first method, wounded roots of seedlings are immersed in serially diluted inoculum, and then maintained in moist growth pouches (Hawes et al 1989). Under those conditions, inoculated roots consistently developed discrete rounded tumors whose weight was proportional to inoculum concentration. Tumor development in response to inoculation with mutant strains was indistinguishable from response to the parent strain (Hawes and Smith, unpublished). These data indicate that the effect of Tn5 insertion in the mutants is relatively specific, with no apparent effect on basic functions required to induce tumors.

Chemotaxis and motility mutants exhibited wild type virulence on roots of plants grown in sand, suggesting that even under semi-natural conditions directed movement toward roots is not always required. In contrast, however, mutant strains were avirulent on seedlings grown in silt-loam soil. Chemotaxis is therefore apparently a conditional determinant of virulence that is crucial in some circumstances and superfluous in others. Particularly intriguing was the discovery that the "wound-specific" mutant was also avirulent on plants grown in soil, despite the presence of wounds which release substances that are chemoattractants to the mutant in in vitro assays (Ashby et al 1988, Hawes et al 1988). Only one wound-specific mutant has been isolated to date, and generalizations therefore have to be made with caution. Nevertheless, the data suggest that, on pea plants grown under the test conditions, attraction to molecules released from SRC cells appear to be more important in pathogenicity than molecules released from wounds (Hawes and Smith, unpublished).

It is interesting to note that the phenotype of infection by A. tumefaciens on roots of plants grown in soil or sand was distinct from that of plants in growth pouches (Hawes & Smith, unpublished). Most inoculation sites developed a profusion of roots rather than tumors, with symptoms that were indistinguishable from hairy root disease induced by A. rhizogenes. Apparently the expression of plant or T-DNA hormone genes in roots of soil-grown plants is distinct from that of plants grown in pouches.

Preliminary experiments indicate that the separation of SRC cells from the root is developmentally regulated, and that some environmental factors that have been reported to influence "exudation" of carbon-rich compounds actually influence the number of SRC cells that are released from the root (Hawes, unpublished). These experiments suggest that treatments that modify levels of root exudates in the rhizosphere could do so in part by regulating the expression of genes or gene products that control root cap cell sloughing. The isolation and characterization of such genes would make it possible to determine the impact of the phenomenon on plant health, and perhaps also to modify the rhizosphere in order to manipulate the balance of rhizosphere inhabitants.

REFERENCES

Ashby AM, Watson MD, Loake GJ, Shaw CH (1988) Ti plasmid specified chemotaxis of Agrobacterium tumefaciens toward vir inducing phenolic compounds and soluble factors from plants. J Bacteriol 170: 4181-4187

Caetano-Anolles G, Wall LG, DeMicheli AT, Macchi EM, Bauer WD, Favelukes G (1988) Role of motility and chemotaxis in efficiency of nodulation by Rhizobium meliloti. Plant Physiol 86: 1228-1235

Curl EA, Truelove B (1986) The Rhizosphere. Springer-Verlag, Berlin Heidelberg New York Tokyo

Douglas CJ, Halperin W, Nester E (1982) Agrobacterium tumefaciens mutants affected in attachment to plant cells. J Bacteriol 152: 1265-1275

Goldberg NP, Hawes MC, Stanghellini ME (1989) Specific attraction to and infection of cotton root cap cells by zoospores of Pythium dissotocum. Can J Bot (in press)

Hawes MC (1983) Detection of effects of nuclear genes on susceptibility to Helminthosporium maydis race T by a root cap cell bioassay for HMT-toxin. Physiol Plant Pathol 24: 163-168

Hawes MC (1989) Sloughed root cap cells: A regulator of microbial populations in the rhizosphere? Plant & Soil (in press)

Hawes MC, Pueppke SG (1985) Relationship between binding of root cap cells to Agrobacterium tumefaciens and susceptibility of plants to crown gall. Current Topics in Plant Biochemistry and Physiology, Vol. 4. D. R. Randall,

D. Blevins, R. L. Larson, and B. J. Rapp Eds.
Hawes MC, Pueppke SG (1986) Sloughed peripheral root cap cells: yield from different species and callus formation from single cells. Amer J Bot 73: 1466-1473
Hawes MC, Pueppke SG (1987) Correlation between binding of Agrobacterium tumefaciens by root cap cells and susceptibility of plants to crown gall. Plant Cell Rep 6: 287-290
Hawes MC, Pueppke SG (1989) Reduced rhizosphere colonization ability of Agrobacterium tumefaciens chromosomal virulence mutants. Plant Soil 113: 129-132
Hawes MC, Pueppke SG (1989a) Variation in binding and virulence of Agrobacterium tumefaciens chromosomal virulence mutants on different plant species. Plant Physiol (in press)
Hawes MC, SL Robbs, SG Pueppke (1989) Use of a root tumorigenesis assay to detect genotypic variation in susceptibility of pea to crown gall. Plant Physiol 90: 180-184
Hawes MC, Smith LY, Howarth AJ (1988) Agrobacterium tumefaciens mutants deficient in chemotaxis to root exudates. Molec Plant Microbe Interactions 1: 182-186
Hawes MC, Wheeler HE (1982) Factors affecting victorin-induced cell death: temperature and plasmolysis. Physiol Plant Pathol 20: 137-144
Oades JM (1978) Mucilages at the root surface. J Soil Sci 29: 1-16
Vermeer J, ME McCully (1982) The rhizosphere in Zea: new insight into structure and development. Planta 156: 45-61

ROLE OF ATTACHMENT OF RHIZOBIUM LEGUMINOSARUM CELLS TO PEA ROOT HAIR TIPS IN TARGETING SIGNALS FOR EARLY SYMBIOTIC STEPS

Ben J.J. Lugtenberg, Gerrit Smit, Clara Díaz and Jan W. Kijne
Leiden University, Department of Plant Molecular Biology,
Nonnensteeg 3, 2311 VJ Leiden,
The Netherlands

INTRODUCTION

Attachment of Rhizobium and Bradyrhizobium bacteria to the tip of the root hairs of their host plants is supposed to be an essential step of the usual nodulation pathway of these organisms (e.g. Rolfe and Shine 1981). Attachment is a prerequisite for marked curling of root hairs (Yao and Vincent 1969; Van Batenburg et al 1986). Attachment is often assumed to be a determinant of host plant specificity (Bohlool and Schmidt 1974; Dazzo and Hubbell 1975) and it was hypothesized that host specificity at this level is the result of specific binding of the homologous rhizobial partner to the cell surface of the host plant's root hair. Evidence was provided that the molecular basis for this recognition mechanism is the binding of a host plant lectin to a bacterial sugar residue, since attachment was largely reduced by the presence of lectin-haptenic monosaccharides (Dazzo et al 1976). Moreover, lectin-mediated attachment to clover root hairs was reported to be dependent on the presence of the Sym plasmid (Dazzo et al 1985), the genetic element that was known to determine host specificity (Johnston et al 1978). Consistent with this notion was also that heterologous rhizobial strains usually fail to form infection threads.

However, several laboratories reported results which apparently were inconsistent with the hypothesis that a lectin-mediated attachment step is a major determinant of host specificity (for review see Smit 1988). (i) Several groups were unable to demonstrate specific lectin-mediated binding of

NATO ASI Series, Vol. H 36
Signal Molecules in Plants and Plant-Microbe Interactions
Edited by B. J. J. Lugtenberg

homologous rhizobia to legume root hairs. (ii) Heterologous rhizobia were reported to adhere equally well to host and non-host plant surfaces in comparison with homologous rhizobia. (iii) Absence of the Sym plasmid did not affect rhizobial attachment. (iv) If attachment is a prerequisite for marked root hair curling, the lectin-recognition hypothesis is inconsistent with the reported induction of root hair curling by heterologous rhizobia. It should be noted that the experiments mentioned under (i) through (iv) have been carried out under conditions which largely differed (e.g. with respect to medium composition, growth phase, criteria for attachment etc.).

In an attempt to systematically elucidate the role of attachment in nodulation we decided to define attachment relevant to nodulation as adhesion at the site of infection. Therefore attachment was microscopically monitored and ranked in classes of increasing degree of bacterial attachment to the tips of pea root hairs (Smit et al 1986). Using this assay we asked the following questions relevant to attachment. (i) What is the influence of bacterial growth medium and growth phase? (ii) Is attachment a host-specific process? (iii) Which adhesins play a role in attachment? Is lectin one of these adhesins? (iv) What is the role of the Sym plasmid in attachment? (v) Is attachment a prerequisite for nodulation, e.g. by generating a molecular signal or targeting nod genes to a site for optimal interaction with signal molecules?

ATTACHMENT ASSAY

Attachment to tips of developing root hairs was ranked from class 1 (no attached bacteria) to class 4 (many attached bacteria forming a cap-like aggregate on the tip of the root hair (Smit et al 1986). Since attachment was quantified at the site of infection, this assay must be superior over a number of other assays in which attachment was quantified without taking the precize infection site into account. Since prior to examination the roots are washed several times, only irreversible attachment is measured. This assay showed that attachment can be dissected into two steps, namely (i) binding of single bacteria to the root hair surface and (ii) formation of bacterial aggregates at the root hair tip (cap formation) (Smit et al 1986).

ROLE OF BACTERIAL GROWTH MEDIUM AND GROWTH PHASE ON ATTACHMENT AND INFECTION ABILITY

Bacteria were either grown in TY medium (rich in tryptone, yeast extract and calcium ions) (Smit et al 1986) or in YM medium (rich in mannitol and poor in yeast extract) (Kijne et al 1988). Logarithmic growth of rhizobia in TY-medium is inhibited by carbon-limitation, whereas manganese ions are the primary growth-limiting nutrient in YM-medium. Optimal cap formation appeared to be dependent on the growth phase, and coincided with nutrient-limitation. Addition of glucose to TY medium shifted optimal cap formation to a higher bacterial density (Smit et al 1986). Similarly, it was observed that in YM medium Mn^{2+} limitation is related to optimal attachment (Kijne et al 1986). In conclusion, different types of growth limitation result in optimal attachment. It should be noted that several attachment characteristics of bacteria grown in the two mentioned media also differ (see later on). Interestingly, later stages of the infection process also appeared to be dependent on the growth conditions of the rhizobia. In contrast to Mn^{2+}-limited cells, carbon-limited bacteria harbouring a Sym plasmid appeared to be unable to induce marked root hair curling and infection thread formation (Kijne et al 1988). _nod_ genes of carbon-limited _R.leguminosarum_ cells can readily be induced by appropriate plant flavonoids, suggesting that carbon-limited cells are unable to produce the factor(s) required for marked curling. It can further be speculated that this is related to the observation that carbon-limited cells are not encapsulated and produce greatly reduced amounts of high molecular weight EPS. Given these results it is important to know the factor that limits bacterial growth in the rhizosphere.

ATTACHMENT, HOST SPECIFICITY AND LECTIN

Cap formation on pea root hair tips by several TY-grown heterologous rhizobia (e.g. R.trifolii, R.phaseoli) is practically equally efficient as that by homologous R.leguminosarum cells. Cap formation and also adhesion of rhizobia to the plant root surface was poor for R.meliloti, R.lupini and B.japonicum. Apparently attachment under these conditions is not host-specific. It should be noted that - since the pea lectin hapten 3-O-methyl-D-glucose did not decrease attachment - lectin is apparently not involved in attachment under these conditions (Smit et al 1986).

Comparison of the attachment characteristics of Mn^{2+}- and C-limited R.leguminosarum cells revealed a number of remarkable differences. Cap formation by Mn^{2+}-limited cells, (i) is a much faster process, (ii) is less H^+-sensitive, (iii) is less sensitive to inhibition by salts and (iv) is slowed down by the presence of the pea lectin hapten 3-O-methyl-D-glucose to a rate comparable with the attachment rate of C-limited rhizobia, and (v) is less sensitive to cellulase pretreatment than cap formation by C-limited cells. The latter results suggest that cellulose fibrils (see further on) are not the only adhesins involved in the accumulation of rhizobia on pea root hair tips. Pea lectin may have a similar effect. Preliminary results indicate that also heterologous R.trifolii bacteria (grown in a third growth medium, B^-) are capable of pea lectin-enhanced accumulation on pea root hair tips. This observation is consistent with the notion that extracellular polysaccharides of R.trifolii bind pea lectin (Kamberger 1979). Lectin-mediated accumulation of R.trifolii cells is accompanied by hair curling but not by initiation of infection threads. In conclusion, lectin-enhanced accumulation of bacteria is not sufficient for initiation of infection threads.

MOLECULAR NATURE OF ADHESINS INVOLVED IN ATTACHMENT

So far 3 components have been identified to play a role in the attachment of Rhizobium bacteria to pea root hair tips (see also Fig. 1).

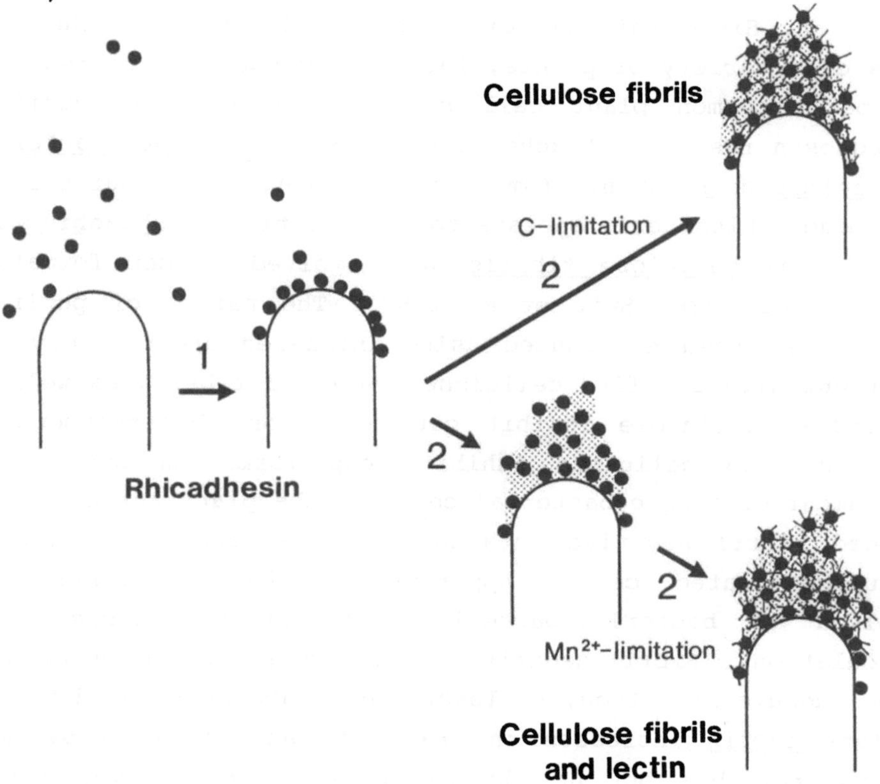

Fig. 1. Attachment of Rhizobium cells to pea root hair tips. Under both C- and Mn^{2+}-limitation rhicadhesin mediates the attachment of single Rhizobium cells to the root hair tips whereas cellulose fibrils mediate the second step, namely aggregate (cap) formation. Under C-limitation the second step occurs slower than under Mn^{2+}-limitation. Increased rate of cap formation in the latter case is correlated with involvement of plant lectin and with infectivity.

Rhicadhesin is a rhizobial surface protein termed rhicadhesin because it is common for Rhizobiaceae. It has been purified as a factor that inhibits attachment of (both C- and Mn^{2+}-limited) bacteria to the surface of pea root hairs (Smit et al 1989b). It has the following properties. 1. It is sensitive to boiling and to protease treatment. 2. It is released from cells by mild blending or growth under Ca^{2+}-limitation. 3. It has a high affinity for hydroxyapatite, which was the basis of a major purification step. 4. Purified rhicadhesin has a IEP of 5.1 and a Mr of 14 kD. 5. It has a high affinity for Ca^{2+} with approx. 5

binding sites per monomer. 6. Ca^{2+} is required for binding to the bacterium, not to the plant. 7. Mutants have not been isolated so far. 8. Since rhicadhesin inhibits attachment to the root hairs of a variety of plants, including monocots, its receptor may be a common plant cell surface molecule. 9. Purified rhicadhesin restores attachment and virulence of Agrobacterium tumefaciens chvB mutants (Smit et al 1989c). 10. To our present knowledge, rhicadhesin appears to be specific for Rhizobiaceae.

Bacterial cellulose fibrils are involved in cap formation (Smit et al 1986, Smit et al 1988). The nature of purified fibrils has been elucidated using infra-red spectroscopy and treatment with purified cellulose. Purified fibrils as well as commercial cellulose inhibit cap formation. Pretreatment of bacteria with cellulose inhibits cap formation but allows attachment of single bacterial cells to the plant surface. Tn5-induced fibril negative mutants have the same properties as cellulose-treated cells. Apparently cellulose fibrils are involved in bacterium-bacterium interactions (aggregation, flocculation). Fibril-negative mutants are able to apparently normal nodule formation, at least under laboratory conditions.

Plant lectin increases the rate of cap formation by Mn^{2+}-limited rhizobia but not by C-limited rhizobia. The latter observation is consistent with the inability of C-limited R.leguminosarum cells to be agglutinated by pea lectin (Díaz 1989). The nature of the bacterial receptor involved in lectin-mediated accumulation/agglutination is not known. More details on the role of the plant lectin, including results with transgenic plants, will be reported by Kijne et al in this volume.

Future research will be directed to characterizing the receptors of rhicadhesin and lectin. Subsequently genetic tools will be used to determine whether these adhesins and their receptors play an essential role in nodulation.

ATTACHMENT, SYM PLASMID AND SIGNALS FOR NODULATION

Comparison of the behaviour of Sym plasmid-containing and

cured strains has shown that the Sym plasmid is not required for the production of rhicadhesin (Smit et al 1986, Smit et al 1988), cellulose fibrils (Smit et al 1986) or the pea lectin receptor (Kijne et al 1988).

There are no indications that attachment as such is absolutely required for or sufficient for an early response in the plant like the production of early nodulins. It is clear, however, that attachment at the site of secretion of flavonoids is very helpful for the optimal induction of nod genes (see also Dazzo et al 1988). Moreover, attachment is responsible for the delivery of the resulting Rhizobium signals (e.g. those in which one or more of the nod genes nodA, B or C play a role) at the correct sites on the target cells. Furthermore it is likely - but not proven - that attachment, in combination with one or more signal molecules, is a prerequisite for root hair curling (Van Batenburg et al 1986), which in turn is likely to be a prerequisite for nodulation. The observation of Yao and Vincent (1969) that curling of clover root hairs is induced by the heterologous R.leguminosarum bacteria, which are not able to interact with clover lectin, shows that lectin-mediated accumulation in general is not a prerequisite for the correct targeting of curling signals.

REFERENCES

Bohlool BB and Schmidt EL (1974) Lectins: a possible basis for specificity in the Rhizobium root nodule symbiosis. Science 185: 269-271

Dazzo FB and Hubbell DH (1975) Cross-reactive antigens and lectins as determinants of symbiotic specificity in the Rhizobium-clover association. Appl Microbiol 30: 1018-1033

Dazzo FB, Napoli CA and Hubbell D (1976) Adsorption of bacteria to roots as related to host specificity in the Rhizobium-clover symbiosis. Appl Environ Microbiol 32: 166-171

Dazzo FB, Hollingsworth RI, Sherwood JE, Abe M, Hrabak EM, Gardiol AE, Pankratz HS, Smith KB and Yang H (1985) Recognition and infection of clover root hairs by Rhizobium trifolii. In: Evans HJ, Bottomley PE and Newton WE (eds) Nitrogen fixation research progress p 239-245. Martinus Nijhoff Publishers BV Dordrecht, The Netherlands

Dazzo F, Hollingsworth R, Philip-Hollingsworth S, Robeles M, Olen T, Salzwedel J, Djordjevic M and Rolfe B (1988) Recognition process in the Rhizobium trifolii - white clover symbiosis. In: Nitrogen fixation: hundred years after. Bothe H, de Bruijn FJ and Newton WE (eds). Gustav Fischer, Stuttgart, New York p 431-435

Díaz CL (1989) Root lectin as a determinant of host-plant specificity in the Rhizobium-legume symbiosis. PhD Thesis, Leiden University

Johnston AWB, Beynon JL, Buchanan-Wollaston AV, Setchell SM, Hirsch PR and Beringer JE (1978) High frequency of transfer of nodulating ability between strains and species of Rhizobium. Nature 276: 635-636

Kamberger W (1979) An Ouchterlony double diffusion study on the interaction between legume lectins and rhizobial cell surface antigens. Arch Microbiol 121: 83-90

Kijne JW, Smit G, Díaz C and Lugtenberg BJJ (1988) Lectin-enhanced accumulation of manganese-limited Rhizobium leguminosarum cells on pea root hair tips. J Bacteriol 170: 2994-3000

Rolfe BG and Shine J (1981) Rhizobium-Leguminosae symbiosis: the bacterial point of view. In: Verma DPS and Hohn TH (eds) Genes involved in microbe-plant interactions p 95-128. Springer-Verlag, Wien, New York

Smit G (1988) Adhesins from Rhizobiaceae and their role in plant-bacterium interactions. Ph D Thesis, Leiden University, The Netherlands

Smit G, Kijne JW and Lugtenberg BJJ (1986) Correlation between extracellular fibrils and attachment of Rhizobium leguminosarum to pea root hair tips. J Bacteriol 168: 821-827

Smit G, Kijne JW and Lugtenberg BJJ (1988) Involvement of both cellulose fibrils and a Ca^{2+}-dependent adhesin in the attachment of Rhizobium leguminosarum to pea root hair tips. J Bacteriol 169: 4294-4301

Smit G, Logman TJJ, Kijne JW and Lugtenberg BJJ (1989a) Roles of flagella, lipopolysaccharide and a Ca^{2+}-dependent cell surface protein in attachment of Rhizobium leguminosarum biovar viciae to pea root hair tips. J Bacteriol 171: 569-572

Smit G, Logman TJJ, Boerrigter MB, Kijne JW and Lugtenberg BJJ (1989b) Purification and partial characterization of the Rhizobium leguminosarum biovar viciae Ca^{2+}-dependent adhesin, which mediates the first step in attachment of cells of the family Rhizobiaceae to plant root hair tips. J Bacteriol 171: 4054-4062

Smit G, Kijne JW and Lugtenberg BJJ (1989c) Restoration of attachment and virulence of Agrobacterium tumefaciens chvB mutants by rhicadhesin, in preparation

Van Batenburg FHD, Jonker R and Kijne JW (1986) Rhizobium induces marked root hair curling by redirection of tip growth: a computer simulation. Physiol Plant 66: 476-480

Yao PY and Vincent JM (1969) Host specificity in the root hair "curling factor" of Rhizobium spp. Aust J Biol Sci 22: 413-423

REGULATION OF NOD GENE EXPRESSION : THE ROLE OF NOD D PROTEIN

Carel Wijffelman, Herman Spaink, Helmi Schlaman, Bas Zaat, Kees Recourt, Ruud de Maagd, Rob Okker and Ben Lugtenberg
Department of Plant Molecular Biology, Leiden University
Nonnensteeg 3, 2311 VJ Leiden, The Netherlands

INTRODUCTION

Bacteria of the genus Rhizobium are able to establish a symbiosis with leguminous plants resulting in the formation of root nodules in which atmospheric nitrogen is fixed. The nodulation of each bacterial species is restricted to its specific group of host plants (cross-inoculation group). Several nodulation genes, which are involved in determining the host-range, have been identified. In contrast to these host specific nod genes, the five nod genes ABCIJ, which constitute one operon, are functionally interchangeable, i.e. common, between Rhizobium species. Both these common and host specific nod genes are present on large Sym(biose) plasmids in the fast growing Rhizobium species and are regulated at the transcriptional level as one regulon. For the activation of the transcription of this regulon three factors are required (i) a nod box, (ii) an activating flavonoid factor secreted by the roots of leguminous plants and (iii) the nodD gene product.

THE NOD BOX

A so-called nod box is present upstream all inducible nod operons and seems to be essential for inducible promoter activity (Rostas et al 1986; Schofield et al 1986; Spaink et al 1987a). The nod box is a strongly conserved regulatory domain. Comparison of the nod boxes of several Rhizobium species lead to a consensus sequence of 49 base pairs. In Rhizobium leguminosarum biovar viciae four nod boxes have been identified in front of the nodA, nodF, nodM and nodO (previously nolR) genes. Deletion analysis of the

NATO ASI Series, Vol. H 36
Signal Molecules in Plants and Plant-Microbe Interactions
Edited by B. J. J. Lugtenberg

nod box in front of nodA showed that the activity of this promoter is located within a stretch of 72 base pairs, whose sequence contains the complete nod box and in addition 21 base pairs downstream of the nod box consensus. Removal of part of these 72 base pairs strongly reduces the promoter activity of the nod promoter, indicating that the nod box is an integral part of the nod promoter. The transcriptional start of this promoter is located 23 base pairs downstream of the nod box.

THE NATURAL INDUCER

Recent work in several laboratories has shown that leguminous plants exude flavonoids, that can activate the transcription of the nod genes (for a review see Rolfe 1988) and that the inducers can be different, even for plants of the same cross-inoculation group (Zaat et al 1988). The major inducers in extract of *Trifolium repens* seedlings are DHF (7,4'-dihydroxyflavone) and geraldone (7,4'-dihydroxy-3'-methoxyflavone) (Redmond et al 1986). Seed extract of alfalfa contains luteolin as the major inducer of *R.meliloti* (Peters et al 1986). In root exudate of *Vicia sativa* at least 7 different inducers of *R.leguminosarum* biovar *viciae* are present, six of which are flavanones. The major inducer, exuded by this plant is 3,5,7,3'-tetrahydroxy-4'methoxyflavanone (Zaat et al 1989).

The plant roots not only secrete substances which act as an activator of the transcription of the nod genes, but also substances which act as inhibitors (Firmin et al 1986). For example the roots of *T.repens* exude umbelliferone, which is a powerful inhibitor even in the presence of inducing substances (Djordjevic et al 1987). Presently it is uncertain whether these inhibitors play an (essential) role in the nodulation process.

THE NodD GENE

The nodD gene is the only nod gene in *Rhizobium* which is constitutively transcribed and it plays a key role in the process of nodulation. The *nodD* genes of the various *Rhizobium* species are very well conserved and play a role in the activation of the other *nod* genes. In this respect the *nodD* genes are common, i.e. functionally interchangeable.

For two reasons flavonoids do not seem likely candidates for functioning as host-specific signals towards *Rhizobium*. Firstly flavonoids are not unique for the roots of leguminous plants, and secondly, it has been shown that several leguminous plants exude flavonoids which can also activate NodD proteins of heterologous rhizobia. However, it has been shown that the *nodD* gene products of the fast-growing *Rhizobium* species differ in their flavonoid induction pattern. The sets of flavonoids which can activate the *nodD* genes is different for the *nodD* genes of *R.leguminosarum* biovars *viciae* and *trifolii* and for the three *nodD* genes of *R.meliloti*. Furthermore it has been shown that some host plants exude flavonoids which exclusively activate particular *nodD* gene products. This led to the conclusion that the *nodD* gene can also be an important determinant of host specificity (Györgypal et al 1988; Spaink et al 1987b).

INTRACELLULAR LOCALIZATION OF NodD PROTEIN

Several laboratories have studied the mechanism by which the *nod* genes are activated. Evidence has been presented by Recourt et al (1989) that the flavonoid inducer molecules are accumulated in the cytoplasmic membrane of *Rhizobium*. Furthermore, the NodD protein binds to *nod* box sequences even in the absence of flavonoids, as was deduced from DNA retardation studies (Hong et al 1987; Fisher et al 1988). Therefore an important aspect of the transcription activating properties mediated by the NodD protein is unclear. How can the NodD protein, bound to the *nod* box DNA in the cytoplasm be activated by the flavonoid inducer molecule in the cytoplasmic membrane? A solution for this controversy is given by the intracellular localization of NodD. We found that the NodD protein in wild type *Rhizobium* is located exclusively in the cytoplasmic membrane of the bacterium (Schlaman et al 1989). On the basis of the mentioned results a model is proposed (Fig. 1), in

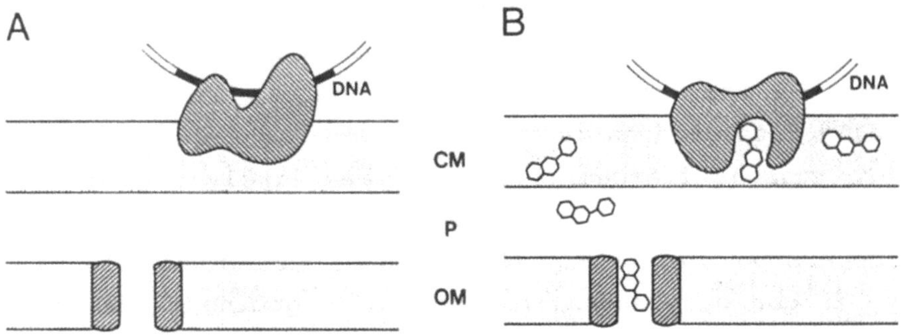

Fig.1 A model for the NodD protein-mediated activation of transcription of the inducible nod genes. The interaction between the NodD protein and the nod box (black string) in the absence (A) and in the presence (B) of a flavonoid inducer is shown.
CM, cytoplasmic membrane; OM, outer membrane; P, periplasmic space.

which NodD protein is partly located in the cytoplasmic membrane, where it interacts with the flavonoid inducer present in the cytoplasmic membrane. A cytoplasmic part of NodD is bound to nod box DNA. We suggest that, by binding to specific flavonoids present in the cytoplasmic membrane, the NodD protein undergoes a conformational change resulting in an activated form of the NodD protein which is capable of activating transcription.

RHIZOBIA CONTAINING A FITA nodD GENE

Using mutagenesis of the nodD gene (Burn et al 1987) or the construction of hybrid nodD genes (Horvath et al 1987; Spaink et al 1989) derivatives of the nodD gene of Rhizobium were isolated which affect both the inducing and autoregulating properties of nodD. One main conclusion from these studies is that NodD protein is a very complicated protein, in which several parts play a role in determining the flavonoid specificity. In our lab a set of hybrid nodD genes were constructed with very peculiar properties. One class of these hybrid nodD genes consists of 75% of the nodD1 gene of R.meliloti at the 5′ end and 25% of the nodD gene of R.leguminosarum biovar trifolii at the 3′ end. These nodD genes are called FITA (Flavonoid Independent Transcription Activation), because they activate the nodA promoter of R.leguminosarum biovar viciae to a maximum level, even in the absence of flavonoid inducer. Moreover, this activation of the *nod* genes is not influenced by known inhibitors like daidzein and umbelliferone.

The capacities of wild type *nodD* genes and of the FITA *nodD* gene, *nodD604*, to activate the promoters *nodA*, *nodF*, *nodM* and *nodO* of *R.leguminosarum* biovar *viciae*, the *nodA* and *nodF* promoter of *R.leguminosarum* biovar *trifolii* and the *nodA* promoter of *R.meliloti*, were compared. The level of activation by *nodD604* in the absence of inducer with all promoters was at least as high, and in most cases much higher, than the levels reached in the presence of the homologous *nodD* gene and the universal inducer luteolin. The transcriptional start of the *nodA* and *nodF* promoters of *R.leguminosarum* biovar *viciae* in the presence of *nodD604* in the presence and in the absence of inducer was identical to the start found with the homologous *nodD* gene in the presence of inducer. Also the intracellular localization of NodD604 protein was identical to that of the homologous NodD, namely exclusively in the cytoplasmic membrane (Schlaman, unpublished). In conclusion, except their flavonoid independent activation of *nod* promoters, the studied characteristics of FITA *nodD604* seem to be identical to those of a wild type *nodD* gene.

SYMBIOTIC PROPERTIES OF RHIZOBIA CONTAINING FITA *nodD*

To investigate the influence of a completely deregulated *nodD* gene on the symbiotic process we have analyzed the properties of *Rhizobium* strains containing *nodD604*. For that purpose an IncP plasmid containing either *nodD604* or the *nodD* gene from *R.leguminosarum* biovar *viciae* or *trifolii* was introduced in *nodD* mutants of *R.leguminosarum* biovar *viciae* (RBL5561) and *trifolii* (ANU851), and subsequently these strains were used to inoculate several plant species (Table 1). Firstly, the results clearly demonstrate that replacement of the homologous *nodD* by a wild type *nodD* of another cross-inoculation group narrows the host range. Microscopic analysis of the roots of the infected plants showed that the limitation of the host range was not at the level of root hair curling, but at the level of infection thread formation. Secondly, the *Rhizobium* strains containing *nodD604* were able to

TABLE 1. NODULATION OF PLANTS BY RHIZOBIUM STRAINS CONTAINING WILD TYPE OR FITA nodD GENES

Rhizobium leguminosarum on test plant	Nodulation characteristics[a] and average number of nodules per nodulated plant		
	R.leguminosarum nodD	R.trifolii nodD	FITA nodD604
biovar viciae RBL5561 on			
Vicia sativa	+ (2)	+ (2)	+ (2)
V.vicoides	+ (8)	+/-(2)	+ (7)
Lathyrus nissolia	+ (3)	-	+ (5)
biovar trifolii ANU851 on			
Trifolium repens	+ (4)	+ (4)	+ (4)
T.pratense	-	+ (4)	+ (5)
Macroptilium atropurpureum	n.d	-	+ (6)
Lablab purpureus	n.d	-	+ (4)

a +, more than 80% of plants nodulated; +/-, 10 to 30% of plants nodulated; -, no plants nodulated; n.d, not determined.

nodulate all plants, indicating that nodD604 is not only a good substitute for wild type nodD, but that nodD604 even extends the host range of R.leguminosarum biovar trifolii strain ANU851 to tropical legumes like Macroptilium and Lablab. However, these latter nodules were not able to fix nitrogen.

Finally, we compared the nitrogen fixing ability of some of the above mentioned plant-Rhizobium combinations. The nitrogen fixing ability of V.sativa plants inoculated with RBL5561 containing nodD604 or nodD wild type (biovar viciae) was determined during 40 days. Nitrogen fixation started about 10 days after inoculation. Surprisingly, the plants inoculated with the strain containing nodD604 reduced acetylene more efficiently between 10 and 20 days after inoculation than the control strain. At later times with both strains the same level of nitrogen fixation was measured. The results with T.repens inoculated with ANU851 containing nodD604 or nodD wild type of R.leguminosarum biovar trifolii were even more surprising. When the nitrogen fixing ability of the plants infected with the nodD604 containing strain was compared to the control, the nitrogen fixing ability of the nodD604 containing strain was about twice as high during the whole period of 40 days.

CONCLUDING REMARKS

The interaction between the symbiotic partners - plant and bacterium - is guided by a delicate exchange of molecular signals that regulate the expression of genes essential for infection and nodule development. In fact one of the earliest steps of symbiosis is signal-exchange, before any visible sign of the nodulation process can be observed. The nodD gene plays a key role in this process by activating the nod promoters in response to the first plant signal i.e. the flavonoids secreted by the roots. The host specific aspects of nodD suggest that the nodD genes are adapted to the flavonoids secreted by the roots of plants of their cross-inoculation group to assure a sufficiently high level of transcription of the nod genes. On the other hand the presence of a FITA nodD gene, which constitutively activates the transcription of all inducible nodulation genes at a maximum level, leads to a normal nodulation, demonstrating (i) that no fine-regulation of the nodulation process is required at the level of transcription of the nod genes and (ii) that nodD604 can be considered as a real common nodD gene. In addition, the insensitivity of FITA nodD for inhibitors like daidzein and umbelliferone suggests that natural inhibitors secreted by plant roots do not play a role in the nodulation process. The observation that the level of nitrogen fixation can be improved by the presence of a FITA nodD gene suggests that the presence of an activated NodD protein is also of crucial importance during later stages of symbiosis and offers a tool for improving the nitrogen fixing ability of Rhizobium strains.

REFERENCES

Burn J, Rossen L and Johnston AWB (1987) Four classes of mutations in the nodD gene of Rhizobium leguminosarum biovar. viciae that affect its ability to autoregulate and/or activate other nod genes in the presence of flavonoid inducers. Genes and Development 1: 456-464

Djordjevic MA, Redmond JW, Batley M and Rolfe BG (1987) Clovers secrete specific phenolic compounds which either stimulate or repress nod gene expression in Rhizobium trifolii. EMBO J 6: 1173-1179

Firmin JL, Wilson KE, Rossen L and Johnston AWB (1986) Flavonoid activation of nodulation genes in Rhizobium reversed by other compounds present in plants. Nature(London) 324: 90-92

Györgypal Z, Iyer N and Kondorosi A (1988) Three regulatory nodD alleles of divergent flavonoid-specificity are involved in host-dependent nodulation by Rhizobium meliloti. Mol Gen Genet 212: 85-92

Horvath B, Bachem CW, Schell J and Kondorosi A (1987) Host-specific regulation of nodulation genes in Rhizobium is mediated by a plant-signal, interacting with the nodD gene product. EMBO J 6: 841-848

Peters NK, Frost JW and Long SR (1986) A plant flavone, luteolin, induces expression of Rhizobium meliloti genes. Science 233: 977-980

Recourt K, Van Brussel AAN, Driessen AMJ and Lugtenberg BJJ (1989) Accumulation of a nod gene inducer, the flavonoid naringenin, in the cytoplasmic membrane of Rhizobium leguminosarum biovar viciae by the pH-dependent hydrophobicity of naringenin. J Bacteriol 171: in press

Redmond JW, Batley M, Djordjevic MA, Innes RW, Kuempel PL and Rolfe BG (1986) Flavones induce expression of nodulation genes in Rhizobium. Nature(London) 323: 632-635

Rolfe BG (1988) Flavones and isoflavones as inducing substances of legume nodulation. Biofactors 1: 3-10

Rostas K, Kondorosi E, Horvath B, Simoncsits A and Kondorosi A (1986) Conservation of extended promoter regions of nodulation genes in Rhizobium. Proc Natl Acad Sci USA 83: 1757-1761

Schlaman H, Spaink HP, Okker RJH and Lugtenberg BJJ (1989) The subcellular localization of the nodD gene product in Rhizobium leguminosarum. J Bacteriol 171 : in press

Schofield PR and Watson JM (1986) DNA sequence of Rhizobium trifolii nodulation genes reveals a reiterated and potentially regulatory sequence preceding nodABC and nodFE. Nucleic Acids Res 14: 2891-2903

Spaink HP, Okker RJH, Wijffelman CA, Pees E and Lugtenberg BJJ (1987a) Promoters in the nodulation region of the Rhizobium leguminosarum Sym plasmid pRL1JI. Plant Mol Biol 9: 27-39

Spaink HP, Wijffelman CA, Okker RJH and Lugtenberg BJJ (1989) Localization of functional regions of the Rhizobium nodD product using hybrid nodD genes. Plant Mol Biol 12: 59-73

Spaink HP, Wijffelman CA, Pees E, Okker RJH and Lugtenberg BJJ (1987b) Rhizobium nodulation gene nodD as a determinant of host specificity. Nature(London) 328: 337-340

Zaat SAJ, Schripsema J, Wijffelman CA, Van Brussel AAN and Lugtenberg BJJ (1989) Analysis of the major inducers of the Rhizobium nodA promoter from Vicia sativa root exudate and their activity with different nodD genes. Plant Mol Biol in press

Zaat SAJ, Wijffelman CA, Mulders IHM, Van Brussel AAN and Lugtenberg BJJ (1988) Root exudates of various host plants of Rhizobium leguminosarum contain different sets of inducers of Rhizobium nodulation genes. Plant Physiol 86: 1298-1303

NODULATION GENES AND THEIR REGULATION IN RHIZOBIUM MELILOTI

Sharon R. Long, Julie Schwedock, Thomas Egelhoff, Melanie Yelton, John Mulligan, Melanie Barnett, Brenda Rushing, and Robert Fisher
Department of Biological Sciences
Stanford University
Stanford CA 94305-5020
U.S.A.

INTRODUCTION

Our long term goals are to study the interaction of Rhizobium and plants at the molecular and cellular levels. Our system of study is Rhizobium meliloti and its host plants such as alfalfa (Medicago sativa L.). The approaches we use are primarily to combine genetic analysis of nodulation with microscopic observation of normal and mutant bacteria during interaction with plants. Our recent work includes the definition of two new nodulation genes, and the further characterization of the network for regulation of nodulation genes in Rhizobium meliloti. One theme found in both these studies is the complexity arising from multiple gene copies in R. meliloti.

NOD GENE EXPRESSION, THE nodD FAMILY, AND syrM

In previous work, we reported that the R. meliloti nodC locus was induced by exposure to exudates of alfalfa plants, and that the nodD gene was required for induction to occur (Mulligan and Long 1985). We found that the flavone, luteolin, present in alfalfa seed exudates, was a very active inducer molecule (Peters et al 1986). Inducibility by plant flavonoids, dependent on nodD, is generally observed for Rhizobium nod genes (reviewed recently by Downie and Johnston 1989; Long 1989).

The role of the nodD protein (NodD) is of interest. It has been observed that extracts of Rhizobium cells can interact with the promoters of inducible nod genes, to the extent that the Rhizobium cells contain copies of nodD (Hong et al 1987; Fisher

NATO ASI Series, Vol. H 36
Signal Molecules in Plants and Plant-Microbe Interactions
Edited by B. J. J. Lugtenberg

et al 1988; Kondorosi et al 1989). Purified NodD binds to these promoters as assayed by a mobility shift gel (Fisher et al 1988; Kondorosi et al 1989). The inducible nod promoters are characterized by the presence of an extensive repeat sequence, the nod box (Rostas et al 1986; Spaink et al 1987), which is a likely cis-acting control element.

In R. meliloti, there are three copies of nodD: nodD1, linked closely to nodABC, induces nod genes in the presence of luteolin. NodD2, linked about 45 kb away, probably induces nod genes in response to other inducers present in alfalfa exudates, and displays differential response to different plants in the R. meliloti host range (Honma and Ausubel 1987; Göttfert et al 1986). As first reported in a previous volume in this NATO series (Long et al 1986), we have found that the third gene, nodD3 (referred to in the previous paper as nodD15), is a powerful constitutive activator of nod gene expression. When present on a clone, pRmJT5, which includes the host specific nod genes nodH, nodFEG, and others, nodD3 causes more than 3-4 times the nod gene expression normally found in cells displaying maximal induction by luteolin and nodD1. The transcription start site is the same for nod gene expression caused by nodD1 plus luteolin, and by nodD3 (Mulligan and Long 1989).

The sequence of nodD3 has been established, and is interesting to compare with partially constitutive nodD mutants reported from other systems. Burn et al (1987) and Djordevic and Rolfe (personal communication) have isolated nodD mutants which cause elevated levels of nod gene expression in the absence of inducer. However, all of these mutants also display at least some inducibility. They differ in this respect from nodD3, which is unaffected by inducer. The positions of the nodD-constitutive mutant codon changes have been determined by sequencing. Comparing these with the sequence of nodD3, it is striking to note that all of the positions which have been found to confer some NodD3-like behavior in point mutants of nodD, nonetheless are conserved between nodD3 and the normally inducer-dependent nodD1 of R. meliloti. This suggests that an overall protein configuration, rather than a change in a single domain, is responsible for the inducer-independence of the constitutive allele nodD3.

If such a powerful transcriptional activator is present in R. meliloti, why is any induction by nodD1 seen? This appears to be due to the fact that nodD3 is typically not expressed in culture conditions. Rather, it is under the control of another locus, syrM (symbiotic regulator). Using antibody to NodD protein as an assay, we have found that the syrM locus is important for production of NodD3 protein, not for its activity: when the nodD3 gene is placed under control of another promoter, there is no requirement for syrM to obtain high transcription of nod genes. We have determined the nucleotide sequence of syrM, and have found that it, too, appears to be a regulator of the nodD family! By mobility-shift assay, we have found that proteins in R. meliloti extracts are able to bind the nodD3 5' region, according to their genetic content of syrM; however, the apparent binding pattern is complex, and work with purified proteins will be necessary in order to establish a mechanism of action for the SyrM protein and other regulatory proteins which may be interacting with SyrM at the nodD3 promoter. The physiological relevance of the multiple nodD's and of syrM is not known. We have approached this by following the expression of these loci during infection, by means of nod-uidA (glucuronidase) assays and in situ staining of nodules. Preliminary tests indicate expression of the nodD genes occurs throughout infection and possibly later in the symbiotic process. One possible clue to syrM function may lie in the observation that this regulatory locus affects not only nod gene expression, but also the expression of extracellular polysaccharide synthesis (exo) genes. This syrM effect is mediated by a locus, syrA, which may act analogously to nodD3 (Mulligan and Long 1989). Studies by others on the phenotype of Exo$^-$ mutants have indicated that the need for exo gene function begins later than that for nod genes, and continues at least through bacteroid release (Finan et al 1985). That syrM controls both nod and exo suggests the possibility that this system takes over control from "inducer-signal" control of nod genes, and keeps nod and exo function together during later stages of infection. Thus, signals may vary, or may fade away or become limiting during infection. R. meliloti may have built in the syrM hierarchy to deal with developmental signal variation, just as it is hypothesized to use the "nodD1-nodD2" difference to deal with ecological or evolutionary variation.

The properties of NodD1 and NodD3 have been compared at the biochemical level, and at least one interesting difference has been observed. Both forms of the protein are able to bind the promoters of nodA, nodF and nodH with or without inducer present (Fisher et al 1988). The "footprints" of each protein on each of the three promoters were also determined, both by DNAse I protection, and by phenanthrolene-Cu^{++} protection. The extent of the footprint in each case was a total of about 55-60 bases, and the position was equivalent, beginning at about 25 bases upstream from the transcript start site of each nod gene mRNA. However, an additional hypersensitive site is seen in the footprint of the NodD3 protein on one strand of each promoter (Fisher and Long, submitted). This would be consistent with an alteration in the strain, degree of bending, or overall geometry of the DNA-protein interaction during binding with NodD3 compared with NodD1, but there may also be other explanations.

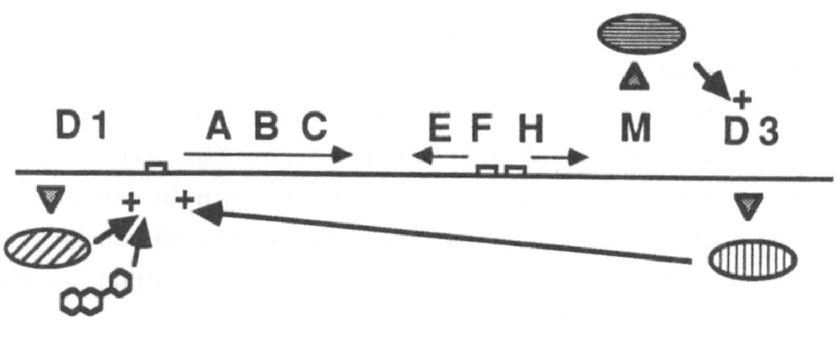

Figure 1. A model for nodD1 and nodD3 regulation of R. meliloti nod genes. Activation by nodD (and by nodD2, not shown) depends on presence of inducers such as flavonoids. A second circuit involves the nodD-like activator syrM, which controls expression of nodD3. The NodD3 protein activates transcription of other nod operons in the absence of inducer.

MORE MULTIPLE GENES: nodP AND nodQ

In a previous study (Swanson et al 1987) we found that there appeared to be nodulation "efficiency" genes downstream of nodG. Some such mutations occurred in the location identified as region "IIb" by Debellé et al 1986 and one,designated mutant nod-216, was in a location not previously observed to have an effect on nodulation. We have sequenced this region (Schwedock and Long 1989; also, see a parallel study by Cervantes et al in press) and have determined that nod-216 corresponds to a new gene, nodP, and that the downstream DNA from this encodes a second gene, nodQ. We found that nodP is expressed in R. meliloti in vitro transcription-translation extracts off of its own promoter, which suggests it may be to some extent constitutive. However, a lacZ fusion in nodP carried on a plasmid with other linked nod genes shows some inducibility; whether this represents physiologically relevant expression, or is an artifact of high-level plasmid expression, is not currently known. The sequences of nodP and nodQ were examined for homology to other known gene products by computer search. While nodP shows no sequence similarity to any known protein (however, see below), nodQ is interesting in that it is homologous to Elongation Factor Tu, and to some initiation factors. The homology is highest in the protein domains used in GDP binding. This may be a clue that NodQ is itself a GDP binding protein. The possibility is open that NodQ is involved in translational control of nod gene expression; alternatively, NodQ may carry out some other function involving GDP or GTP, such as synthesis or signalling.

Filter hybridization was used to determine whether copies of nodP and nodQ are present in other Rhizobium. We found that nodQ was detectable in ANU265, a pSym deleted derivative of NGR234, but was not observable in other bacteria. The nodP homology test, however, indicated that nodP is present in all other Rhizobium tested, and is present in E. coli as well. We are attempting to locate the region of the E. coli chromosome in which the nodP homology is found, in the hopes that some clues as to its function may emerge. It has previously been reported that Azospirillum brasilense shows homology to the region now known to encode nodP (Vieille et al 1988).

In the process of carrying out the filter hybridization studies to look for homology to nodP and nodQ, we included on the filters the DNA from several known deletion strains of R. meliloti. To our surprise, we found that the nodP and nodQ bands were still present in the lanes representing deletion-strain DNA. We confirmed that the deletion strains, which were independently constructed by the Toulouse group and the MGH group, behaved exactly as predicted when probes from linked genes such as nodC or syrM were tested. We concluded, then, that the nodPQ DNA region is present in a second copy in the R. meliloti genome, so highly conserved that most of the restriction sites are exactly identical, and the fragments representing the first and second copies are indistinguishable in most restriction digests. We have confirmed this by establishing two restriction digests which distinguish the first and second copies, and we are in the process of cloning and mapping the second set of nodPQ genes. It will be of interest to determine the phenotype of mutant strains lacking both copies of nodP and nodQ.

REFERENCES

Burn J, Rossen L, Johnston, AWB (1987) Four classes of mutations in the nodD gene of Rhizobium leguminosarum biovar viciae that affect its ability to autoregulate and/or activate other nod genes in the presence of flavonoid inducers. Genes Dev 1:456-464

Cervantes E, Sharma SB, Maillet F, Vasse J, Truchet G, Rosenberg C (to be published) The Rhizobium meliloti host-range nodQ gene encodes a protein which shares homology with translation elongation and initiation factors. Molecular Microbiol

Debellé F, Rosenberg C, Vasse J, Maillet F, Martinez E, Dénarié J, Truchet, G (1986) Assignment of symbiotic developmental phenotypes to common and specific nodulation (nod) genetic loci of Rhizobium meliloti. J Bacteriol 168:1075-1086

Downie JA, Johnston AWB (1988) Nodulation of legumes by Rhizobium. Plant, Cell and Environ 11:403-412

Finan TM, Hirsch AM, Leigh JA, Johansen E, Kuldau GA, Deegan S, Walker GC, Signer ER (1985) Symbiotic mutants of Rhizobium meliloti that uncouple plant from bacterial differentiation. Cell 40:869-877

Fisher RF, Long SR (submitted) DNA footprint analysis of the transcriptional activator proteins NodD1 and NodD3 on inducible nod gene promoters. J Biol Chem

Fisher RF, Egelhoff TT, Mulligan JT, Long SR (1988) Specific binding of proteins from Rhizobium meliloti cell-free extracts containing NodD to DNA sequences upstream of inducible

nodulation genes. Genes Dev 2:282-293
Fisher RF, Swanson J, Mulligan JT, Long SR (1987) Extended region of nodulation genes in Rhizobium meliloti 1021. II. Nucleotide sequence, transcription start sites, and protein products. Genetics 117:191-201
Göttfert M, Horvath B, Kondorosi E, Putnoky P, Rodriguez-Quinones F, Kondorosi A (1986) At least two nodD genes are necessary for efficient nodulation of alfalfa by Rhizobium meliloti. J Mol Biol 191:411-420
Honma MA, Ausubel FM (1987) Rhizobium meliloti has three functional copies of the nodD symbiotic regulatory gene. Proc Natl Acad Sci USA 84:8558-8562
Kondorosi E, Gyuris J, Schmidt J, John M, Duda E, Hoffmann B, Schell J, Kondorosi A (to be published) Positive and negative control of nod gene expression in Rhizobium meliloti is required for optimal nodulation. EMBO J
Long SR (1989) Rhizobium-legume nodulation: Life together in the underground. Cell 56:203-214
Long SR, Peters NK, Mulligan JT, Dudley ME (1986) Genetic analysis of Rhizobium-plant interactions. In: Lugtenberg B (ed) Recognition in microbe-plant symbiotic and pathogenic interactions, vol 4. Springer-Verlag Berlin, p 1-15
Mulligan JT, Long SR (1985) Induction of Rhizobium meliloti nodC expression by plant exudate requires nodD. Proc Natl Acad Sci USA 82:6609-6613
Mulligan JT, Long SR (1989) A family of activator genes regulates expression of Rhizobium meliloti nodulation genes. Genetics 122:7-18
Peters NK, Frost JW, Long SR (1986) A plant flavone, luteolin, induces expression of Rhizobium meliloti nodulation genes. Science 233:917-1008
Rostas K, Kondorosi E, Horvath B, Simoncsits A, Kondorosi A (1986) Conservation of extended promoter regions of nodulation genes in Rhizobium. Proc Natl Acad Sci USA 83:1757-1761
Schwedock J, Long SR (1989) Nucleotide sequence and protein products of two new nodulation genes of Rhizobium meliloti, nodP and nodQ. Mol Plant-Microbe Int 2:181-194
Spaink HP, Okker RJH, Wijffelman CA, Pees E, Lugtenberg BEJ (1987) Promoters in the nodulation region of the Rhizobium leguminosarum Sym plasmid pRL1JI. Plant Mol Biol 9:29-37
Vieille C, Onyeocha I, Baca B-E, Palacios R, Flores M, Elmerich C (1988) Genetic analysis of a 90 Md plasmid of Azospirillum brasilense sp7. In: Palacios R, Verma DPS (eds) Molecular Genetics of Plant Microbe Interactions. APS Press, St Paul, MN p 209

REGULATION OF THE AGROBACTERIUM VIRULENCE GENES BY PLANT FACTORS

P.J.J. Hooykaas, L.S. Melchers, C.W. Rodenburg & R.J.M. van Veen
Department of Plant Molecular Biology, Leiden University
Wassenaarseweg 64, 2333 AL Leiden,
The Netherlands

The soil bacterium Agrobacterium tumefaciens causes the plant tumour crown gall at wound sites on dicotyledonous plants. During crown gall induction, which is only possible during a limited period of time (less than 72 hours) after wounding, agrobacteria transfer a segment of their tumour-inducing (Ti)-plasmid to plant cells at the wound sites. This transferred (T)-DNA becomes integrated in one or more of the chromosomes of the host plant cells usually either in one copy or in a limited number of copies. The T-DNA contains genes for enzymes capable of mediating the production of an auxin (IAA) and a cytokinin (isopentenyl-AMP), which are responsible for the tumorous character of crown gall cells. It is not yet known how exactly the T-DNA is transferred from agrobacteria to plant cells, but numerous genes have been identified in the bacterium which are involved in this process. Some of these genes are located in the chromosome (chromosomal virulence (chv)-genes), whereas others are Ti plasmid borne (vir-genes). Mutants with defects in either of the two chv-genes do not produce extracellular β1,2-glucan and show a loss of the ability to attach to the plant cells. The role of vir-genes in crown gall induction will be described below. A number of recent reviews describe several aspects of the Agrobacterium-plant interaction in detail (Hooykaas and Schilperoort 1984; Nester et al 1984; Koukolikova-Nicola et al 1987; Melchers and Hooykaas 1988).

THE VIRULENCE GENES

The vir-genes are located in one particular segment of the Ti plasmid, the Vir(ulence) region, which is about 40 kbp in size.

NATO ASI Series, Vol. H 36
Signal Molecules in Plants and Plant-Microbe Interactions
Edited by B.J.J. Lugtenberg

For the octopine type Ti plasmid 22 vir-genes have been identified in seven operons, which together belong to one regulon (see below). The operons may contain from one up to 11 genes, and have been given the names virA-virG (see also Table). The Vir-regions of other types of Ti plasmids may be slightly different in their gene composition. For instance the host range gene virF, is absent from the nopaline type Ti plasmid. The virF gene is necessary for full tumour induction on plant species such as Nicotiana glauca (unpublished). The exact role of most of the products determined by the vir-genes is still unknown. However, we know that the virA and virG genes together constitute a two-factor regulatory system which is responsible for the induction of the vir-operons through plant factors (Stachel and Zambryski 1986). Proteins encoded by the first two genes of the virD-operon mediate the production of single stranded copies of the T-region (T-strands), which probably are the intermediates that are transferred to plant cells during tumour induction (Stachel et al 1987).
The second gene of the virE-operon encodes a protein which has single stranded DNA-binding properties (Gietl et al 1987), and is capable of forming a protein-DNA complex with the T-strands (Citovsky et al 1989). Finally, the virB-operon codes for a set of proteins that mostly either have a potential signal peptide or are hydrophobic. For some of these it was already shown that they are located in the cytoplasmic membrane of the bacterium (Engström et al 1987; Thompson et al 1988) and, therefore, it may be that these proteins together form a structure (a pilus?) capable of mediating the transfer of T-strands between the bacterium and the plant cell. The finding that incQ plasmids as such (and possibly any mobilizable plasmid) can be transferred to plant cells via the Agrobacterium delivery system - provided that it has an intact oriT site and the genes for the Mob-proteins - also points to the possibility that plant cells are transformed via a conjugation-like mechanism (Buchanan-Wollaston et al 1987). That plant cells can be transformed by single stranded DNA was shown by Rodenburg et al (1989), who used electroporation to achieve transformation and who found that single stranded DNA was even a significantly better substrate

for plant cell transformation than the homologous double stranded form.

REGULATION OF THE VIR-GENES

It was shown earlier that the _vir_-genes are silent during growth in the standard laboratory media for bacteria, but become induced in media that had contained plant cultures or in which plant (root) exudate had been incorporated. Further research has shown that the responsible inducers are plant phenolics, and that the most important natural inducer present in tobacco is the compound acetosyringone (Stachel et al 1985). This latter compound is released from tobacco tissue especially after wounding, which by itself is a prerequisite for tumour induction by _Agrobacterium_. Besides acetosyringone a variety of other phenolic compounds can be used for the induction of _vir_-genes, the most simple inducer being guiacol, a derivative of phenol with a methoxyl-group in the _ortho_-position (Spencer and Towers 1988; Melchers et al 1989b). Substitutions in the _para_-position are tolerated to some extent, and can lead to compounds with either enhanced or reduced inducing activity.

Even in the presence of an inducer particular conditions have to be met in order to obtain an optimal induction of the _vir_-genes: 1) the pH of the medium must be adjusted to about 5.3, (2) the temperature must be below 30°C, (3) the presence of yeast extract in the medium must be avoided, (4) a high sugar (preferably sucrose) concentration must be present in the medium. Although the _vir_-genes of different types of Ti and Ri plasmids can be induced in similar acetosyringone containing media, there are small differences in their requirement for optimal induction. For instance, while octopine and leucinopine Ti plasmids require pH5.3 for maximal _vir_-gene induction, for the _vir_-genes of nopaline Ti and agropine Ri plasmids the preferred medium is at pH5.8. This probably reflects small differences in the VirA regulatory proteins determined by these plasmids. A different situation applies to strains containing the limited host range (LHR)-Ti plasmid pTiAG57. Here, the virulence genes are not induced at all in the presence of acetosyringone or any of the other phenolic compounds tested sofar. It may therefore be that in this particular case induction needs a different class of plant compounds. This may also be one of the reasons for the restricted host range of bacteria with this Ti plasmid for tumour formation (in nature: Vitis vinifera).

THE REGULATORY SYSTEM

The virA and virG gene products together are necessary for the induction of the other vir-operons (Stachel and Zambryski 1986). Sequence analysis in our laboratory (Melchers et al 1986; 1987) as well as by others showed that these proteins have homology with proteins comprising other two-factor regulatory systems such as envZ / ompR, ntrB / ntrC, phoR / phoB (Ronson et al 1987).
The model is that the virA-like component of the system is a sensor (for phenolic compounds in the case of the VirA protein), and the virG-like component a DNA-binding activator protein capable of specifically activating a particular set of genes (the other vir-operons in the case of the VirG protein). How the signal is transmitted from the sensor to the activator protein is not known, but it might be that the activator becomes active via phosphorylation by the signalling domain of the sensor which becomes unlocked only after the receptor domain of the sensor has received a stimulating signal. Indeed this has been shown to be the case for the ntrB / ntrC (Keener and Kustu 1988; Weiss and Magasanik 1988) and the cheA / cheY (Hess et al 1988) two component systems.

For the virA / virG system data have been obtained which are in line with this model. A particular sequence ("vir-box") has been found with only little variation in front of all the vir-operons sequenced sofar as well as in front of the tzs-gene of the nopaline Ti plasmid, which is co-regulated with the vir-genes. The VirG protein has been shown to have aspecific DNA-binding capacity (Powell et al 1989), although it is still up to future research to show that it binds with higher affinity to the *vir*-promoters at least in the presence of an inducer. For the VirA protein the sequence data show that it has two long stretches of hydrophobic amino acids capable of forming membrane-crossing α-helices. Using antibodies against the C-terminal part of the VirA protein we could show (after membrane separation) that the VirA protein is present in the cytoplasmic membrane of *Agrobacterium* indeed (Melchers et al 1987). Further work in which we used *in vitro* constructed fusions between different N-terminal parts of VirA and PhoA devoid of its signal peptide revealed that the topology of the VirA protein is as predicted from the DNA-sequence, *i.e.* having in the N-terminus a periplasmic domain followed by a membrane spanning domain (TM2) and a cytoplasmic C-terminal domain (Melchers et al 1989a). On the basis of this it was postulated that - like with other types of receptor proteins - the periplasmic (outside) domain would form the sensor (signal receiver) domain and the cytoplasmic (inside) domain the signal transmitter domain. In order to find out whether this is indeed the case, we made hybrids between VirA and the *E.coli* chemoreceptor Tar, which activates the chemotaxis apparatus after sensing the presence of aspartate or loaded maltose binding protein. The Tar-protein has a basic structure similar to that of the VirA protein, and it has been shown that the periplasmic domain of Tar is the sensor domain, while the cytoplasmic domain contains the signal transmitter region. When replacing the periplasmic part of VirA by that of Tar, we obtained sensor proteins that were active as *vir*-regulators. Unexpectedly, however, they responded still to phenolic compounds such as acetosyringone and not to aspartate or maltose (Melchers et al 1989a). Certain of these hybrid proteins were even more active in mediating the activation of *vir*-genes than the wild-type VirA protein, and rendered the system less sensitive to the negative influence of a relatively high pH or temperature. Significant induction took place even at pH6.8 and at temperatures as high as 33°C and 37°C. Thus our data show that 1) the sensor domain is not located in the periplasmic region of VirA 2) the amino acid composition of the periplasmic domain is one element conferring pH-and temperature-sensitivity on the system. In line with this we found that the deletion of the periplasmic domain from the VirA protein did not render the protein inactive. In fact such deleted VirA proteins behaved like the wild-type VirA protein in acetosyringone-dependent signal transduction. Further work showed that the deletion of the second transmembrane domain (TM2) from VirA or the replacement of this domain by the corresponding domain of Tar led to proteins that no longer responded to plant phenolics, but rather mediated a low, signal-independent level of vir-gene induction. We conclude that the second transmembrane domain of VirA is an essential element of VirA, which possibly contains the sensor (signal receiver) function of this protein. That the location of the sensor domain in VirA is different from that in the more classically studied chemoreceptors does not have to be too surprising in light of the fact that the signals in the case of VirA are formed by relatively apolar phenolic compounds that probably have a relatively high affinity for membranes. The only other example, where the signal receiver function is located in a membrane domain of a chemoreceptor protein is given by the β-adrenergic receptors of eukarotics (Strader et al 1987). The ligands in this case are formed by catecholamines, again compounds which have affinity for membranes. Deletion analysis of the cytoplasmic C-terminal domain of VirA showed that small deletions already render the protein inactive as a signal transducer, which is in line with the proposed function of this part of the protein in signal transmission to the VirG protein. Future research will hopefully shed more light on the exact molecular details of the signal communication between plant cells and agrobacteria as mediated by the proteins VirA and VirG.

REFERENCES

Buchanan-Wollaston V, Passiatore JE & Cannon F (1987) The mob and oriT mobilization functions of a bacterial plasmid promote its transfer to plants. Nature 328:172-175

Citovsky V, Wong ML & Zambryski P (1989) Cooperative interaction of Agrobacterium VirE2 protein with single-stranded DNA: implications for the T-DNA transfer process. Proc Natl Acad Sci USA 86:1193-1197

Engström P, Zambryski P, Van Montagu M & Stachel S (1987) Characterization of Agrobacterium tumefaciens virulence proteins induced by the plant factor acetosyringone. J Mol Biol 197:635-645

Gietl C, Koukolikova-Nicola Z & Hohn B (1987) Mobilization of T-DNA from Agrobacterium to plant cells involves a protein that binds single-stranded DNA. Proc Natl Acad Sci USA 84:9006-9010

Hess JF, Oosawa K, Kaplan N & Simon MI (1988) Phosphorylation of three proteins in the signaling pathway of bacterial chemotaxis. Cell 53:79-87

Hooykaas PJJ & Schilperoort RA (1984) The molecular genetics of crown gall tumorigenesis. Advances in Genetics 22:209-283

Keener J & Kustu S (1988) Protein kinase and phosphoprotein phosphatase activities of nitrogen regulatory proteins NTRB and NTRC of enteric bacteria: roles of the conserved amino-terminal domain of NTRC. Proc Natl Acad Sci USA 85:4976-4980

Koukolikova-Nicola Z, Albright L & Hohn B (1987) The mechanism of T-DNA transfer from Agrobacterium tumefaciens to the plant cell. In: Hohn T, Schell J (eds) Plant-DNA Infections Agents. Springer Verlag, Vienna, pp 109-148

Leroux B, Yanofsky MF, Winans SC, Ward JE, Ziegler SF & Nester EW (1987) Characterization of the virA locus of Agrobacterium tumefaciens: a transcriptional regulator and host range determinant. EMBO J 6:849-856

Melchers LS & Hooykaas PJJ (1987) Virulence of Agrobacterium. In: Miflin BJ (ed) Oxford Surveys of Plant Molecular and Cell Biology, Vol 4, Oxford University Press, Oxford UK, pp 167-220

Melchers LS, Thompson DV, Idler KB, Schilperoort RA & Hooykaas PJJ (1986) Nucleotide sequence of the virulence gene virG of the Agrobacterium tumefaciens octopine Ti plasmid: significant homology between virG and the regulatory genes ompR, phoB and dye of E.coli. Nucl Ac Res 14:9933-9942

Melchers LS, Thompson DV, Idler KB, Neuteboom STC, De Maagd RA, Schilperoort RA & Hooykaas PJJ (1987) Molecular characterization of the virulence gene virA of the Agrobacterium tumefaciens octopine Ti plasmid. Plant Mol Biol 9:635-645

Melchers LS, Regensburg-Tuink TJG, Bourret RB, Sedee NJA, Schilperoort RA & Hooykaas PJJ (1989a) Membrane topology and functional analysis of the sensory protein VirA of Agrobacterium tumefaciens. EMBO J 8, in press

Melchers LS, Regensburg-Tuink AJG, Schilperoort RA & Hooykaas PJJ (1989b) Specifity of signal molecules in the activation of Agrobacterium virulence gene expression. Mol Microbiol 3, in press

Nester EW, Gordon MP, Amasino RM & Yanofsky MF (1984) Crown gall: a molecular and physiological analysis. Ann Rev Plant Physiol 35:387-413

Nixon BT, Ronson CW & Ausubel FM (1986) Two-component regulatory systems responsive to environmental stimuli share strongly conserved domains with the nitrogen assimilation regulatory genes ntrB and ntrC. Proc Natl Acad Sci USA 83:7850-7854

Powell BS, Rogowsky PM & Kado CI (1989) virG of Agrobacterium tumefaciens plasmid pTiC58 encodes a DNA-binding protein. Mol Microbiol 3:411-419

Rodenburg CW, De Groot MJA, Schilperoort RA & Hooykaas PJJ (1989) Single stranded DNA used as an efficient new vehicle for plant protoplast transformation. Plant Mol Biol 13, in press

Ronson CW, Nixon BT & Ausubel FM (1987) Conserved domains in bacterial regulatory proteins that respond to environmental stimuli. Cell 49:579-581

Spencer PA & Towers GHN (1988) Specificity of signal compounds detected by Agrobacterium tumefaciens. Phytochemistry 27: 2781-2785

Stachel SE & Zambryski PC (1986) virA and virG control the plant-induced activation of the T-DNA transfer process if A.tumefaciens. Cell 46:325-333

Stachel SE, Messens E, Van Montagu M & Zambryski P (1985) Identification of the signal molecules produced by wounded plant cells that activate T-DNA transfer in Agrobacterium tumefaciens. Nature 318:624-629

Stachel SE, Timmerman B & Zambryski P (1987) Activation of Agrobacterium tumefaciens vir gene expression generates multiple single stranded T-strand molecules from the pTiA6 T-region: requirement for 5′ virD gene products. EMBO J 6:857-863

Strader CD, Sigal IS, Register RB, Candelore MR, Rands E, Dixon RAF (1987) Identification of residues required for ligand binding to the β-adrenergic receptor. Proc Natl Acad Sci USA 84:4384-4388

Thompson DV, Melchers LS, Idler KB, Schilperoort RA & Hooykaas PJJ (1988) Analysis of the complete nucleotide sequence of the Agrobacterium tumefaciens virB operon. Nucl Ac Res 16:4621-4636

Weiss V & Magasanik B (1988) Phosphorylation of nitrogen regulator I(NR_I) of Escherichia coli. Proc Natl Acad Sci USA 85:8919-8923

VIRUS-INDUCED GLYCANHYDROLASES AND EFFECTS OF OLIGOSACCHARIDE SIGNALS ON PLANT-VIRUS INTERACTIONS

B. Fritig, J. Rouster, S. Kauffmann, A. Stinzi, P. Geoffroy, M. Kopp and M. Legrand
Institut de Biologie Moléculaire des Plantes du CNRS
12 rue du Général Zimmer
67000 Strasbourg
France

INTRODUCTION

The hypersensitive response is one of the most efficient mechanisms of defence that is induced by the infection itself. Even though the triggering of the hypersensitivity reaction results from a very specific gene-for-gene recognition between plant and pathogen, what the plant does thereafter to defend itself appears to be specific only to the host. The alterations of host metabolism believed to participate in active defence include cell wall thickening resulting from production and deposition of various macromolecules, and the production of defence enzymes and proteins. Among the cell wall macromolecules are lignins and other phenolic polymers, polysaccharides such as callose, and proteins such as the hydroxyproline rich glycoproteins. Defence enzymes fall into two classes : enzymes that catalyze the production of various metabolites participating in resistance (ethylene, phytoalexins, aromatic compounds, oxidized metabolites), and direct defence enzymes (hydrolases such as chitinases and glucanases). The defence proteins include inhibitors of microbial proteases or of polygalacturonases and "pathogenesis-related" (PR) proteins.

Most of the metabolic changes that have been reported as associated to active defence in incompatible plant-fungi and plant-bacteria interactions have also been found during the hypersensitive reaction of plants to viruses (Fritig et al 1987). This is particularly the case for the PR-proteins which were initially discovered in tobacco reacting hypersensitively to tobacco mosaic virus (TMV) (Van Loon and Van Kammen 1970, Gianinazzi et al 1970), but have now been found in more than 20 plant species in various circumstances of plant stress (Van Loon 1985).

THE PATHOGENESIS-RELATED PROTEINS OF TOBACCO

The PR-proteins have several characteristic properties : they are selectively extractable at low pH ; highly resistant to proteolytic enzymes ; localized predominantly in the intercellular spaces ; and easily resolved by electrophoresis on polyacrylamide gels under native conditions.

NATO ASI Series, Vol. H 36
Signal Molecules in Plants and Plant-Microbe Interactions
Edited by B. J. J. Lugtenberg

Despite the great wealth of information on their characteristic properties and their occurrence in large amounts in various circumstances, no biological function has been found for any of these proteins for a long time after their discovery.

The highest number of PR-proteins was found in Solanaceae such as tomato (Granell et al 1987) and tobacco (Van Loon 1985). In Samsun NN tobacco, ten major PR proteins could be detected (Van Loon 1982) and were referred to as PR proteins 1a, 1b, 1c, 2, N, O, P, Q, R and S in order of decreasing mobility on electrophoresis in native gels. The three more mobile PR proteins, 1a, 1b and 1c had already been extensively studied and shown to be highly homologous (Cornelissen et al 1986). Nevertheless, their biological function is still unknown.

Since very little information was available on the physicochemical properties of the other PR proteins, we designed a preparative purification procedure using a combination of conventional and high-performance liquid chromatography. The procedure, which is summarized in the right part of Figure 1, enabled us to purify to homogeneity (Jamet and Fritig 1986, Kauffmann 1988) all PR-proteins initially detected in tobacco (Van Loon 1982) and to detect new PR-proteins of acidic isoelectric point designated Q', r1 and r2, s1 and s2, because of similar migrations upon electrophoresis on native basic gels with the known PR-proteins Q, R and S, respectively (Kauffmann 1988). By using antibodies raised against the proteins we confirmed the serological relationships between proteins 1a, 1b and 1c, and proteins PR-2, -N and -O (Fortin et al 1985), and demonstrated new groups of serologically-related proteins : PR-P and -Q (Hooft van Huijsduijnen et al 1987), and PR-R and -S (Kauffmann 1988).

FUNCTION OF SOME OF THE PR-PROTEINS

Previous reports have mentioned highly stimulated activities of chitinase in cucumber leaves infected by tobacco necrosis virus (Métraux and Boller 1986) and of 1,3-ß-glucanase in *Nicotiana glutinosa* infected by TMV (Moore and Stone 1972). We found that these enzyme activities were also strongly enhanced in the Samsun NN-TMV combination, with time course curves that were strikingly similar to those for the induction of several PR proteins (Legrand et al 1987). It turned out that four chitinases - two acidic forms with low isoelectric points and two basic ones with high pI values- were produced upon infection. These four enzymic proteins have been purified to homogeneity (Figure 1) and characterized. Their relative molecular masses (M_r) were estimated at 27500 and 28500 for the acidic isoforms and at 32000 and 34000 for the basic isoforms of chitinase.

Several lines of evidence demonstrated that the two acidic chitinases from tobacco were, in fact, proteins PR-P and PR-Q (Legrand et al 1987). This was the first demonstration of a biological function for proteins of the PRtype. Furthermore, the acidic chitinases, PR-P and -Q, are serologically closely related to the two other chitinases, which can be considered to be new basic pathogenesis-related proteins. The two basic chitinases exhibit higher specific

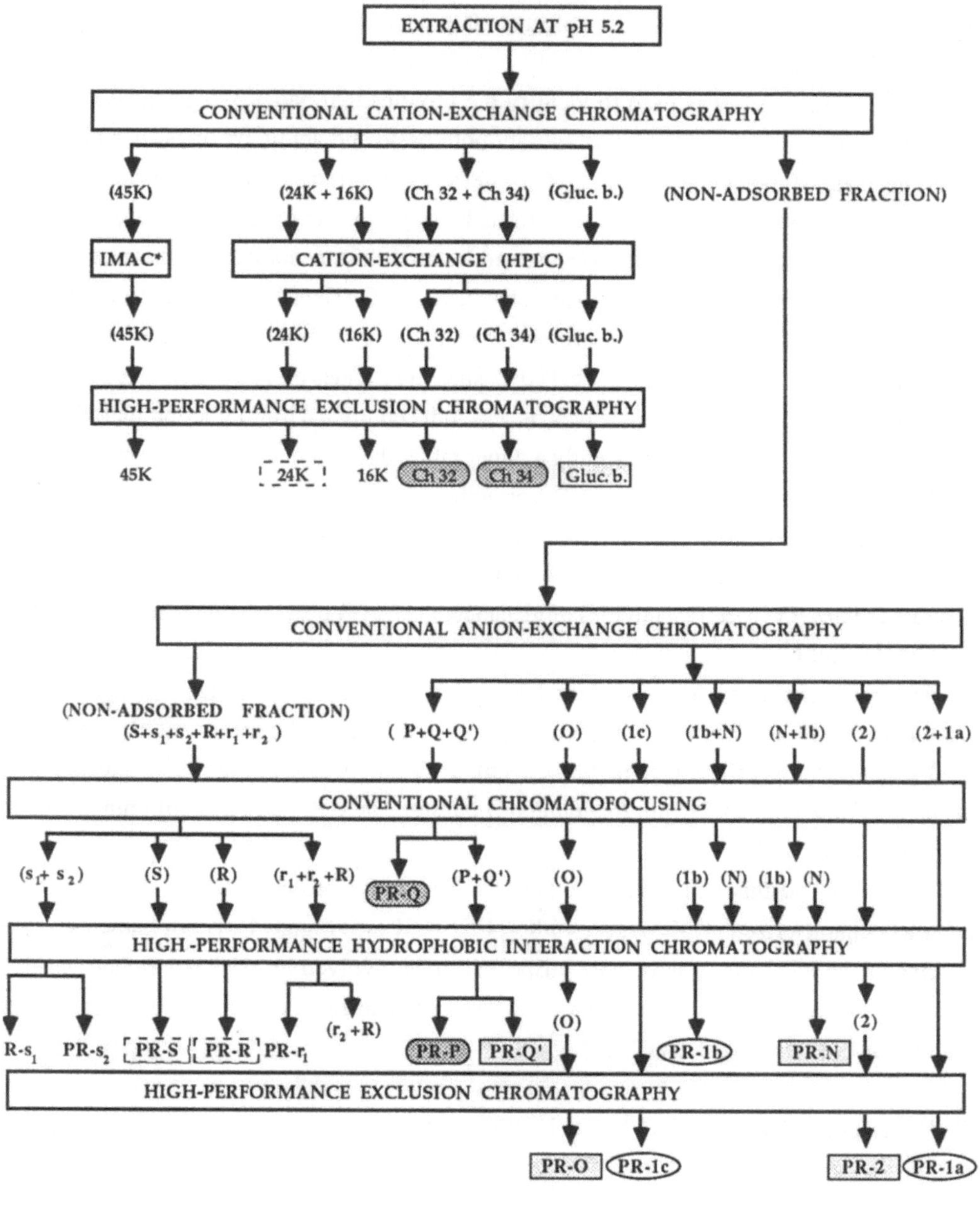

Figure 1. Purification procedure of the tobacco PR ("pathogenesis-related") proteins. *(IMAC = immobilized metal (Zn) affinity chromatography). Serologically related proteins are enclosed by boxes of similar shape. Shaded boxes indicate proteins that are glycanhydrolases.

enzyme activity than the two acidic isoforms. These two basic chitinases were found to be produced abundantly in cultured tobacco callus (Shinshi et al 1987).

A similar approach was used to identify the proteins that had 1,3-ß-glucanase activity. Enzymatic activity was found associated with both basic and acidic proteins. A major basic 1,3-ß-glucanase of molecular weight 33000 (referred to as Gluc.b. in Figure 1) was identified whose features resemble those described for a 1,3-ß-glucanase produced in tobacco tissues cultured on auxin-containing medium (Felix and Meins 1985). Among the acidic glucanases, 3 were identified as the well known PR-proteins 2, N and O (Kauffmann et al 1987). Another acidic glucanase was identified as protein Q' that has a typical behaviour of PR-protein (Kauffmann 1988). A sixth glucanase, called O' since it migrates on native gels like PR-protein O, is readily detectable in healthy leaves, is less induced upon TMV infection than the other glucanases and PR-proteins, is apparently a dimer rather than a monomer, and therefore has not all of the properties that are typical for PR-proteins (C. Michel et al, unpublished results). All acidic glucanases, except O', are serologically related to the basic 33000 glucanase that can be considered as a new tobacco PR-protein with basic isoelectric point. When laminarin is used as substrate *in vitro*, the specific activities of glucanases O and Gluc.b. are much higher than those of 2, N', Q' and O'.

At almost the same time as several PR-proteins induced in tobacco leaves reacting hypersensitively to TMV were identified as chitinases (Legrand et al 1987) and 1,3-ß-glucanases (Kauffmann et al 1987), a similar finding was made in potato leaves during the incompatible interaction with *Phytophthora infestans* or after treatment with fungal elicitor (Kombrink et al 1988).

The biological function(s) of the other well known PR-proteins, 1a, 1b, 1c, R, S and of the newly characterized r1, r2, s1, s2, 45 K, 24 K, 16 K (see Figure 1), is not known yet. The proteins of the PR-1 group have been suspected to be involved in localized and/or systemic resistance to virus infection, but there are also arguments difficult to reconcile with such a role (Van Loon 1985, Linthorst et al 1989).

GLYCANHYDROLASES AND ACTIVE DEFENCE

The lack of specificity of the metabolic alterations which confer resistance contrasts with the high specificity (gene-for-gene recognition) required to initiate the hypersensitive response. As shown in Figure 1, a significant proportion of major proteins induced in hypersensitively reacting tobacco are glycanhydrolases, and similar results have been found in potato (Kombrink et al 1988). The glycanhydrolases may represent very efficient "direct" defence enzymes against fungi and bacteria (Boller et al 1983, Schlumbaum et al 1986) and a combination of chitinases and 1,3-ß-glucanases may increase their efficiency (Mauch et al 1988).

But in addition to their role of direct anti-microbial enzymes, glycanhydrolases might also be involved in the production of chemical signals of pathogen and/or plant origin. Such regulatory molecules, called "elicitors", were obtained first from various pathogens (for a review see Darvill and Albersheim 1984) and, later, from plant cell walls (for reviews see Darvill and Albersheim 1984, McNeil et al 1984). Many are carbohydrates or contain carbohydrates. Synergistic effects between elicitors of pathogen or host origin were observed in some cases (Darvill and Albersheim 1984, Davis et al 1986). All these observations led to a hypothetical scheme involving these regulatory molecules in plant-fungi interactions (Albersheim and Darvill 1985, Fritig et al 1987, Lamb et al 1989, Hahn et al 1989). Some of the hydrolases would be able to attack the host cell wall and release endogenous elicitors while others would attack the cell wall of the fungal pathogen and release exogenous elicitors which in turn could activate host hydrolases and thus amplify the host response. Furthermore, the pathogens themselves contain hydrolases that are likely to attack the host cell wall and also release endogenous elicitors. Therefore, it appears that a complex network of elicitors of both pathogen and host origin would act with synergistic and/or cascade effects and that this could explain the amplified efficiency of the defence responses.

Can carbohydrates interfere with a virus infection ? There has been a preliminary report (Modderman et al 1985) suggesting that polysaccharides from plant origin may modulate hypersensitive resistance to virus infection. These authors partially digested isolated cell walls from tobacco leaves with a commercial fungal cellulase. The released carbohydrate material was fractionated by gel filtration and the resulting fractions were injected into the intercellular spaces of tobacco leaves reacting hypersensitively to TMV. A large reduction in mean lesion diameter was observed in the areas injected with carbohydrate containing fractions. However, the active components have not been characterized.

Plant viruses do not contain polysaccharidic cell wall material and hydrolases, so the scheme of elicitation (Albersheim and Darvill 1985, Fritig et al 1987, Lamb et al 1989, Hahn et al 1989) will need to be simplified considerably since it should involve exclusively elicitors of host origin. The occurrence and activity of such elicitors have been reviewed recently (Ryan 1987, 1988, Hahn et al 1989). It is certainly worthwile to search for putative elicitors that might be involved in hypersensitivity to viruses. We intend to characterize such elicitors, and have already shown that pure oligogalacturonides of DP 8-14 are not good candidates for such a role (J. Rouster, unpublished results).

EFFECTS OF OLIGOSACCHARIDE SIGNALS ON PLANT-VIRUS INTERACTIONS

In collaboration with P. Albersheim and A. Darvill (CCRC Russel Laboratories, Agricultural Research Center, Athens, Georgia, USA) we tried an approach using oligosaccharidic fractions known as elicitors of defence reactions in order to tentatively prestimulate typical defence pathways and examine if the pretreatment would result in any interference with a subsequent virus infection. This approach led to very surprising results. A glucan preparation obtained from the mycelial walls of the fungus *Phytophthora megasperma* f.sp. *glycinea* and known as an active elicitor of phytoalexins in soybean (Darvill and Albersheim 1984) was shown to be a very efficient inducer of resistance against viruses in tobacco (Kopp et al 1989). The glucan preparation protected against mechanically transmitted viral infections on the upper and lower leaf surfaces. Whether the glucan preparation was applied by injection, inoculation, or spraying, it protected the plants if applied before, at the same time as, or not later than 8 hours after virus inoculation. At concentrations ranging from 0.1 to 10 µg/ ml, the glucan preparation induced protection ranging from 50 to 100 % against both symptom production (necrotic local lesions, necrotic rings, or systemic mosaic) and virus accumulation in all *Nicotiana*-virus combinations examined. However, no significant protection against some of the same viruses was observed in bean or turnip. The host plants successfully protected included *N. tabacum* (9 different cultivars), *N. sylvestris*, *N. glutinosa*, and *N. clevelandii*. The viruses belonged to several taxonomic groups including tobacco mosaic virus, alfalfa mosaic virus, and tomato black ring virus. The glucan preparation did not act directly on the virus and did not interfere with virus disassembly ; rather, it appeared to induce changes in the host plant that prevented infections from being initiated or recently established infections from enlarging.

Preliminary experiments aiming to find a relationship between the degree of protection and prestimulated defence reactions led also to surprising results (Rouster et al, unpublished results) : the glucan-mediated protection did not depend on induction of the known pathogenesis-related proteins (including glycanhydrolases) or on enzymes of the phenyl-propanoid pathway. We believe the induced protection results from a mechanism that has yet to be described, and that does perhaps not involve typical defence mechanisms. Other oligosaccharides from various sources containing 3-, 6-, and 3,6-linked glucosyl residues, or only 3-linked glucosyl residues protected similarly *Nicotianae* from virus infections, though with less efficiency than the *Phytophthora megasperma* glucan preparation. Other oligosaccharides known as elicitors (Ryan 1987, 1988) have similar protecting activity against virus infections, but with apparently broader host specificities (Rouster et al, unpublished results). Biochemical studies are under way to tentatively correlate, here too, these effects with typical defence reactions.

CONCLUSION

The hypersensitive reaction is believed to be triggered by the very specific recognition between the pathogen's avirulence gene product and the host's resistance gene product. However, active defence of plants is associated with metabolic alterations that are apparently only host specific and confer a universal type of resistance. It is noteworthy that some of the tobacco PR-proteins, which were first detected in viral infections and were thought to be related to antiviral defence, are chitinases and glucanases and are therefore defence enzymes directed against insects and fungi. It follows that controlling the production of regulatory molecules such as elicitors could perhaps confer resistance to various pathogens. Hypersensitivity to viruses might be a simplified model in which to identify elicitors of host origin. These elicitors could then be used to induce metabolic alterations, and the elicitors that proved most effective in inducing resistance against challenge infection by various pathogens could be selected. The gene(s) of the corresponding hydrolase(s) releasing the most efficient elicitor(s) would be useful for plant genetic engineering. If adequate regulatory sequences were used, the expression of these genes could be targeted to the sites of preferential attack by pathogens.

The finding that various oligosaccharides known as elictors of defence reactions can protect plants against virus infections is intriguing. More has to be discovered about the generality of these oligosaccharide-induced protections, their host specificities, their efficiency against modes of transmission of virus infection other than mechanical, their relationship with defence or stress reactions, before it can be concluded that oligosaccharides may be of practical use in protecting plants against virus infections.

REFERENCES

Albersheim P, Darvill AG (1985) Oligosaccharins. Sci Am 253:44-50

Boller T, Gehri A, Mauch F, Vögeli U (1983) Chitinase in bean leaves: induction by ethylene, purification, properties, and possible function. Planta 157:22-31

Cornelissen BJC, Hooft van Huijsduijnen RAM, van Loon LC, Bol JF (1986) Molecular characterization of the messenger RNAs for 'pathogenesis-related' proteins 1a, 1b and 1c, induced by TMV infection of tobacco. EMBO J 5:37-40

Darvill AG, Albersheim P (1984) Phytoalexins and their elicitors- a defense against microbial infection in plants. Annu Rev Plant Physiol 35:243-275

Davis KR, Darvill AG, Albersheim P (1986) Host-pathogen interactions XXI Several biotic and abiotic elicitors act synergistically in the induction of phytoalexin accumulation in soybean. Plant Mol Biol 6:23-32

Felix G, Meins J (1985) Purification, immunoassay and characterization of an abundant, cytokinin-regulated polypeptide in cultured tobacco tissues. Evidence the protein is a ß-1,3-glucanase. Planta 164:423-428

Fortin MG, Parent JG, Asselin A (1985) Comparative study of two groups of b proteins (pathogenesis-related) from the intercellular fluid of *Nicotiana* leaf tissue infected by TMV. Can J Bot 63:932-93

Fritig B, Kauffmann S, Dumas B, Geoffroy P, Kopp M, Legrand M (1987) Mechanism of the hypersensitivity reaction of plants. In : Evered D, Harnett S (eds) Ciba Foundation Symposium 133 : Plant Resistance to Viruses. John Wiley & Sons, Chichester, pp 92-108

Gianinazzi S, Martin C, Vallée JC (1970) Hypersensibilité aux virus, température et protéines solubles chez le *Nicotiana* Xanthi nc. Apparition de nouvelles macromolécules lors de la répression de la synthèse virale. C R Acad Sci (Paris) D270:2383-2386

Granell A, Bellés JM, Conejero V (1987) Induction of pathogenesis-related proteins in tomato by citrus exocortis viroid, silver ion and ethephon. Physiol Mol Plant Pathol 31:83-90

Hahn MG, Bucheli P, Cervone F, Doares SH, O'Neil RA, Darvill A, Albersheim P (1989) The roles of cell wall constituents in plant pathogen interactions. In : Nester E, KosugeT. (eds) Plant-Microbe Interactions, Volume 3, Macmillan Press, (in the press).

Hooft van Huijsduijnen RAM, Kauffmann S, Brederode FT, Cornelissen BJC, Legrand M, Fritig B, Bol J (1987) Homology between chitinases that are induced by TMV infection of tobacco. Plant Mol Biol 9:411-420

Jamet E, Fritig B (1986) Purification and characterization of 8 of the pathogenesis-related proteins in tobacco leaves reacting hypersensitively to tobacco mosaic virus. Plant Mol Biol. 6:69-80

Kauffmann S (1988) Les protéines PR (Pathogenesis-Related) du Tabac : Des protéines impliquées dans les réactions de défense aux agents phytopathogènes. Isolement, propriétés sérologiques et activités biologiques. Thèse de l'Université Louis Pasteur, Strasbourg

Kauffmann S, Legrand M, Geoffroy P, Fritig B (1987) Biological function of 'pathogenesis-related' proteins : four PR proteins of tobacco have 1,3-ß-glucanase activity. EMBO J 6:3209-3212

Kombrink E, Schröder M, Hahlbrock K (1988) Several "pathogenesis-related" proteins in potato are 1,3-ß glucanases and chitinases. Proc Natl Acad Sci USA 85:782-786

Kopp M, Rouster J, Fritig B, Darvill A, Albersheim P (1989) Host-pathogen interactions XXXII. A fungal glucan preparation protects Nicotianae against infection by viruses. Plant Physiol 90:208-216

Lamb CJ, Lawton MA, Dron M, Dixon RA (1989) Signals and transduction mechanisms for activation of plant defenses against microbial attack. Cell 56:215-224

Legrand M, Kauffmann S, Geoffroy P, Fritig B (1987) Biological function of "pathogenesis-related" proteins : four tobacco PR-proteins are chitinases. Proc Natl Acad Sci USA 84:6750-6754

Linthorst HJM, Meuwissen RLJ, Kauffmann S, Bol J (1989) Constitutive expression of pathogenesis-related proteins PR-1, GRP, and PR-S in tobacco has no effect on virus infection. The Plant Cell 1:285-291

McNeil M, Darvill AG, Fry SC, Albersheim P (1984) Structure and function of the primary cell walls of plants. Annu Rev Biochem 53:625-663

Mauch F, Mauch-Mani B, Boller T (1988) Antifungal hydrolase in pea tissue. II. Inhibition of fungal growth by combinations of chitinase and ß-1,3-glucanase. Plant Physiol 88: 936-942

Métraux JP, Boller T (1986) Local and systemic induction of chitinase in cucumber plants in response to viral, bacterial and fungal infections. Physiol Mol Plant Pathol 28:161-169

Modderman PW, Schot CP, Klis FM, Wieringa-Brants DH (1985) Acquired resistance in hypersensitive tobacco against tobacco mosaic virus, induced by plant cell wall components. Phytopathol Z 113:165-170

Moore AE, Stone BA (1972) Effect of infection with TMV and other viruses on the level of a ß-1,3-glucan hydrolase in leaves of *Nicotiana glutinosa*. Virology 50:791-798

Ryan CA (1987) Oligosaccharide signalling in plants. Annu Rev Cell Biol 3:295-317

Ryan CA (1988) Oligosaccharides as recognition signals for the expression of defensive genes in plants. Biochemistry 27:8879-8883

Schlumbaum A, Mauch F, Vögeli U, Boller T (1986) Plant chitinases are potent inhibitors of fungal growth. Nature 324:365-367

Shinshi H, Mohnen D, Meins F (1987) Regulation of a plant pathogenesis-related enzyme : inhibition of chitinase and chitinase mRNA accumulation in cultured tobacco tissues by auxin and cytokinin. Proc Natl Acad Sci USA 84:89-93

Van Loon LC (1982) Regulation of changes in proteins and enzymes associated with the active defence against virus infection. In : Wood RKS (ed) Active defence mechanisms in plants. Plenum, NY, p. 247-274

Van Loon LC (1985) Pathogenesis-related proteins. Plant Mol Biol 4:111-116

Van Loon LC, van Kammen A (1970) Polyacrylamide disc electropheresis of the soluble leaf proteins from *Nicotiana tabacum* var. "Samsun" and "Samsun NN". Changes in protein constitution after infection with TMV. Virology 40:199-201

CHARACTERIZATION OF TWO VIRUS-INDUCIBLE PLANT PROMOTERS

J.F. Bol, M.D. van de Rhee, J.A.L. van Kan, M.T. González Jaén and H.J.M. Linthorst
Leiden University , Gorlaeus Laboratories, Einsteinweg 5, 2333 CC Leiden, The Netherlands

INTRODUCTION

The hypersensitive response of plants to infection with pathogens is accompanied by the induction of host genes encoding a diversity of proteins such as enzymes from the phenylpropanoid patway, peroxidases, hydrolytic enzymes, proteinase inhibitors, etcetera (for references see Bol and Van Kan 1988, Collinge and Slusarenko 1987). Studies on the induction of host proteins by virus infection were initiated by Van Loon and Van Kammen (1970) and Gianinazzi et al (1970) in the early seventies. Their model system consisted of tobacco plants reacting hypersensitively to infection with tobacco mosaic virus (TMV). The induction of at least 10 acidic proteins was observed which are all excreted by the plant cells into the intercellular space of the leave (see Bol and Van Kan 1988). These were named "pathogenesis-related" (PR) proteins 1a, 1b, 1c, 2, N, O, P, Q, R and S. Proteins 1a, 1b and 1c (15 kDa) have a highly homologous amino acid sequence; their function is not known (Cornelissen et al. 1987). Proteins 2, N and O (40 kDa) and proteins P and Q (29 kDa) were shown to be β-1,3-glucanases and chitinases, respectively (Kauffmann et al. 1987, Legrand et al. 1987). The function of protein R (13-15 kDa) is not known. Protein S (23 kDa) is homologous to the sweet-tasting protein thaumatin and to a maize protein that inhibits enzymes in the digestive tract of insects (Cornelissen et al 1986, Pierpoint et al 1987, Richardson et al 1987). It has been speculated that the hydrolytic enzymes (PR proteins 2, N, O, P and Q) play a role in the TMV-induced resistance of tobacco to fungal infection while PR-S could be involved in an induced resistance to insect attack (see Bol and Van Kan 1988).

Basic isoforms of all acidic PR proteins, with the exception of PR-R, have been identified in tobacco. Only minor amounts of these basic proteins are found in the extracellular fluid of infected tobacco. By analogy to the

NATO ASI Series, Vol. H 36
Signal Molecules in Plants and Plant-Microbe Interactions
Edited by B. J. J. Lugtenberg

basic chitinase of bean plants (Boller and Vögli 1984) they may accumulate in vacuoles. Proteins related to the acidic or basic tobacco PR's by amino acid sequence homology or serology have been observed in many plant species, including both monocots and dicots (see Bol et al 1989). In addition to virus infection, PR proteins are also known to accumulate after treatment of plants with a variety of chemicals. An inducer that has been widely used in salicylate. The observation that spraying of plants with salicylate also induces a resistance to virus infection has led to the suggestion that some PR's may be involved in an antiviral response of the plant (White 1979, Hooft van Huijsduijnen 1986b). Here we report on the characterization of regulatory elements involved in the induction of tobacco genes by TMV-infection or salicylate treatment.

INDUCTION OF TOBACCO GENES BY SALICYLATE AND TMV

Figure 1 shows seven Northern blots probed with cDNA clones corresponding to different classes of tobacco mRNAs that are induced by TMV-infection. The blots were probed with cDNA clones corresponding to acidic and basic PR-1 proteins (Cornelissen et al 1987), acidic β-1,3-glucanases (Linthorst et al , to be published), basic β-1,3-glucanases (kindly provided by Dr. J. Memelink), acidic and basic chitinases (Hooft van Huijsduijnen et al 1987) and the acidic thaumatin-like protein PR-S (Cornelissen et al 1986). A basic isoform of PR-S has been identified that is induced by salt stress of tobacco (Singh et al 1987) but cDNA clones to this tobacco protein were not yet available.

All mRNAs shown in Fig. 1 are present in relatively low amounts in healthy plants (lanes H) and are strongly induced after TMV-infection (lanes T). In addition, the mRNAs encoding acidic and basic PR-1 proteins, acidic glucanases and basic chitinases are also strongly induced by salicylate (lanes S) while the other mRNAs are only weakly induced by this treatment. The mRNAs encoding acidic and basic isoforms of a given class of PR proteins may differ somewhat in size and so do the encoded proteins (see Bol et al 1989). A significant difference is that the genes encoding the basic isoforms are constitutively expressed in the roots of healthy tobacco plants whereas the genes for acidic proteins are not (Memelink 1988).

In addition to the mRNAs shown in Fig. 1, we have cloned cDNA to two

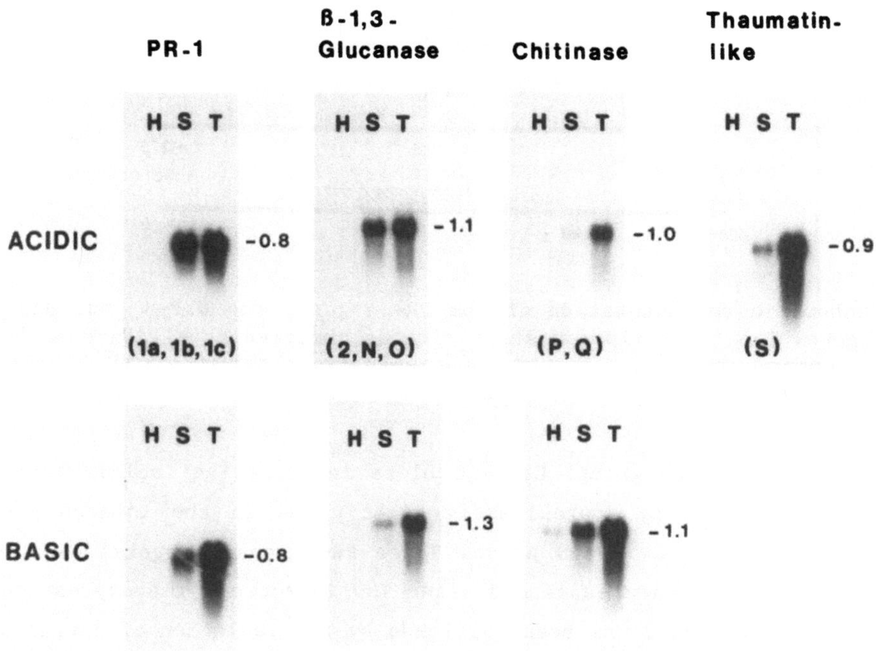

Fig. 1. Induction of mRNAs in Samsun NN tobacco by salicylate-treatment or TMV-infection. Seven Northern blots were loaded with RNA from healthy plants (lanes H), plants sprayed with 5 mM salicylate (lanes S) and plants infected with TMV (lanes T). The blots were hybridized to ^{32}P-labeled cDNA clones corresponding to the acidic and basic isoforms of PR-1 proteins, β-1,3-glucanases and chitinases and the acidic isoform of PR-S. The estimated size of the mRNAs (kb) is indicated in the right margin of each blot.

other TMV-inducible tobacco mRNAs which were initially called "A" and "C" (Hooft van Huijsduijnen et al 1986a). As mRNA C was found to encode a glycine-rich protein it was renamed GRP (Van Kan et al 1988). The GRP mRNA is also strongly induced within two days after spraying of tobacco with a salicylate solution.

STRUCTURE OF PR-1, GRP AND PR-S GENES

The cDNA clones of the TMV-inducible mRNAs have been used to isolate corresponding genes from a Samsun NN tobacco genomic library. So far, we have reported the sequencing of three genes encoding acidic PR-1 proteins (Cornelissen et al 1987), two GRP genes (Van Kan et al 1988) and two PR-S genes (Van Kan et al 1989). Fig. 2 shows a schematic representa-

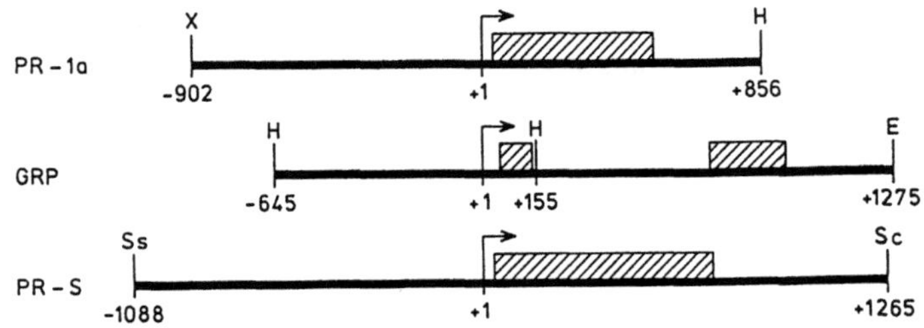

Fig. 2. Schematic representation of the PR-1a gene, the GRP-8 gene and the PR-S E22 gene. The transcription start site is numbered as +1. Hatched boxes show the coding regions of the genes. Restriction sites: X, XhoI; H, HindIII, E, EcoRI; Ss, SstI; sc, ScaI.

tion of agene from each group. Genomic blots indicate that acidic PR-1 and GRP are each encoded by approximately eight genes in the tobacco genome while there are about two PR-S genes. Three PR-1 genes, several GRP genes and the two PR-S genes are expressed after TMV-infection. The expression of the genes shown in Fig. 2 has been confirmed by the isolation of corresponding cDNA clones or by S1-nuclease mapping. Interestingly, a putative silent GRP gene was found to have a 2 kb intron with a mutated splice-donor site while in an active gene a 555 bp intron was flanked by correct splice-sequences (Van Kan et al 1988).

CIS-ACTING ELEMENTS OF PR-1 AND GRP GENES

The cloned PR-1a and GRP-8 genes shown in Fig. 2 were used to study regulatory elements involved in the induction of these genes by TMV-infection and salicylate-treatment of tobacco. The transcription start sites, designated as +1 in Fig. 2, were mapped by primer-extension studies done with the corresponding mRNAs. The sequence of -902 to +29 of the PR-1a gene was fused to the GUS reporter gene while the sequence of -645 to +8 of the GRP-8 gene was fused to the CAT reporter gene. These fusions were used to transform tobacco plants by an Agrobacterium tumefaciens mediated transformation procedure. Transformants were assayed for induction of the reporter gene after salicylate treatment or TMV-infection.

About half of the primary transformed plants containing the PR-1a upstream sequence showed a significant induction of GUS activity by salicylate and TMV. This property was reproducibly retained in the progeny of these plants. Induction by TMV occurred also in systemically induced virus-free leaves indicating that the 902 bp promoter fragment contains elements that respond to the signal molecules that migrate in the plant to activate PR genes at a distance from the site of infection. Results obtained with a nested set of deletions of the PR-1a promoter indicated that an upstream sequence of 689 bp was sufficient for induction by TMV or salicylate while constructs with 633 bp or shorter sequences were inactive (Van de Rhee et al , manuscript in preparation).

Similar results were obtained with the GRP promoter. The upstream sequence of 645 bp was sufficient to induce the CAT gene in response to salicylate. Reduction of this sequence to 400 bp or less resulted in a very low inducibility of the reporter gene with salicylate. The data indicated that salicylate responsive elements are present immediately upstream from the transcription start site, but that enhancer-like sequences further upstream are required for an efficient induction of the gene by salicylate (Van de Rhee et al , manuscript in preparation).

CONCLUSIONS

The observation that virus-inducible tobacco genes respond differently to various chemical inducers, show differences in their tissue-specific expression and are expressed in a light-dependent manner, indicates that the promoter regions of these genes are composed of a complex mosaic of various cis-acting elements interacting with different trans-acting factors. The results reported here offer a possibility to identify some of these factors as a first step towards the characterization of the "mobile compounds" that are believed to be involved in the systemic induction of PR genes by a local necrotic infection.

REFERENCES

Bol JF, Van Kan JAL (1988) The synthesis and possible functions of virus-induced proteins in plants. Microbiol Sci 5: 47-52

Bol JF, Linthorst HJM, Meuwissen RLJ (1989) Molecular responses of tobacco to virus infection. In: Smith CJ (ed) Biochemistry and molecular biology of plant-pathogen interactions. Oxford University Press, Oxford, in press

Boller T, Vögeli U (1984) Vacuolar localization of ethylene-induced chitinase in bean leaves. Plant Physiol 74: 442-444

Collinge DB, Slusarenko AJ (1987) Plant gene expression in response to pathogens. Plant Molec Biol 9: 389-410

Cornelissen BJC, Hooft van Huijsduijnen RAM, Bol JF (1986) A tobacco mosaic virus-induced tobacco protein is homologous to the sweet-tasting protein thaumatin. Nature 321: 531-532

Cornelissen BJC, Horowitz J, Van Kan JAL, Goldberg RB, Bol JF (1987) Structure of tobacco genes encoding pathogenesis-related proteins from the PR-1 group. Nucl Acids Res 15: 6799-6811

Gianinazzi S, Martin C, Vallee JC (1970) Hypersensibilite aux virus, temperatures et proteines solubles chez le Nicotiana Xanthi-nc. Apparition de nouvelles macromoles lors de la repression de la synthese virale. C R Acad Sci Paris D 270: 2383-2386

Hooft van Huijsduijnen RAM, Van Loon LC, Bol JF (1986a) cDNA cloning of six mRNAs induced by TMV infection of tobacco and a characterization of their translation products. EMBO J 5: 2057-2061

Hooft van Huijsduijnen RAM, Alblas SW, De Rijk RH, Bol JF (1986b) Induction by salicylic acid of pathogenesis-related proteins and resistance to alfalfa mosaic virus infection in various plant species. J Gen Virol 67: 2135-2143

Hooft van Huijsduijnen RAM, Kauffmann S, Brederode FTh, Cornelissen BJC, Legrand M, Fritig B, Bol JF (1987) Homology between chitinases that are induced by TMV infection of tobacco. Plant Mol Biol 9: 411-420

Kauffmann S, Legrand M, Geoffroy P, Fritig B (1987) Biological function of "pathogenesis-related" proteins. Four PR-proteins of tobacco have 1,3--glucanase activity. EMBO J 6: 3209-3212

Legrand M, Kauffmann S, Geoffroy P, Fritig B (1987) Biological function of "pathogenesis-related" proteins: four tobacco PR-proteins are chitinases. Proc Natl Acad Sci USA 84: 6750-6754

Memelink J (1988) Altered gene expression in T-DNA transformed tobacco tissues. Thesis, Leiden University

Pierpoint WS, Tatham AS, Pappin DJC (1987) Identification of the virus-induced protein of tobacco leaves that resembles the sweetprotein thaumatin. Physiol Mol Plant Pathol 31: 291-298

Richardson M, Valdes-Rodriguez S, Blanci-labra A (1987) A possible function for thaumatin and a TMV-induced protein suggested by homology to a maize inhibitor. Nature 327: 432-434

Singh NK, Bracker CA, Hasegawa PM, Handa AK, Buckel S, Hermodson MA, Pfankoch E, Regnier FE, Bressan RA (1987) Characterization of osmotin; a thaumatin-like protein associated with osmotic adaptation in plant cells. Plant Physiol 85: 529-536

Van Kan JAL, Cornelissen BJC, Bol JF (1988) A virus-inducible tobacco gene encoding a glycine-rich protein shares putative regulatory elements with the ribulose bisphosphate carboxylase small subunit gene. Molec Plant-Microbe Int 1: 107-112

Van Kan JAL, Van de Rhee MD, Zuidema D, Cornelissen BJC, Bol JF (1989) Structure of tobacco genes encoding thaumatin-like proteins. Plant Molec Biol 12: 153-155

Van Loon LC, Van Kammen A (1970) Polyacrylamide disc electrophoresis of the soluble leaf proteins from Nicotiana tabacum var "Samsun" and "Samsun NN". II. changes in protein constitution after infection with tobacco mosaic virus. Virology 40: 199-221

White RF (1979) Acetylsalicylic acid (aspirin) induces resistance to tobacco mosaic virus in tobacco. Virology 99: 410-412

A GENE FOR AVIRULENCE FROM PSEUDOMONAS SYRINGAE PV. PHASEOLICOLA

Carol Jenner, Ed Hitchin, Ian Brown, Karen Walters and John Mansfield,
Department of Biochemistry and Biological Sciences, Wye College, Wye, Ashford, Kent, UK (for correspondence)
Peter Betteridge,
Shell Research Ltd., Sittingbourne, Kent, UK

Analysis of the virulence of isolates of *P. syringae* pv. *phaseolicola* and the susceptibility of varieties and genotypes of *Phaseolus vulgaris* and related *Phaseolus* spp. from Africa is producing new insights into specificity in the halo-blight disease of bean. Taylor et al (1987) described the identification of race 3 of *P.s.* pv. *phaseolicola* and subsequent unpublished studies indicate that at least six races may be differentiated on certain genotypes. In our work we have focused on isolates of race 3 which were found to be differentiated into two pathotypes based on their reaction on the tepary bean *Phaseolus acutifolius* as summarized in Table 1.

TABLE 1 Summary of the interactions between *Phaseolus* and isolates of *Pseudomonas syringae* pv. *phaseolicola*

		Reaction[a] on pods of French bean cultivars and *P. acutifolius* (tepary bean)			
Isolate	Race	Canadian Wonder	Red Mexican	Tendergreen	*P. acutifolius*
52	1	S	HR	S	HR
882	2	S	S	S	HR
1301A	3A	S	S	HR	S
1302A	3B	S	S	HR	HR
Transconjugant in 52	1+avirulence gene[b]	S	HR	HR	HR
	1+virulence gene[b]	S	S	S	HR

[a] S, water-soaked lesion; HR, hypersensitive reaction, small brown lesion
[b] In cosmids from the genomic libraries of races 3A or 3B

NATO ASI Series, Vol. H 36
Signal Molecules in Plants and Plant-Microbe Interactions
Edited by B. J. J. Lugtenberg

The cloning strategy we have adopted is based on the cosmid vectors pLAFR1 and pLAFR3 which have a broad host range being derived from the IncP plasmid RP4 (Friedman et al 1982; Staskawicz et al 1987). Genomic libraries of races 3A (in pLAFR1) and 3B (in pLAFR3) were conjugated into race 1 using the helper plasmid pRK2013 and transconjugants tested for virulence on cvs. Canadian Wonder, Red Mexican and Tendergreen. In this way we were able to test for the presence of genes controlling avirulence or virulence in isolates of race 3 as indicated in Table 1. Young bean pods were used for the initial pathogenicity tests (Harper et al 1987; Hitchin et al 1989). Shortage of seed of the line of P. acutifolius differentiating races 3A and 3B, and very low efficiency of conjugation into race 3A (isolate 1301A) precluded screening the race 3B library for a gene for avirulence interacting with the tepary bean and differentiating between 3A and 3B.

PROPERTIES OF DETERMINANTS OF AVIRULENCE

Isolation of a determinant of avirulence from race 3A was described by Hitchin et al (1989). None of the 842 pLAFR1-based genomic clones tested altered the race 1 phenotype from avirulent to virulent on cv. Red Mexican, but two clones (p3A1000 and p3A2000) with a common 19.2 kb EcoR1 fragment conferred avirulence in cv. Tendergreen.

Leaves and pods of cv. Tendergreen inoculated with race 1 (p3A1000) or race 1 (p3A2000) underwent a hypersensitive reaction apparently identical to that caused by race 3A. Similarity between the reactions was indicated by symptom appearance, phytoalexin accumulation and the restricted multiplication of bacteria as a result of the HR (Hitchin et al 1989). The response of leaves was also examined by electron microscopy using the quantitative techniques described by Brown and Mansfield (1988). Results of an ultrastructural comparison between the responses of Tendergreen leaves to race 1 (p3A1000) and race 1 containing p3A001 (a clone which did not affect the virulence of race 1) are given in Fig. 1.

Plant cells adjacent to colonies of virulent and avirulent transconjugants of race 1 in leaves of cv. Tendergreen underwent the same sequence of responses as reported for the wild-type strains race 1 and race

3A respectively (Brown and Mansfield 1988). Both types of transconjugant caused an initial, localized convolution of the plasma membrane in adjacent plant cells during the first 3 h after inoculation. Paramural papillae were subsequently deposited at sites of membrane convolution. Formation of papillae was most noticeable during the compatible interaction to race 1 (p3A001) but they were less conspicuous and eventually dispersed in cells undergoing the HR. The first sign of an obvious differential response and the onset of the HR was the rupture of the tonoplast in cells adjacent of race 1 (p3A1000). Visible damage to the tonoplast is thought to follow closely earlier irreversible damage to the plasma membrane (Brown and Mansfield 1988). Once the tonoplast has ruptured, cytoplasmic disorganization and collapse occur very rapidly (Fig. 1).

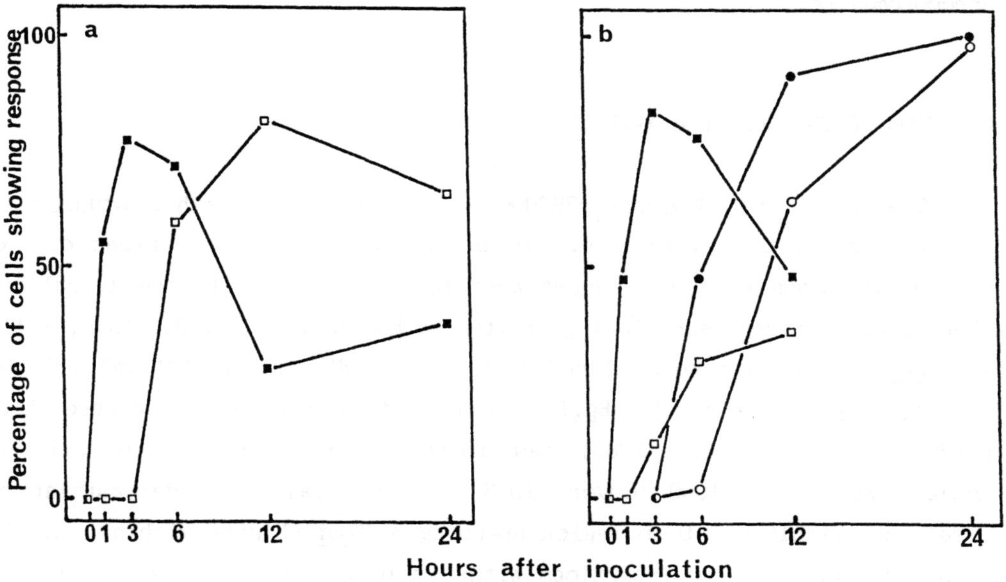

Fig. 1 Responses of mesophyll cells in leaves of bean cv. Tendergreen inoculated with a) P.s. pv. phaseolicola race 1 (p3A001) which causes a susceptible reaction and b) race 1 (p3A1000) which induces the HR. Cells adjacent to bacterial colonies were examined and the percentage of plant cells showing localized convolution of the plasma membrane (■), deposition of papille (□), tonoplast rupture (●) and cytoplasmic disorganization (○) recorded. At least 30 colonies of each transconjugant were recorded for the calculation of each percentage.

The transconjugants used for this work are effectively "isogenic lines" and therefore allow comparisons to be made without the influence of differing genetic background. Our findings therefore confirmed earlier proposals that one of the primary functions of the avirulence gene is to control the occurrence of irreversible damage to the plant plasma membrane which is the irrevocable event during the onset of the HR.

Genomic clones conferring avirulence to race 1 in cv. Tendergreen were also recovered from the race 3B library (p3B1000 and p3B2000), again no clones from the 950 tested conferred virulence to race 1 on cv. Red Mexican. Initial examination of clones from races 3A and 3B by restriction digests indicated that they contained different regions of DNA. However a common c. 10 kb region was identified within different flanking regions of chromosomal DNA.

LOCATION OF THE GENE FOR AVIRULENCE

The clones p3A1000 and p3B2000 were chosen for further studies to locate gene(s) for avirulence. In order to ensure that regions of DNA outside the common 10 kb fragment were not contributing to the avirulence phenotype, different sub-cloning strategies had to be adopted. Thus, BglII and BamHI fragments were cloned from insert DNA in p3A1000 and p3B2000 respectively. The 8.5 kb BglII fragment from p3A1000 sub-cloned into pRK290 (Ditta et al 1980) was found to be fully active and was mutagenized with Tn1000 (Guyer 1978). This located the determinant of virulence within a 0.9 kb region spanning a HindIII site within a 3.4 kb BamHI fragment; Tn1000 insertions within this region abolished ability to cause the HR. Sub-cloning from p3B2000 revealed that the same fragment contained the gene(s) for avirulence from race 3B. Restriction maps and the activity of sub-cloned DNA from within the common 3.4 kb BamHI fragment are given in Fig. 2. An HR identical to that caused by wild type race 3 was induced by transconjugants of race 1 containing the 1.4 kb BglII/HindIII fragment.

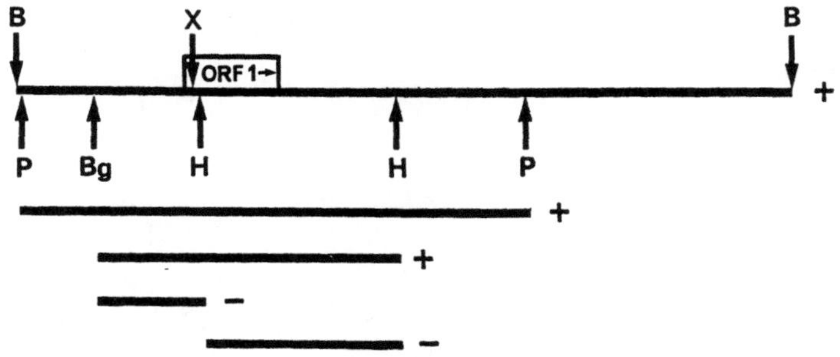

Fig. 2 Restriction maps of sub-cloned fragments common to p3A1000 and p3B2000, and their ability to cause the HR (+ or -). Sites for BamHI (B), BglII (Bg), HindIII (H), PstI (P) and XbaI (X) are given. Note that the 1.4 kb BglII/HindIII fragment cloned into pLAFR3 caused transconjugants of race 1 to cause the HR in plants with the gene for resistance in cv. Tendergreen. The position of the open reading frame (ORF1) identified as the gene for avirulence in race 3B is indicated.

GENE FOR GENE INTERACTION

Sub-cloning and transposon mutagenesis therefore indicated that a single gene may control the avirulence of race 3A and 3B towards cv. Tendergreen. Transconjugants with genomic clones and sub-cloned fragments were also found to cause the HR on other cultivars and genotypes of bean resistant to races 3A and 3B (Hitchin et al 1989). The implication of these observations was that a gene for gene relationship controls interactions between pathotypes of P.s. pv. phaseolicola and P. vulgaris. In order to test this hypothesis the virulence of wild type races 1, 3A and 3B and tranconjugants of race 1 containing the cloned avirulence gene from 3A and 3B was examined on the F_2 progeny of crosses between cv. Canadian Wonder (susceptible to race 3) and cv. Tendergreen (resistant to race 3), both cultivars are susceptible to race 1 (see Table 1). To date, analysis of 121 F_2 progeny from 10 families has indicated co-segregation for resistance to races 3A and 3B and reaction to transconjugants. Numbers of resistant and susceptible progeny were 95 and 26 respectively giving a good fit to the 3:1 ratio expected if resistance to races 3A and 3B in cv. Tendergreen was controlled by a single dominant gene. Our results indicate that race 3B probably contains a second gene for avirulence corresponding to a gene for resistance in P. acutifolius which is not present in cv. Tendergreen.

SEQUENCING AND HOMOLOGY

The nucleotide sequence of the 2.2 kb PstI fragment from race 3B containing the determinant of avirulence was determined. The fragment was sub-cloned into pBR322 mutagenized with Tn1000 and sequenced by dideoxy methods from $\gamma\delta$ primers (Liu et al 1987). An opening reading frame (ORF1) of 384bp spanning the HindIII site and unique XbaI site was located within the active 1.4 kb BglII/HindIII fragment (Fig. 2). The 2.2 kb PstI fragment cloned in pLAFR3 (p3B2302) was digested with XbaI, ends filled with Klenow fragment and the blunt ends religated. This procedure introduced a frameshift into ORF1. Following the frameshift, p3B2302 no longer caused the HR indicating that ORF1 is the gene for avirulence. The predicted amino acid sequence of ORF1 contained no hydrophobic, potentially membrane-spanning regions and was not highly homologous to any sequences in data banks, including the avirulence genes avr A, B and C from P.s. pv. glycinea (Napoli and Staskawicz 1987; Tamaki et al 1988).

The 3.4 kb BamH1 and internal 0.86 kb HindIII fragments (Fig. 2) were used as probes in Southern blot analyses of digests of DNA recovered from various isolates of P.s. pv. phaseolicola and other pathovars of P. syringae. Results indicated that all race 3 isolates of P.s. pv. phaseolicola contained single copies of an homologous sequence located in different regions of DNA. Hybridization was also observed to an isolate of P.s. pv. glycinea but not to four isolates of races 1 and 2 of P.s. pv. phaseolicola from diverse geographical origins, to P.s. pv. coronafaciens, P.s. pv. pisi or to P.s. pv. syringae.

Our results strongly suggest that a gene for gene relationship Flor, 1971) controls interactions between races of P.s. pv. phaseolicola and Phaseolus. Similar conclusions have been reached for P.s. pv. glycinea and soybean (Staskawicz et al 1984; Tamaki et al 1988) although the genetics of resistance in soybean is less clear (Crute 1986). The avirulence gene (ORF1) from race 3B would encode a much smaller protein (16.6 kDa) than has been described for the avr genes from P.s. pv. glycinea; the smallest, encoded by avrB, being 36 kDa. Determination of the function of the encoded protein is the next step in our programme to investigate the control of cultivar and race specificity in P.s. pv. phaseolicola; the protein may itself act as a cultivar specific elicitor of the HR.

We wish to acknowledge support from the Agricultural and Food Research Council and Science and Engineering Research Council of the United Kingdom.

REFERENCES

Brown IR, Mansfield JW (1988) An ultrastructural study, including cytochemistry and quantitative analyses, of the interactions between pseudomonads and leaves of Phaseolus vulgaris L. Physiol & Molec Plant Pathol 33: 351-376

Crute IR (1986) Investigations of gene-for-gene relationships: the need for genetic analyses of both host and parasite. Plant Pathology 35: 15-17.

Ditta G, Stanfield S, Corbin D, Helinski DR (1980) Broad host range cloning system for gram-negative bacteria: construction of a gene bank of Rhizobium meliloti. Proc Natl Ac Sci USA 77: 7347-7351

Flor H (1971) Current status of the gene-for-gene concept. Ann Rev Phytopathol 9: 275-296

Friedman AM, Long SR, Brown SE, Buikema WJ, Ausubel F (1982) Construction of a broad host range cosmid cloning vector and its use in the genetic anaysis of Rhizobium meliloti. Gene 18: 289-296

Guyer MW (1978) The $\gamma\delta$ sequence of F is an insertion sequence. J Mol Biol 126: 347-365

Harper S, Zewdie N, Brown IR, Mansfield JW (1987) Histological, physiological and genetical studies of the responses of leaves and pods of Phaseolus vulgaris to three races of Pseudomonas syringae pv. phaseolicola and to Pseudomonas syringae pv. coronafaciens. Physiol & Molec Plant Pathol 31: 153-172

Hitchin FE, Jenner CE, Harper S, Mansfield JW, Barber CE, Daniels MJ (1989) Determinant of cultivar specific avirulence cloned from Pseudomonas syringae pv. phaseolicola race 3. Physiol & Molec Plant Pathol 34: 309-322

Liu L, Whalen W, Das A, Berg CM (1987) Rapid sequencing of cloned DNA using a transposon for bidirectional priming: sequence of the Escherichia coli K-12 avtA gene. Nucl Acids Res 15: 9461-9469

Napoli C, Staskawicz BJ (1987) Molecular characterization and nucleic acid sequence of an avirulence gene from race 6 of Pseudomonas syringae pv. glycinea. J Bacteriol 169: 572-578

Staskawicz BJ, Dahlbeck D, Keen NT (1984) Cloned avirulence gene of Pseudomonas syringae pv. glycinea determines race-specific incompatibility on Glycine max (L.) Merr. Proc Natl Acad Sci USA 81: 6024-6028

Tamaki S, Dahlbeck D, Staskawicz B, Keen NT (1988) Characterization and expression of two avirulence genes cloned from Psudomonas syringae pv. glycinea. J Bacteriol 170: 4846-4854

Taylor JD, Teverson DM, Davies JHC (1987) Halo-blight of Phaseolus bean. Report of the National Vegetable Research Station for 1986, p. 63-64. N.V.R.S. Wellesbourne Warwick

CHARACTERIZATION AND FUNCTION OF AVIRULENCE GENES FROM *PSEUDOMONAS SYRINGAE* PV. *TOMATO*

N.T. Keen, S. Tamaki, D. Kobayashi, M. Stayton, D. Gerhold, H. Shen, S. Gold, J. Lorang and H. Thordal-Christensen
Department of Plant Pathology,
University of California
Riverside, California 92521, U.S.A.

INTRODUCTION

Avirulence genes in plant pathogens did not evolve for that purpose but are a consequence of the occurrence in plants of surveillance mechanisms that detect highly specific features of only certain pathogen biotypes. Thus, while the functions of avirulence genes in pathogens are generally not known, it is assumed that their role in recognition by plants is gratuitous. If recognition of an avirulent pathogen occurs, the plant undergoes a complex active defense reaction called the hypersensitive response (HR) that restricts further pathogen development. The plant surveillance system is controlled by plant disease resistance genes that are thought to encode specific receptors. They recognize pathogen metabolites called specific elicitors. The formation of these elicitors is poorly understood, but recent evidence suggests that they are either the primary protein products of pathogen avirulence genes or cellular metabolites arising from their catalytic activity. We also know little about the structure and function of plant disease resistance genes, although recent progress has occurred in attempts to clone them (e.g., Bennetzen et al 1988). In contrast, several pathogen avirulence genes have been cloned and sequenced (Keen and Staskawicz 1988).

RECOGNITION LEADING TO THE HYPERSENSITIVE REACTION

Following initial recognition of a pathogen to elicit the HR, 30 or more plant genes are depressed and this eventually leads to inhibition of pathogen development. Several of these genes control biochemical events that have been associated with expression of the HR, including the production of phytoalexins, chitinases, antiviral proteins, β-1,3-glucanases, hydroxyproline rich glycoproteins and lignin. All of these agents are potential antagonists to certain classes of pathogens. However, since expression of the hypersensitive response is not the major topic of this paper, the reader is referred to Lamb et al (1989) for a recent review.

NATO ASI Series, Vol. H 36
Signal Molecules in Plants and Plant-Microbe Interactions
Edited by B.J.J. Lugtenberg

Although the HR involves varied and highly specific surveillance systems that detect a range of incompatible pathogens (but, alas, fail to recognize compatible pathogens), we know surprisingly little about the chemistry of these recognitional mechanisms. It is clear that the HR operates on various levels in plant-pathogen interactions. One of these is termed general resistance, in which an entire plant species is uniformly resistant to an entire pathogen taxon. Another level, called specific resistance, involves the resistance of certain cultivars of a single plant species to only certain biotypes or races of a single pathogen taxon. Specific resistance is generally controlled by single plant disease resistance genes and single complementary avirulence genes in the pathogen (Flor 1942). Thus, for plant recognition and the HR to occur, both the plant and the invading pathogen must carry dominant alleles of at least one pair of complementary resistance genes/avirulence genes. This gene-for-gene relationship has been known for some 50 years but only recently have we begun to understand the underlying mechanisms.

MODELS TO EXPLAIN GENE-FOR-GENE SPECIFICITY

Several models with varying degrees of complexity have been devised to explain recognition in gene-for-gene systems. One of them is the elicitor-suppressor model (Doke et al 1979) in which pathogens are thought to produce general elicitors that initiate host defense unless a specific suppressor substance is also produced by the pathogen. While this model has found only limited experimental support to date in gene-for-gene systems (e.g. Doke et al 1979), it may function in those rare cases where virulence rather than avirulence is dominant in the pathogen or when inhibitors of avirulence genes are present.

The dimer model (Ellingboe 1982) proposed that the primary gene products of pathogen avirulence genes and plant disease resistance genes interacted in a a ligand-receptor fashion to form a dimer which caused disease resistance *per se* in an unspecified manner. This model ignored the probable role of the hypersensitive reaction and phytoalexins as conferral mechanisms of disease resistance. To account for this shortcoming, the elicitor-receptor model has been advanced (Dixon 1986; Gabriel et al 1988; Keen 1982). In this scheme, either the primary avirulence gene protein product or a second metabolite resulting from its catalytic activity is recognized by a specific plant receptor encoded by a disease resistance gene.

SPECIFIC ELICITORS

Specific elicitors that mirror the specificity of certain pathogen races have been detected in several systems (Bruegger and Keen 1979; De Wit et al 1985; Keen and Legrand 1980; Mayama et al 1986; Tepper and Anderson 1986), but relatively few of them have been isolated and characterized. Of these, the classic host selective toxin, victorin, was shown to behave as a specific elicitor in *Pc*-2 oat plants (Mayama et al 1986). A small linear peptide from *Cladosporium fulvum* has also been identified and was found to be a race specific elicitor in tomato plants carrying the *Cf*9 resistance gene (Schottens-Toma and De Wit 1988). While the relationship of specific elicitors to avirulence gene function has not been fully established, it is important to note that a specific elicitor may or may not be the primary gene product of an avirulence gene.

AVIRULENCE GENES

Avirulence genes have been cloned by several laboratories from various *P. syringae* and *Xanthomonas campestris* pathovars [e.g. J. Leach and co-workers from *X. campestris* pv. *oryzae* (personal communication); S. Patil and colleagues with *P. syringae* pv. *phaseolicola* (personal communication); J. Mansfield and collaborators with *P. s. phaseolicola* (this book); Vivian et al (1989) with *P.s. pisi*; Gabriel et al (1986) with *X. c. malvacearum*; Staskawicz, Keen and co-workers from various *P. syringae* and *X. campestris* pathovars (see Keen and Staskawicz 1988)].

All of the avirulence (*avr*) genes thus far sequenced consist of single open reading frames. These proteins have not exhibited significant homology to those in data banks and, with a few exceptions, are unrelated to each other. To date, therefore, the sequence data have not answered the questions of (i) how do *avr* genes function to cause plant recognition of the pathogen? and (ii) what is the function of *avr* genes in the pathogen? Recent advances with avirulence genes from *P. syringae* pv. *tomato* have iluminated these questions.

AVIRULENCE GENE D

We have cloned four different avirulence genes from *P. syringae* pv. *tomato* (*Pst*) that function in the related bacterium, *P.s. glycinea (Psg*), causing it to elicit a hypersensitive reaction (HR) on some but not all soybean cultivars (Kobayashi et al 1989; Shen and Keen, unpublished). Surprisingly, one of these genes was identical to *avr*A, previously cloned from *Psg* race 6. A second *Pst* avirulence gene, called *avr*D, has been studied in attempts to determine how it initiates the soybean HR. A functional copy of this gene is not present in any tested race of *Psg*, but we were surprised to find that all *Psg* races contained a DNA sequence with high homology to *avr*D from *Pst*. These genes were located on *ca.* 80 kb indigenous plasmids in both *Pst* and *Psg* (Kobayashi and Keen, manuscript in preparation). Sequence analysis showed that the *avr*D genes from both *Pst* and *Psg* were 92% homologous and encoded colinear proteins of 311 amino acids, of which 86% were identical. However, two downstream ORFs contained 98 and 99% identical amino acids. When the *avr*D homologue from *Psg* was cloned and reintroduced into *Psg* or *E. coli* cells on expression plasmids, no HR was observed in soybean leaves. However, expression of the *Pst avr*D gene in either *E. coli* cells or in *Psg* resulted in hypersensitive reactions on certain soybean cultivars. The *avr*D protein extracted from overproducing *E. coli* cells did not cause a HR when infiltrated into soybean leaves, but culture fluids from *E. coli* or *Psg* cells expressing *avr*D contained a low molecular weight race specific elicitor (Keen et al., in preparation). This molecule has been purified and partially characterized, (Stayton, Tamaki, Gerhold and Keen, in preparation), but its structure has not yet been determined. The *avr*D elicitor is relatively potent, producing HR on sensitive soybean cultivars in the nM range, but not on insensitive cultivars at 1000x or higher concentrations. The elicitor also appears to move systemically through soybean leaves when infiltrated in relatively high concentrations.

The *avr*D gene was found to be the first of five open reading frames organized in tandem and possibly comprising an operon (Kobayashi et al in preparation). It is therefore possible that *avr*D encodes an enzyme in a secondary metabolic pathway. Elucidation of the structure of the *avr*D elicitor should permit tests of the hypothesis that the *avr*D protein catalyzes conversion of a normal bacterial metabolite present in *P. syringae* and *E. coli* cells to the *avr*D elicitor. It is also hoped that the structure will offer an important clue to the bacterial function of *avr*D and the other open reading frames in the putative *avr*D operon. Significantly, our preliminary observations using RNA slot blot experiments with a gene specific *avr*D probe as well as a probe to two of the downstream ORFs in the putative *avr*D operon have indicated that their expression is plant inducible. Experiments in which the *Vibrio lux* genes were used as a promoter probe confirmed that bacteria grown in soybean leaves give *ca.* 100x greater expression of *avr*D than those grown on several laboratory culture media (Shen, Kobayashi and Keen, unpublished).

It would have been impossible to detect the presence of the putative soybean disease resistance gene complementing *avr*D before availability of the cloned *avr*D gene. Since *avr*D was cloned from *Pst* rather than *Psg*, it also was not clear if this avirulence gene interacted with a classic single disease resistance gene. We therefore analyzed soybean cross progeny in order to test whether a classical resistance gene in fact interacts with *avr*D. The experiment also permitted testing whether resistance to *Psg* carrying *avr*D and sensitivity to the isolated *avr*D elicitor co-segregated. Crosses of appropriate soybean cultivars showed that a single dominant character, assumed to represent a disease resistance gene, indeed segregated in a 3:1 ratio. Further, all of 552 F_2 and F_3 progeny that were resistant to *Psg* race 4 cells carrying *avr*D were also sensitive to the isolated *avr*D elicitor. All of the 190 segregants that were susceptible to Psg race 4 carrying *avr*D were also insensitive to the *avr*D elicitor (Keen and Buzzell, unpublished). The data therefore prove that the *avr*D elicitor mediates the phenotype of *avr*D and indicate that the elicitor interacts with a single disease resistance gene to elicit the soybean HR.

SUMMARY AND CONCLUSIONS

The cloning of avirulence genes has greatly aided our understanding of plant-pathogen specificity. It has proven that the gene-for-gene relationship first noted by Flor (1942) is correct--single avirulence genes encoding single protein products indeed are the entities that interact with plant disease resistance genes. Furthermore, genetic evidence now links one cloned avirulence gene, *avr*D, with initiation of the plant hypersensitive response by causing bacterial hosts to produce a specific elicitor. A consequence of these results has also been the establishment of firm genetic and biochemical evidence supporting the elicitor-receptor model for recognition of incompatible pathogen races by plants.

REFERENCES

Bennetzen JL, Qin MM, Ingels S, Ellingboe AH (1988) Allele-specific and mutator-associated instability at the *Rp1* disease-resistance locus of maize. Nature 332:369-370

Bruegger BB, Keen NT (1979) Specific elicitors of glyceollin accumulation in the *Pseudomonas glycinea*-soybean host-parasite system. Physiol Plant Pathol 15:43-51

De Wit PJGM, Hofman AE, Velthuis GCM, Kuc JA (1985) Isolation and characterization of an elicitor of necrosis isolated from intercellular fluids of compatible interactions of *Cladosporium fulvum* (syn. *Fulvia fulva*) and tomato. Plant Physiol 77:642-647

Dixon RA (1986) The phytoalexin response: elicitation, signalling and control of host gene expression. Biol Rev 61:239-291

Doke N, Garas NA, Kuc J (1979) Partial characterization and aspects of the mode of action of a hypersensitivity inhibiting factor (HIF) isolated from *Phytophthora infestans*. Physiol Plant Pathol 15:127-140.

Ellingboe AH (1982) Genetical aspects of active defence. In: Wood, RKS (ed) Active defense mechanisms in plants. Plenum Press, New York, p. 179

Flor HH (1942) Inheritance of pathogenicity in *Melampsora lini*. Phytopathology 32:653-669

Gabriel DW, Burges A, Lazo GR (1986) Gene-for-gene interactions of five cloned avirulence genes from *Xanthomonas campestris* pv. *malvacearum* with specific resistance genes in cotton. Proc Nat Acad Sci USA 83:6415-6419

Gabriel DW, Loschke DC, Rolfe BG (1988) Gene-for-gene recognition: the ion channel defense model. In: Palacios R, Verma DPS (eds) Molecular genetics of plant-microbe interactions 1988. APS Press, St. Paul, Minn., p. 3

Keen NT (1982) Specific recognition in gene-for-gene host-parasite systems. Adv Plant Pathol 1:35-82

Keen NT, Legrand M (1980) Surface glycoproteins: evidence that they may function as the race specific phytoalexin elictors of *Phytophthora megasperma* f.sp. *glycinea*. Physiol Plant Pathol 17:175-192

Keen NT, Staskawicz B (1988) Host range determinants in plant pathogens and symbionts. Annu Rev Microbiol 42:421-440

Kobayashi DY, Tamaki S, Keen NT (1989) Cloned avirulence genes from the tomato pathogen, *Pseudomonas syringae* pv. *tomato*, confer cultivar specificity on the non-host, soybean. Proc Nat Acad Sci USA 86:157-161

Lamb CJ, Lawton MA, Dron M, Dixon RA (1989) Signals and transduction mechanisms for activation of plant defenses against microbial attack. Cell 56:215-224

Mayama S, Tani T, Ueno T, Midland SL, Sims JJ, Keen NT (1986) The purification of victorin and its phytoalexin elicitor activity in oat leaves. Physiol Molec Plant Pathol 19:1-18

Schottens-Toma IMJ, De Wit PJGM (1988) Purification and primary structure of a necrosis-inducing peptide from the apoplastic fluids of tomato infected with *Cladosporium fulvum* (Syn *Fulva fulvia*). Physiol Molec Plant Pathol 33:59-67

Tepper CS, Anderson AJ (1986) Two cultivars of bean display a differential response to extracellular components from *Colletotrichum lindemuthianum*. Physiol Molec Plant Pathol 29:411-420

Vivian, A, Atherton GT, Bevan JR, Crute IR, Mur LAJ, Taylor JD (1989) Isolation and characterization of cloned DNA conferring specific avirulence in *Pseudomonas syringae* pv. *pisi* to Tea (*Pisum sativum*) cultivars, which possess the resistance allele, R2. Physiol Molec Plant Pathol 34:335-344

REGULATION IN *XANTHOMONAS* - PLANT INTERACTIONS

M J Daniels, A E Osbourn and J-L Tang
The Sainsbury Laboratory
John Innes Institute
Colney Lane
Norwich NR4 7UH
UK

INTRODUCTION

Current concepts of the nature and significance of intercellular signals between plants and microbes have evolved almost entirely from work on *Rhizobium* and *Agrobacterium*. The relationship between these bacteria and plants is of a highly specialised nature, and progress along the developmental pathway or life cycle of the microbe involves specific structural changes in the plant, and in the case of *Agrobacterium*, genetic changes also. It is not surprising therefore that interorganismal regulatory systems have evolved which control gene expression in both partners in the interactions.

In contrast, most bacterial pathogens are often regarded as rather crude destructive agents whose interactions with plants display little of the finesse which might require complex signalling systems. This view certainly represents an oversimplification, and in this paper we summarise some recent experiments in our laboratory on regulatory genes in *Xanthomonas campestris* pathovar *campestris* and interpret them in the context of the interaction of the bacterium with its *Brassica* host. Three approaches to the identification and characterisation of regulatory genes are described, and the effect of specific mutations in these genes on pathogenicity is discussed. In the first example, a mutation which abolished pathogenicity was subsequently found to lie in a cluster of "global" regulatory genes controlling the synthesis of several pathogenicity factors. In the second example, genes encoding products with amino acid sequence homology to conserved prokaryotic regulatory elements were isolated, and in the third case we used a technique to identify plant-inducible genes with a promoter-probe plasmid.

A CLUSTER OF "GLOBAL" POSITIVE REGULATORY GENES

Some years ago we isolated mutants of *X c campestris* defective in pathogenicity by direct screening on turnip seedlings (Daniels *et al* 1984 a). One of these mutants, designated 8237, was subsequently found to be defective in production of extracellular protease and polygalacturonate lyase, and a recombinant plasmid pIJ3020 from a pLAFR1 wild type DNA library was able to complement 8237 to restore enzyme production and pathogenicity (Daniels *et al* 1984 b). *X c campestris* also produces endoglucanase, amylase and extracellular polysaccharide (EPS, xanthan), and levels of these are also severely depressed in 8237, and restored to wild type levels by genes in pIJ3020.

NATO ASI Series, Vol. H 36
Signal Molecules in Plants and Plant-Microbe Interactions
Edited by B. J. J. Lugtenberg

Presumably 8237 is avirulent because one or more of the enzymes or EPS is necessary for pathogenicity. A similar phenotype (in terms of pathogenicity) is given by mutations in a cluster of genes, distinct from those cloned in pIJ3020, required for export of extracellular enzymes (Dow *et al* 1987) in which the set of enzymes are still synthesised at near-normal levels, but are retained within the bacterial periplasm. There is no evidence for intracellular or periplasmic accumulation of enzymes in 8237; the mutation affects enzyme synthesis.

The genes in pIJ3020 have been defined by a combination of subcloning, mutagenesis with Tn*5* and Tn*5lac*, promoter cloning and DNA sequencing (Tang 1989 and manuscripts in preparation). All genetic manipulations have strictly coordinate effects on the synthesis of the four enzymes and EPS, and on pathogenicity. Tn*5* mutagenesis initially indicated the presence of five distinct regions, A to E, wherein mutations gave negative phenotypes indistinguishable from one another, and complementation data suggested that these presumed genes acted in a cascade fashion. Subsequent experiments showed that the system is more complex.

A *Bam*H1 fragment of 3.2 kb containing the presumed gene C was sequenced and an open reading frame encoding a 74 kd protein was found, which showed strong amino acid sequence homology to both the "sensor" and "regulator" members of a family of prokaryotic two-component regulatory systems (Ronson *et al* 1987). From amino acid residues 138 to 376 there is strong homology with the conserved C-terminal domains of the EnvZ and PhoR proteins (sensors), and from residues 408 to 563 there is homology to the regulators NtrC, OmpR, SpoOF, Chey, PhoB and Dye (Fig 1). EnvZ and PhoR proteins are transmembrane proteins, and the putative gene C protein has significant hydrophobic domains in its N-terminal portion. It thus appears that *X c campestris* possesses a two-component regulatory system, with both components encoded by the same gene, although we cannot yet rule out the possibility that the gene product is cleaved post-translationally to give independent functional molecules. Another bifunctional regulator gene has recently been described in the human pathogen *Bordetella* (Stibitz *et al* 1989). The nucleotide sequence of the 3.2 kb fragment also revealed part of a second ORF in the opposite orientation to gene C, tentatively designated gene F, which interestingly shows homology to the PhOM protein of *Escherichia coli* (Amemura *et al* 1986) and the HR118 protein of *R leguminosarum* (Khanaka *et al* 1985), a DNA-binding protein. Work is currently in progress to complete the sequencing of gene F and the adjacent presumed gene B. A 5.1 kb *Bam*H1 fragment to the right of gene C (Fig 1), believed to contain gene D (identified from Tn*5* mutagenesis) and gene G (suggested by promoter probing) has recently also been sequenced and found to contain 4 ORF's, of which one has strong homology to regulator proteins.

More than 60 Tn*5lac* insertions throughout the pIJ3020 genomic region have been studied, and from the data minimal (constitutive) transcription units shown in Fig 1 can be deduced. Although the analysis of these genes is not yet complete it appears that *X c campestris* contains a cluster of positively

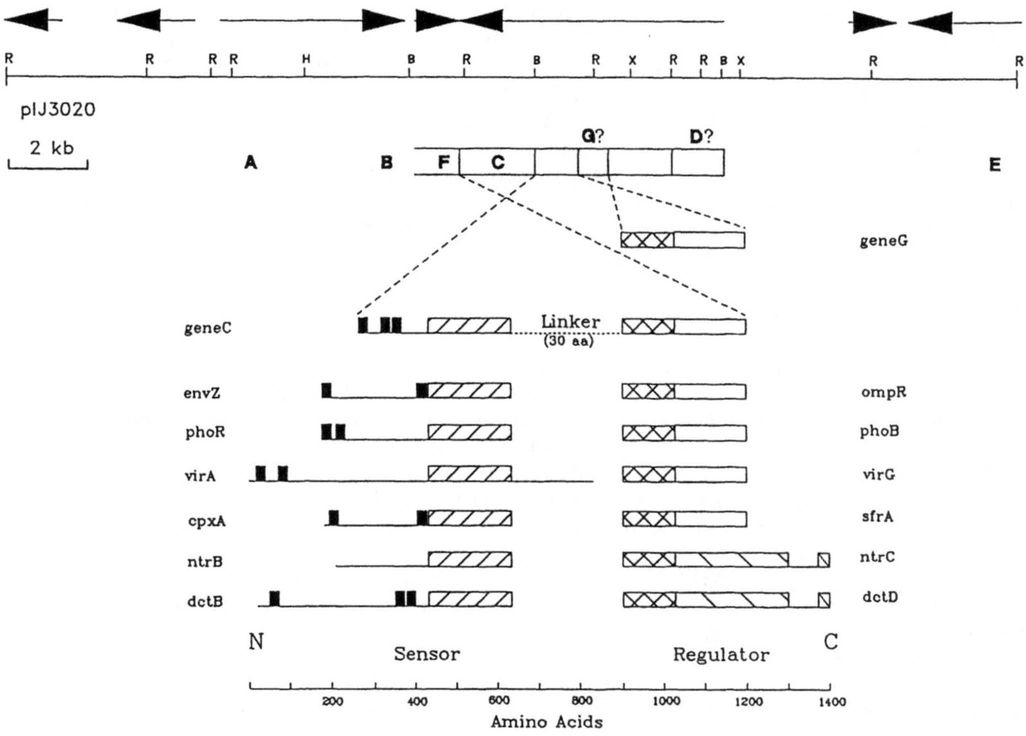

Figure 1. Organisation of genes involved in regulation of synthesis of extracellular enzymes and polysaccharide of *Xanthomonas campestris* pv *campestris*.

The genes were cloned in the cosmid pIJ3020, a partial restriction map of which is shown at the top of the diagram (B = *Bam*H1, R = *Eco*R1, X = *Xho*1). The arrows indicate minimal transcription units deduced from data from more than 60 Tn*5lac* insertions distributed throughout the region . The existence of genes A,B,C,D and E was inferred from Tn*5* mutagenesis, insertions in all these regions giving a similar pleiotropic negative phenotype. The presence of gene G was revealed by promoter cloning, and gene F by sequencing. The relationship between the deduced protein product of gene C and the family of two-component regulatory elements is shown below, adapted in part from Ronson *et al* (1987). The transmembrane domains (black blocks) and conserved sensor domain (diagonal hatching) are found in the N-terminal half of the protein, separated by a linker of about 30 amino acids from a region containing the conserved regulator domain (double hatched). Gene G also has a regulator domain. Two other ORF's are found in the boxes between genes C, G and D, but it is not yet known whether these are part of the regulatory system.

acting genes regulating coordinately the synthesis of extracellular enzymes and EPS. The regulatory genes themselves interact in a complex manner. It is not known whether the system senses an extracellular signal (perhaps of plant origin) or the metabolic state of the bacterial cell. Presumably one or more of the gene products interacts with the structural genes for the enzymes, although the DNA sequence has not yet revealed possible DNA-binding domains. The structural genes for a protease and endoglucanase regulated by the pIJ3020 gene cluster have recently been sequenced (Tang 1989; Gough 1989), and these data should contribute to the understanding of the final stages of the regulatory pathway.

Since the pIJ3020 genes activate the coordinate synthesis of the enzymes and EPS, we asked whether a "balancing" negative regulatory system exists. A candidate gene has been identified in the recombinant pLAFR3 plasmid pIJ3079 which when introduced into the wild type *X c campestris* (thereby increasing the gene dosage) coordinately depresses extracellular enzyme and EPS levels and reduces pathogenicity, but has no effect on growth *in vitro*. Disruption of the gene(s) by marker exchange of Tn*5* mutations from pIJ3079 to the wild-type chromosome gives a pathogenic strain producing elevated levels of enzymes and EPS. Preliminary mutation data indicate that the pIJ3079 and pIJ3020 systems operate in parallel.

Both the pIJ3020 and pIJ3079 systems have coordinate effects on all the extracellular enzymes. However at least some of the enzymes are also subject to individual regulation. Protease synthesis is induced by peptides and repressed by amino acids, and polygalacturonate lyase is induced by polygalacturonate (Tang *et al* 1987; Dow *et al* 1987).

ADDITIONAL ENVIRONMENT-SENSING REGULATORY GENES?

Conserved, environment-sensing, two-component regulatory systems are ubiquitous in bacteria. Because of the importance of genes of this type in pathogenicity of *X c campestris* (see above), we decided to isolate other such regulatory genes and investigate their role in plant-bacteria interactions, using oligonucleotide probes coding for conserved amino acid sequences. The high G+C content of *X c campestris* DNA (65%) imposes strict codon-usage constraints, simplifying the design of oligonucleotides. A 24-mer homologous to the presumed coding sequence for the octapeptide GGTGLGLA, which is highly conserved in several sensor gene products (Nixon *et al* 1986), was allowed to hybridise at various stringencies with Southern blots of *Bam*H1-digested *X c campestris* DNA, and clear hybridisation was obtained with a 4 kb band, together with weaker hybridisation to four other bands. *Bam*H1 DNA fragments in the 3.5 - 6 kb range were cloned and positive clones were identified by hybridisation of the oligonucleotide to colony lysates and Southern blots of isolated plasmid digests. Two types of clone were recovered, containing 4 and 5 kb *Bam*H1 fragments. Both clones also hybridised to a mixed oligonucleotide probe (kindly provided by Dr S Cole) designed to detect the regulator component of two-component systems (Winans *et al* 1986), suggesting that, as in other systems, sensor and regulator genes are linked. As expected from sequencing, the oligonucleotides also hybridised to pIJ3020 DNA, but the clones are otherwise unrelated to pIJ3020.

The *Bam*H1 fragments were also cloned into pLAFR3, mutagenised with Tn*5*, and after mapping the insertions were transferred into the wild-type *X c campestris* genome by marker exchange. All mutants were prototrophic, grew at normal rates, and produced normal levels of extracellular enzymes, and could not be distinguished from the wild-type in several pathogenicity assays. However, one of the clones appears to contain genes involved in regulation of EPS synthesis. DNA sequencing verified that sequences capable of encoding sensor and regulator proteins were present in the clone.

Although the work has not so far revealed any genes involved in plant interactions, it illustrates a simple approach to cloning bacterial regulatory genes, some of which (*vir*A and *vir*G in *A tumefaciens*) are required for plant-microbe signalling. Studies of the other homologous genes in *X c campestris* are in progress.

PLANT-INDUCIBLE BACTERIAL GENES

Osbourn *et al* (1987) described a technique for identifying plant-inducible genes in *X c campestris*. Random DNA fragments were cloned upstream of a promoterless chloramphenicol acetyltransferase gene in a broad host range promoter probe plasmid, and the clones were transferred into the wild-type strain and individual transconjugants were inoculated into chloramphenicol-treated seedlings (only Cm^R bacteria can grow and cause disease), and on to Cm-agar. Two putative plant-inducible promoters were studied further. Genomic regions containing the promoters were isolated from a pLAFR3 library and used as vehicles for Tn5 mutagenesis of the genome by marker-exchange. Mutations in the gene controlled by one of the promoters reduced pathogenicity, but the bacteria showed impaired growth *in vitro*, and the effect on pathogenicity is probably due to the mutant growing more slowly in plant tissues. The gene can be regarded as a "housekeeping" gene, and the apparent plant-inducibility may be a response to a nutritional shift in changing from laboratory medium to plant tissues, rather than to a specific chemical trigger.

Mutation of the gene controlled by the second promoter had no observable effect on pathogenicity (or on any other observable phenotypic property). However, a second closely-linked gene was discovered, controlled from another promoter which is probably not plant-regulated. Mutations in this gene dramatically reduced symptom production in all pathogenicity assays, but did not affect growth *in planta*. Mutants were prototrophic, motile, grew at normal rates and produced normal levels of extracellular enzymes and EPS. The gene has been sequenced but database searches for homology have so far proved fruitless, and we have no information about the function of the gene product (A E Osbourn and M J Daniels, in preparation). Although neither of the promoters studied in detail have yielded important data on chemical communication between *X c campestris* and plants, this approach has led to the discovery of a new pathogenicity gene. However several other plant-inducible promoters have been cloned, and systematic analysis of these and their cognate genes may reveal true signalling interactions.

DISCUSSION

The objective of our research is to identify and characterise genes of *X c campestris* which are necessary for pathogenicity, and an important element has been the study of regulatory genes, in the belief that the bacteria are capable of adaptation to meet the constantly changing physicochemical environment presented by infected plant tissues. The coordinate regulation of extracellular enzyme and polysaccharide synthesis is surprisingly complex, and is superimposed on individual regulatory systems. The nature of the presumed external stimulus to which the "regulon" responds is not known, but as the only common feature of the products of the regulated genes is the ability to degrade plant polymers or otherwise contribute to symptom production, a substance of plant origin is a strong possibility. It is notable that regulation of EPS synthesis appears to involve at least two two-component systems.

The approaches to gene regulation we have described using oligonucleotide probes and promoter cloning are to be regarded as preliminary screening tests for genes which may be important in plant-microbe interactions, and it is recognised that many genes involved in simple "housekeeping" regulation functions may also be cloned. Thus a quick and simple but "noisy" screen is followed by a more careful detailed genetic analysis. The primary screen does not involve plant tests on a large scale, which are either laborious and time consuming, or, if simplified, of low stringency, and thus these alternative approaches may reveal genes which would be overlooked using more conventional direct strategies.

To assess the role of a gene in pathogenicity it is necessary to investigate the behaviour of mutants defective only in the gene of interest. Experience has shown that is not a straightforward matter, and a complementary set of pathogenicity assays have to be employed. Many assays commonly used were devised for the purpose of rapidly screening strains to isolate mutants, and do not mimic the natural infection cycle. We have found that mutants of *X c campestris* deficient in protease (Tang *et al* 1987) and EPS (Barrère *et al* 1986) can give near normal symptoms if introduced into seedlings or intercellular spaces of leaves, but if inoculated at the edges of leaves (hydathode regions, where natural infection begins) they show much reduced virulence. These observations suggest that protease and EPS are necessary only at certain stages of the infection process, or in certain anatomical regions of the plant (manuscript in preparation). It will be interesting to study regulation of expression of *X c campestris* genes in different plant tissues, and techniques are being developed to permit this. Although studies with *X c campestris* and other pathogens are at an early stage of development there are already indications that interorganismal signalling plays an important part in the disease process.

ACKNOWLEDGEMENTS

The Sainsbury Laboratory is supported by the Gatsby Charitable Foundation. J-L Tang received a studentship from the Government of the Chinese People's Republic and Guangxi Agricultural College.

REFERENCES

Amemura M, Makino K, Shinagawa H, Nakata A (1986) Nucleotide sequence of the *pho*M region of *Escherichia coli*: four open reading frames may constitute an operon. J Bacteriol **168**:294-302

Barrère G C, Barber C E, Daniels M J (1986) Molecular cloning of genes involved in the production of the extracellular polysaccharide xanthan by *Xanthomonas campestris* pv *campestris*. Int J Biol Macromol **8**:372-374

Daniels M J, Barber C E, Turner P C, Cleary W G, Sawczyc M K (1984a) Isolation of mutants of *Xanthomonas campestris* pv *campestris* showing altered pathogenicity. J Gen Microbiol **130**:2447-2455

Daniels M J, Barber C E, Turner P C, Sawczyc M K, Byrde R J W, Fielding A H (1984b) Cloning of genes involved in pathogenicity of *Xanthomonas campestris* pv *campestris* using the broad host range cosmid pLAFR1. EMBO J **3**:3323-3328

Dow J M, Scofield G, Trafford K, Turner P C, Daniels M J (1987) A gene cluster in *Xanthomonas campestris* pv *campestris* required for pathogenicity controls the excretion of polygalacturonate lyase and other enzymes. Physiol Mol Plant Pathol **31**:261-271

Gough C L (1989) Molecular genetics of cellulase production by *Xanthomonas campestris*. PhD Thesis, University of East Anglia, UK.

Khanaka H, Laine B, Santiere P, Guillaume J (1985) Characterisation and primary stuctures of DNA-binding HU-type proteins from *Rhizobiaceae*. Eur J Biochem **147**:343-349

Nixon B T, Ronson C W, Ausubel F M (1986) Two-component regulatory systems responsive to environmental stimuli share strongly conserved domains with the nitrogen assimilation regulatory genes *ntr*B and *ntr*C. Proc Natl Acad Sci USA **83**:7850-7854

Osbourn A E, Barber C E, Daniels M J (1987) Identification of plant-induced genes of the bacterial pathogen *Xanthomonas campestris* pathovar *campestris* using a promoter-probe plasmid. EMBO J **6**:23-28

Ronson C W, Nixon B T, Ausubel F M (1987) Conserved domains in bacterial regulatory proteins that respond to environmental stimuli. Cell **49**:579-581

Stibitz S, Aaronson W, Monack D, Falkow S (1989) Phase variation in *Bordetella pertussis* by frameshift mutation in a gene for a novel two-component system. Nature **338**:266-269

Tang J-L (1989) Aspects of extracellular enzyme production by *Xanthomonas campestris*. PhD Thesis, University of East Anglia, UK

Tang J-L, Gough C L, Barber C E, Dow J M, Daniels M J (1987) Molecular cloning of protease gene(s) from *Xanthomonas campestris* pv *campestris*: expression in *Escherichia coli* and role in pathogenicity. Mol Gen Genet **210**:443-448

Winans S C, Ebert P R, Stachel S E, Gordon M P, Nester E W (1986) A gene essential for *Agrobacterium* virulence is homologous to a family of positive regulatory loci. Proc Natl Acad Sci USA **83**:8278-8282

PSEUDOMONAS SPP. WITH MUTATIONAL CHANGES IN THE O-ANTIGENIC SIDE CHAIN OF THEIR LIPOPOLYSACCHARIDE ARE AFFECTED IN THEIR ABILITY TO COLONIZE POTATO ROOTS

Letty A. de Weger[1], Peter A.H.M. Bakker[2], Bob Schippers[2], Mark C.M. van Loosdrecht[3], and Ben J.J. Lugtenberg[1]

[1]Leiden University, Department of Plant Molecular Biology, Nonnensteeg 3, 2311VJ Leiden, [2]Department of Plant Pathology, Phytopathological Laboratory, Baarn, and [3]Department of Microbiology, Agricultural University Wageningen, The Netherlands

INTRODUCTION

Microorganisms that efficiently colonize plant roots are ideal tools for the production of biocontrol agents, since many major and minor root pathogens are present at the root surface. In the past decade Pseudomonas spp. have received much attention as biocontrol agents (Kloepper et al 1980, Schippers et al 1986). The interest in this group of bacteria is mainly due to their antagonistic potentials in vitro towards pathogens combined with their ability to colonize the roots of many plants efficiently.

The mechanism underlying efficient root colonization is essentially unknown. It can be reasonably assumed that processes like survival, growth, attachment and distribution along the root surface are involved in the establishment of a bacterial population on the root. With respect to the latter process, we showed in a previous study that flagella-less mutants of Pseudomonas fluorescens strain WCS374 are impaired in their ability to colonize the deeper root parts of the potato plant (de Weger et al 1987b), suggesting that bacterial motility is important for a good distribution of the cells along the root.

Other bacterial properties that may be involved in the complex process of colonization may be found in cell surface structures, that play a role in the interaction with the plant

NATO ASI Series, Vol. H 36
Signal Molecules in Plants and Plant-Microbe Interactions
Edited by B. J. J. Lugtenberg

root. Cell surface structures may also confer protection against harmful events, like desiccation or predation and thus increase the ability of the bacteria to survive.

This study focuses on the protruding polysaccharide chain (=O-antigenic side chain) of the LPSs (=lipopolysaccharides) of two plant-beneficial Pseudomonas strains. The majority of the pseudomonads isolated from the potato root surface were shown to produce LPSs with O-antigenic side chains (de Weger et al 1987a). Apparently this structure is an conserved trait among these root colonizers. This prompted us to study whether the O-antigen is important for root colonization and/or the interaction with the plant root.

RESULTS AND DISCUSSION

After selection with phages, isolated from soil samples or ditch water, mutant strains LWP358-43b of P.putida strain WCS358 and LWP374-30b of P.fluorescens strain WCS374 were obtained. The LPSs of these strains were characterized using SDS- (=sodium dodecyl sulfate) polyacrylamide gel electrophoresis (Fig.1) and by the chemical analysis of the sugar composition (de Weger et al 1989). The results indicate that the

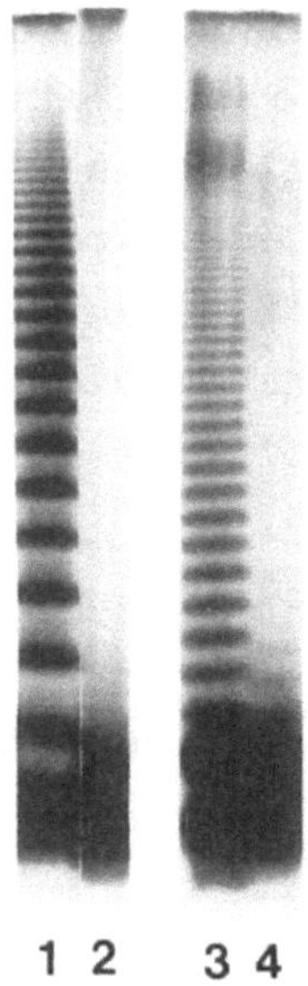

Fig.1 Silver stained patterns of proteinase K-treated cell envelopes obtained after SDS-polyacrylamide gel electrophoresis. Lanes: 1, WCS358; 2, LWP358-43b; 3, WCS374; 4, LWP374-30b.

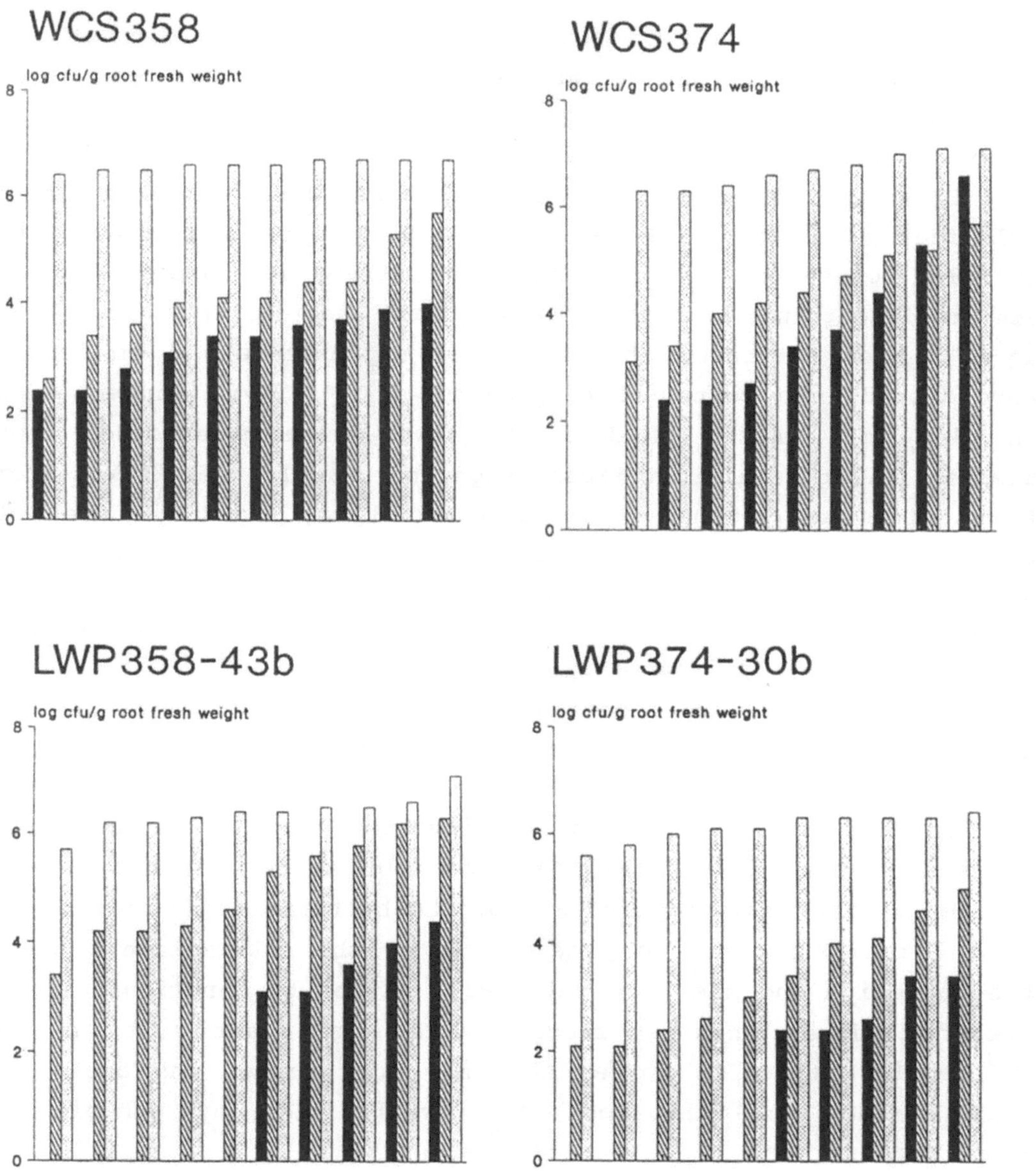

FIG.2. The logarithm of the number of kanamycin and streptomycin resistent CFU of strains WCS358, WCS374 and their mutant strains LWP358-43b and LWP374-30b were determined on root segments taken from 0-1 (□), 4 (▨) and 8 (■) cm depth. The results are shown for each plant individually. Similar results were obtained in two other experiments.

O-antigenic side chain of strain WCS374, containing glucose and fucose, is absent in mutant LWP374-30b and that the quinovosamine- and glucose-containing O-antigenic side chain of strain WCS358 is absent in the mutant strain LWP358-43b.

These mutant and wild type strains were compared in a colonization assay (de Weger et al 1987b) for their ability to colonize potato roots. Tn5-labeled derivatives were used which are resistent to kanamycin (200 µg/ml) and streptomycin (200 µg/ml). Stem cuttings with roots of approximately 1 cm length were dipped in a bacterial suspension of 10^8 CFU/ml and embedded in clay soil. After 14 days of growth in a greenhouse, root samples (0.3 g total root fresh weight per sample) were taken from three different depths (0 to 1, 4 and 8 cm) and the number of kanamycin- and streptomycin-resistent CFU was determined. Both wild type strains were found on the deeper root parts of at least 90 % of the plants, while the LPS-mutant strains colonize the deeper root parts in 50% of the plants (Fig. 2).

This impaired colonization of the deeper root parts by the LPS-mutant strains was not due to reduced abilities in growth, motility or siderophore production.

Another trait that may be involved in the colonization process is the firm adhesion to the root surface. Since the O-antigenic side chain protrudes from the bacterial cell surface it is a likely structure to play a role in the interaction between the bacterium and the root cell wall, either by functioning as a specific signal that is recognized by the plant root, or by influencing aspecific forces that may act between the bacterial cell surface and the root cell wall, like hydrophobic or electrical forces.

The physicochemical cell surface characteristics of wild type and mutant strains were compared. The cell surface hydrophobicity was determined by measuring the contact angle of a drop of water on a layer of the cells and as a measure of the cell surface charge the electrophoretic mobility was determined (van Loosdrecht et al 1987). The mutants lacking the O-antigenic side chain were altered in the hydrophobicity and the electrical charge of the cell surface when compared to their parent strains (Table 1).

Table 1. Electrophoretic mobilities and contact angles of water for parent strains WCS358 and WCS374 and their mutant strains[a]

Strain	Contact angle (°)[b]	Electrophoretic mobility (10^{-8} m. V^{-1}. s^{-1})[c]
WCS358	40	-2.2
LWP58-43b	25	-2.5
WCS374	16	-0.5
LWP74-30b	23	-2.6

[a]For details on these measurements see van Loosdrecht et al 1987, or de Weger et al 1989
[b]Average standard deviation 1.5°
[c]Average standard deviation 0.15 x 10^{-8} $m.V^{-1}.s^{-1}$

These alterations in the cell surface characteristics make these strains ideal tools for studying the influence of the physicochemical cell surface traits on the firm adhesion to potato roots. Sterile potato root pieces were incubated at times varying between 1 to 120 min in a suspension of 5 x 10^8 CFU/ml under agitation at 100 rpm. The bacteria still attached to the roots after thorough washing of the root segments were considered to be firmly bound. The number of firmly bound bacteria, determined by dilution plating after homogenization of the roots, was less than 0.2 % of the bacteria added. In a time course no significant differences in the number of firmly bound cells was observed between the wild type and the O-antigen mutant strains (de Weger et al 1989). These results show that neither these differences in cell surface charge and hydrophobicity between these wild type and mutant strains nor the presence of the O-antigenic side chain of the LPS are relevant for the firm adhesion to the potato plant roots.

For the colonization of the deeper root parts at least a few bacterial cells have to reach the newly formed root parts. This may result in the formation of a few microcolonies on the root surface which again may form a basis for the distribution of the cells along the growing root parts. The results presented here show that the O-antigen mutants of strains WCS358 and WCS374 are impaired in their ability to establish their progeny on the growing roots. At this stage we cannot exclude that the mutations that effect the O-antigenic side chains are pleiotrophic and that another molecule than LPS is in fact responsible for the reduced colonization ability. However, these mutations do not exert their effects on the growth rate, motility, siderophore production or on the firm attachment to the root surface. This suggests that the colonization process of these mutant strains is not affected in the multiplication phase, the distribution phase, or the attachment phase. Presumably, these mutants are affected in their survival properties. The O-antigenic side chain may play a role in the protection against a variety of hazardous events in the rhizosphere like predation, desiccation or fluctuating osmotic pressures.

REFERENCES

de Weger LA, Jann B, Jann K, and Lugtenberg B (1987a) Lipopolysaccharides of Pseudomonas spp. that stimulate plant growth; composition and use for strain identification. J. Bacteriol. 169: 1441-1446

de Weger LA, van der Vlugt CIM, Wijfjes AHM, Bakker PAHM, Schippers B, and Lugtenberg B (1987b) Flagella of a plant growth-stimulating Pseudomonas fluorescens strain are required for colonization of potato roots. J.Bacteriol. 169: 2769-2773

de Weger LA, van Loosdrecht MCM, Klaassen HE, Lugtenberg B (1989) Mutational changes in physicochemical cell surface properties of plant growth-stimulating Pseudomonas spp. do not influence their attachment properties. J. Bacteriol. 171: (in press)

Kloepper JW, Leong J, Teintze M, and Schroth MN (1980) Enhanced plant growth by siderophores produced by plant growth-promoting rhizobacteria. Nature 286: 885-886

Schippers B, Lugtenberg B, and Weisbeek PJ (1987) Plant growth control by fluorescent pseudomonads. In: Chet, I (ed) Innovative approaches to plant disease control. Wiley and Sons, New York, p.19-39

van Loosdrecht MCM, Lyklema J, Norde W, Schraa G, and Zehnder AJB (1987) Electrophoretic mobility and hydrophobicity as a measure to predict the initial steps of bacterial adhesion. Appl. Environm. Microbiol. 53: 1898-1901

ELICITATION OF PHYTOALEXIN SYNTHESIS IN SOYBEAN (*GLYCINE MAX*) BY A FUNGAL PATHOGEN AND A FUNGAL ß-GLUCAN

Jürgen Ebel, Eric G. Cosio, Markus Feger, Dorit Grab and Heide Habereder
Biologisches Institut II
Universität Freiburg
D-7800 Freiburg
Federal Republic of Germany

INTRODUCTION

One inducible defence mechanism of soybean (*Glycine max* L. Merrill) against potential pathogens is the production of isoflavonoid phytoalexins at the site of attempted infection (Ebel 1986; Ebel and Grisebach 1988). Mechanisms underlying phytoalexin formation in this plant and their role in defence have been studied in some detail in two experimental systems. One utilizes the inoculation of soybean primary roots with zoospores of the fungal pathogen *Phytophthora megasperma* f. sp. *glycinea*, the causal agent of root and stem rot, and mimics the natural infection process. The second experimental system uses cell-suspension cultures and a ß-glucan elicitor isolated from *P. megasperma* cell walls and represents a system of reduced complexity. Fungus-induced phytoalexin synthesis in roots has been analyzed in relation to fungal race:plant cultivar-specific resistance and to the regulation of phytoalexin accumulation. The analysis of the elicitor-cell culture interaction may provide valuable information on the molecular basis of the initial recognition event between fungus-derived components and the plant cell surface and the subsequent steps involved in the transduction of the primary signal. Different soybean tissues have been used to characterize plasma membrane-associated ß-glucan binding proteins which could be related to putative elicitor "receptors". These studies may, thus, help to understand the nature of the hypothetical "recognition" event between the whole plant and the live pathogen preceding the activation of plant defence responses.

NATO ASI Series, Vol. H 36
Signal Molecules in Plants and Plant-Microbe Interactions
Edited by B. J. J. Lugtenberg

REGULATION OF PHYTOALEXIN ACCUMULATION

Soybean cultivars are differentially resistant to several *P. megasperma* races (Compendium of soybean diseases 1975). A detailed analysis of a single soybean cultivar has shown marked differences in dynamics, amounts, and sites of accumulation of glyceollin I, the major phytoalexin of soybean, following infection of roots with zoospores of either a virulent (compatible interaction, plant susceptible) or an avirulent race (incompatible interaction, plant resistant) of *P. megasperma* (Hahn et al 1985). Only resistant soybean plants rapidly accumulate glyceollin I at infection sites and fungal growth is arrested early in the immediate vicinity of the inoculation site. Similar race:cultivar-specific differences have also been observed for the increases in the catalytic activities and the mRNAs of several enzymes involved in glyceollin biosynthesis during early stages of infection (2-8 hr) (Bonhoff et al 1986a,b; Habereder et al 1989). In the incompatible interaction, rapid increases in the mRNA amounts of two enzymes, phenylalanine ammonia-lyase and chalcone synthase, occur at the time of penetration of fungal germ tubes into the epidermal cell layers (1-2 hr after inoculation; Beagle-Ristaino and Rissler 1983), reaching highest levels at about 7 hr. In the compatible interaction, only a slight early enhancement of the mRNA levels has been found and no further increase occurs until about 9 hr after inoculation. For one of the enzymes, chalcone synthase, mRNA activity changes have also been analyzed. The kinetics shows again major differences between the incompatible and compatible interactions (Habereder et al 1989). The rapidity of mRNA induction in the incompatible interaction, as compared with the timing of fungal invasion of the soybean roots, might imply that at least some of the components involved in the perception and transduction of the signal produced by the invading fungus in the root tissue are available within a short period after inoculation. Another interesting observation is that the kinetics of mRNA induction in the incompatible root-fungus interaction is very similar to that measured after elicitor treatment of soybean cell-suspension cultures. Thus, in the soybean root, not only phytoalexin accumulation itself but also the induction of phytoalexin biosynthetic enzymes and their mRNAs reflect the known differences in the physiological interaction between *P. megasperma* and the soybean plant (Ebel and Grisebach 1988).

CHARACTERIZATION OF THE BINDING OF FUNGAL ß-GLUCANS TO MEMBRANES

It is assumed that the initial event in the elicitor-mediated stimulation of phytoalexin production is the interaction of the elicitor with primary target sites, or receptors, on host cell surfaces. Although soluble or cell-wall bound components cannot be ruled out as possible targets, the likely candidates for the location of these sites are the cell membranes.

The components of *P. megasperma* mycelial walls which show elicitor activity in soybean are ß-1,3- and ß-1,6-linked glucans (Darvill and Albersheim 1984; Ebel 1986). These glucans are prominent structural polysaccharides of the oomycetes. A number of *Phytophthora* spp. ß-glucans have been used as ligands to demonstrate specific glucan binding to soybean membranes (Schmidt and Ebel 1987; Yoshikawa et al 1983) and to analyze ligand specificity (Cosio et al 1988; Schmidt and Ebel 1987). ^{3}H- and ^{125}I-labeled glucan fractions exhibit high-affinity binding (K_d between 10 and 40 nM) which is both saturable and reversible. Specific glucan binding has been detected in membrane preparations from soybean roots, hypocotyls, cotyledons, and cell-suspension cultures. The binding follows the distribution of a plasma-membrane marker, 1,3-ß-glucan synthase, in sucrose density gradients. This finding is in good agreement with the observation that intact protoplasts isolated from soybean cell cultures show displaceable ß-glucan binding of an affinity identical with that found in membrane preparations. The recent synthesis by us of biotinylated glucans with high binding affinity may further aid the study of the properties of the binding sites *in vivo*, including their presumed role in mediating the elicitor stimulus in the plant cells.

Competition studies with a variety of poly- and oligosaccharides have been carried out to investigate the specifity of the glucan binding (Cosio et al 1988; Schmidt and Ebel 1987). The ability of these alternative ligands to compete with ^3H- and ^{125}I-labeled glucan for binding, is dependent on both their structure and size. Only 1,3-ß-glucan-type poly/oligosaccharides have proved to be competitors. Most interestingly, a strong positive correlation has been found between the ability to compete for the binding sites and the activity of the ligands as elicitors of glyceollin synthesis in a cotyledon bioassay.

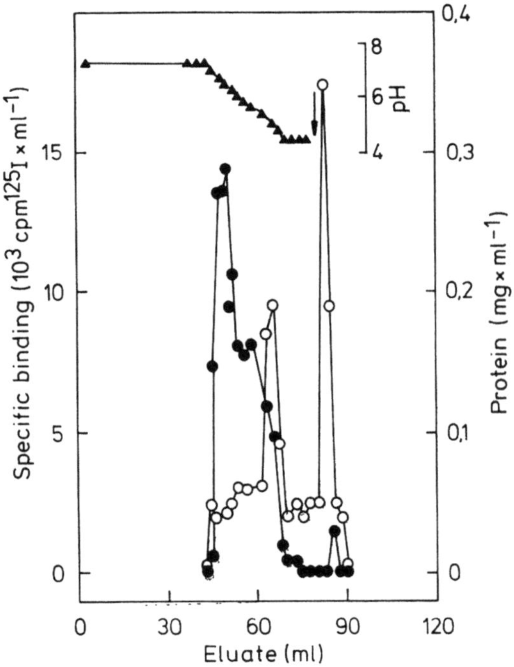

Fig. 1. Chromatofocusing of a detergent-solubilized protein extract from soybean roots. After adsorption to a Mono P column (Pharmacia), proteins were eluted with Polybuffer 74 (pH 4.4), followed by 25 mM Bis-Tris buffer containing 1 M sodium chloride (arrow). Fractions were analyzed for ß-glucan binding activity (●) and for protein (o).

The ß-glucan binding proteins appear to be integrally associated with the membranes. High salt or low detergent concentrations fail to remove them from the membranes. Of a number of nonionic and zwitterionic detergents tested, Zwittergent 3-12 has been chosen for further studies because it offers a good compromise between preservation of binding activity and suitability for protein purification. The Zwittergent-solubilized proteins display similar binding affinity as their membrane-bound counterparts (Cosio and Ebel 1988). The detergent-solubilized binding activity has been fractionated by conventional methods of protein purification including ion exchange followed by gel-permeation chromatography. The latter gave an apparent M_r for the glucan-binding protein-containing micelles of about 300,000 on a Sephacryl S-300 HR column. Ampholyte-displacement chromatography on a Mono P column (Pharmacia) gave an apparent pI of 6.5 for the binding proteins (Fig. 1). Conversely, the bulk of the proteins eluted from the column at a pH below 6.

EFFECTS ASSOCIATED WITH ELICITATION IN CELL-SUSPENSION CULTURES

The existence of plasma membrane-localized binding proteins allows one to investigate the hypothesis whether binding of ß-glucan fragments from *P. megasperma* cell walls to these sites initiates the signal resulting in enhanced transcription of the genes for the enzymes of glyceollin biosynthesis. In soybean cell cultures, the glucan elicitor increases the activities of enzymes of phytoalexin biosynthesis and stimulates phytoalexin accumulation when the culture medium contains 1.2 mM Ca^{2+} (Stäb and Ebel 1987). However, removal of the exchangeable external Ca^{2+} with EGTA (residual Ca^{2+} concentration about 10 µM) or addition of low concentrations of the Ca^{2+} antagonist La^{3+} to the medium containing normal Ca^{2+} concentration, effectively abolished the elicitor-mediated phytoalexin response. These results and the observation that amino acid uptake is rapidly inhibited (Stäb and

Ebel 1987) suggest the possibility that the initial response to elicitor binding to soybean cells is a number of substrate (ion) flux changes across the plasma membrane. This interpretation is supported by the observation that, in the presence of Ca^{2+} in the medium, the Ca^{2+} ionophore A23187 stimulates the phytoalexin response in a dose-dependent manner (Stäb and Ebel 1987). Further analysis has to establish the relationship between the external Ca^{2+} pool and putative Ca^{2+} as well as other ion flux changes in stimulus-response coupling during elicitor action.

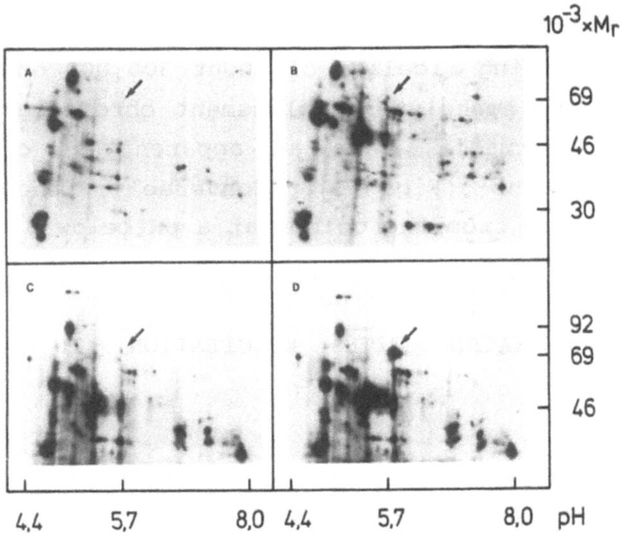

Fig. 2. Effects of Ca^{2+} and of a phosphorylation-stimulating factor (effector) on endogenous protein phosphorylation in extracts from soybean cell cultures. A protein extract was incubated with [γ-^{32}P]ATP in the presence of Ca^{2+} (3 µM) (B, C), or of effector and Ca^{2+} (D), or without any addition (A). The assay mixture was subjected to polyacrylamide gel electrophoresis in the presence of sodium dodecyl sulphate and to autoradiography. A protein with M_r = 69,000 was strongly phosphorylated in the presence of the effector and to a minor degree in the presence of Ca^{2+} (arrows).

Addition of the ß-glucan elicitor to soybean cell-suspension cultures also results in rapid changes in the phosphate turnover of several phosphoproteins (Grab et al 1989). The effect of the elicitor on protein phosphorylation has been tested after labeling of the cells with [^{32}P]orthophosphate. As revealed by gel electrophoretic analysis and autoradiography, decreases as well as increases in the labeling of several phosphoproteins occur rapidly, being detectable within 5 min after elicitor application, and persist for at least 15 min. As judged by their relative molecular masses M_r and isoelectric points pI, a number of proteins which are radioactively labeled *in vivo* are also phosphorylated *in vitro* by endogenous protein kinase activity in the presence of Ca^{2+}. The most pronounced effect has been observed with a protein substrate with M_r = 69,000 and pI = 5.7 whose phosphate labeling decreases in response to elicitor treatment *in vivo*. Phosphorylation *in vitro* of a protein which might be identical with that affected *in vivo* is strongly enhanced by a phosphorylation-stimulating factor (effector) derived from soybean cell cultures (Fig. 2). The effector has been purified and possesses a low apparent M_r, is negatively charged at pH 7.3, and is relatively heat stable. Phosphorylation of the protein with M_r = 69,000 is only slightly stimulated by Ca^{2+} (Fig. 2), but is not enhanced by several other substances known as activators of protein kinases in various systems. Further characterization of the components involved in phosphorylation/dephosphorylation of endogenous soybean phosphoproteins will help to understand the regulation of elicitor-triggered posttranslational metabolic processes.

ACKNOWLEDGEMENTS

This work was supported by the Deutsche Forschungsgemeinschaft (SFB 206), Fonds der Chemischen Industrie, the Alexander von Humboldt Foundation (E.G.C.), and the Studienstiftung des Deutschen Volkes (D.G.).

REFERENCES

Beagle-Ristaino JE, Rissler JF (1983) Histopathology of susceptible and resistant soybean roots inoculated with zoospores of *Phytophthora megasperma* f sp *glycinea*. Phytopathology 73:590-595

Bonhoff A, Loyal R, Ebel J, Grisebach H (1986a) Race:cultivar-specific induction of enzymes related to phytoalexin biosynthesis in soybean roots following infection with *Phytophthora megasperma* f sp *glycinea*. Arch Biochem Biophys 246:149-154

Bonhoff A, Loyal R, Feller K, Ebel J, Grisebach H (1986b) Further investigations of race:cultivar-specific induction of enzymes related to phytoalexin biosynthesis in soybean roots following infection with *Phytophthora megasperma* f sp *glycinea*. Biol Chem Hoppe-Seyler 367:797-802

Compendium of soybean diseases (1975) American Phytopathological Society, St Paul MN USA

Cosio EG, Ebel J (1988) Solubilization of fungal oligoglucan binding proteins from soybean cell membranes. Plant Physiol 86 Suppl:Abstract 129

Cosio EG, Pöpperl H, Schmidt WE, Ebel J (1988) High-affinity binding of fungal ß-glucan fragments to soybean (*Glycine max* L) microsomal fractions and protoplasts. Eur J Biochem 175:309-315

Darvill AG, Albersheim P (1984) Phytoalexins and their elicitors - A defense against microbial infections in plants. Ann Rev Plant Physiol 35:243-275

Ebel J (1986) Phytoalexin synthesis: The biochemical analysis of the induction process. Ann Rev Phytopathol 24:235-264

Ebel J, Grisebach H (1988) Defense strategies of soybean against the fungus *Phytophthora megasperma* f sp *glycinea*: a molecular analysis. Trends Biochem Sci 13:23-27

Grab D, Feger M, Ebel J (1989) An endogenous factor from soybean (*Glycine max* L) cell cultures activates phosphorylation of a protein which is dephosphorylated *in vivo* in elicitor-challenged cells. Planta (in press)

Habereder H, Schröder G, Ebel J (1989) Rapid induction of phenylalanine ammonia-lyase and chalcone synthase mRNAs during fungus infection of soybean (*Glycine max* L) roots or elicitor treatment of soybean cell cultures at the onset of phytoalexin synthesis. Planta 177:58-65

Hahn MG, Bonhoff A, Grisebach H (1985) Quantitative localization of the phytoalexin glyceollin I in relation to fungal hyphae in soybean roots infected with *Phytophthora megasperma* f sp *glycinea*. Plant Physiol 77:591-601

Schmidt WE, Ebel J (1987) Specific binding of a fungal glucan phytoalexin elicitor to membrane fractions from soybean *Glycine max*. Proc Natl Acad Sci USA 84:4117-4121

Stäb MR, Ebel J (1987) Effects of Ca^{2+} on phytoalexin induction by a fungal elicitor in soybean cells. Arch Biochem Biophys 257:416-423

Yoshikawa M, Keen NT, Wang M-C (1983) A receptor on soybean membranes for a fungal elicitor of phytoalexin accumulation. Plant Physiol 73:497-506

STUDIES ON ELICITOR RECOGNITION AND SIGNAL TRANSDUCTION IN HOST AND NON-HOST PLANT/FUNGUS PATHOGENIC INTERACTIONS

Dierk Scheel, Christiane Colling, Harald Keller, Jane Parker, Wolfgang Schulte and Klaus Hahlbrock
Max-Planck-Institut für Züchtungsforschung
Abteilung Biochemie
D-5000 Köln 30
F.R.G.

Plants have developed efficient mechanisms to defend themselves against pathogen attack. Beside a number of preformed barriers they have a whole set of inducible defense reactions at their disposal, which appear to be similar in many diverse plant species and are used against different pathogens in non-host- and host-incompatible interactions (Hahlbrock and Scheel 1987). The induction of these responses requires recognition of the pathogen and consequent signal transduction. Both processes are difficult to study in the intact plant, since successful resistance reactions are usually limited to a few cells. In many plants, however, a typical resistance response is synchronously induced by treatment of cultured cells with plant- or pathogen-derived elicitors (DiCosmo and Misawa 1985). Some of these systems have been used with great success to elucidate the biochemistry and molecular biology of many plant defense responses and therefore represent excellent tools for studies on the mode of elicitor action (Ebel 1986; Hahlbrock and Scheel 1989).

Treatment of cultured cells from potato and parsley with crude elicitor preparations from *Phytophthora infestans* (*Pi*), the causal agent of late blight in potato, or *Phytophthora megasperma* f.sp. *glycinea* (*Pmg*), a soybean pathogen, results in rapid activation of defense responses (Fritzemeier *et al* 1987; Scheel *et al* 1987; Somssich *et al* 1989). Elicitor treatment of potato cells leads to the rapid coordinated induction of phe-

NATO ASI Series, Vol. H 36
Signal Molecules in Plants and Plant-Microbe Interactions
Edited by B. J. J. Lugtenberg

nylalanine ammonia-lyase (PAL) and tyrosine decarboxylase (TDC), which is followed by incorporation of the conjugated products, 4-coumaroyltyramine and feruloyltyramine, into the cell wall (Figure 1). This response is also observed in fungus-infected potato leaves, where tuber-specific sesquiterpenoid phytoalexins are not detected (Rohwer *et al* 1987). Since leaves are the primary infection site of *Pi* and also express complete resistance to specific fungal races (Thurston and Schultz 1981), this cell reaction should be a suitable marker for the response of potato to elicitor.

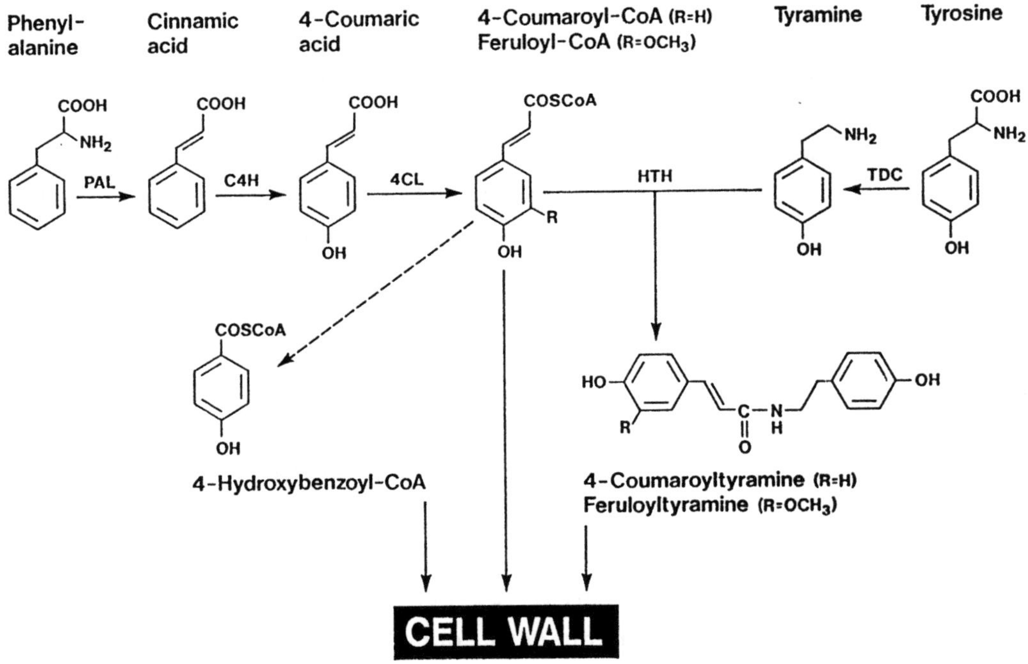

Figure 1. Proposed biosynthetic pathways of phenylpropanoid derivatives incorporated into the cell walls of infected or elicitor-treated potato cells (PAL, phenylalanine ammonia-lyase; C4H, cinnamic acid 4-hydroxylase; 4CL, 4-coumarate:CoA ligase; HTH, hydroxycinnamoyl-CoA:tyramine hydroxycinnamoyl-transferase; TDC, tyrosine decarboxylase).

In elicitor-treated parsley cells autofluorescent furanocoumarin phytoalexins are excreted into the culture medium (Tietjen *et al* 1983; Hauffe *et al* 1986) and a characteristic set of resistance-related genes is activated (Somssich *et al* 1989). All responses of cultured cells to elicitor are expressed with a similar relative timing in parsley leaves inoculated with *Pmg* spores (Scheel *et al* 1986; Jahnen and Hahlbrock 1988). Since, in addition, parsley protoplasts retain their responsiveness to elicitor, these cells are particularly suitable for the detection of pathogen-derived elicitor compounds (Dangl *et al* 1987).

For comparison we included soybean, the host plant of *Pmg*, in our studies on the crude *Pmg* elicitor. In this system we measured glyceollin accumulation which is known to be stimulated by *Pmg*-derived pure glucan molecules (Sharp *et al* 1984).

Table 1. Effect of proteinase treatments and deglycosylation of crude *Pmg* elicitor (Ayers *et al* 1976) prior to its addition to cultured parsley cells and protoplasts, soybean cotyledons and cells, and potato cells.

Sample[a]	Defense response [% of maximum][b]					
	Parsley		Soybean		Potato	
	Cells	Protoplasts	Cotyledons	Cells	PAL	TDC
Elicitor only	100	100	100	100	100	100
Buffer only (PBS)	0	0	3	1	19	18
Pronase only	0	0	3	2	39	22
Pronase+elicitor	5	0	88	63	99	129
Autoclaved pronase +elicitor	98	95	NT	NT	114	84
Trypsin+elicitor	35	12	102	114	119	102
Autoclaved trypsin +elicitor	105	91	NT	NT	109	94
Deglycosylated elicitor	49	86	3	4	18	18

[a] Proteinase digestions and deglycosylations of elicitor were performed as described by Parker *et al.* (1988). Deglycosylated samples contained amounts of protein equivalent to the untreated elicitor.

[b] In parsley and soybean phytoalexin accumulation was measured according to Parker *et al.* (1988). Cell suspension cultures of potato cv. Datura were treated for 6 h with 50 µg/ml elicitor before enzyme activities of PAL and TDC were determined in crude extracts (Palavan and Galston 1982; Kombrink and Hahlbrock 1986). NT, not tested.

Cultured cells of the three plants which represent either host or non-host systems for *Pmg* and *Pi* recognize different components of the crude *Pmg* elicitor (Table 1). Treatments of the cells with proteinase-digested or deglycosylated elicitor preparations identify proteinaceous constituents as active eliciting compounds in parsley, which are inactive in soybean and potato. In cultured cells from potato and soybean the proteinase-treated elicitor is active, whereas the deglycosylated preparation is completely inactive. However, in similar preparations from *Pi*, both parsley and potato recognize carbohydrate constituents (Table 2).

Table 2. Effect of proteinase treatments and deglycosylation of crude *Pi* elicitor (Rohwer *et al* 1987) prior to its addition to cultured cells of parsley and potato cv. Datura.

Sample[a]	Defense response [% of maximum][b]	
	Parsley	Potato
Elicitor only	100	100
Buffer only (PBS)	11	22
Pronase only	16	30
Pronase+elicitor	104	104
Trypsin only	12	22
Trypsin+elicitor	79	114
Deglycosylated elicitor	8	19

[a] All *Pi* elicitor samples were treated as described by Parker *et al*. (1988) for *Pmg* elicitor.

[b] In parsley the response was measured by coumarin accumulation in the culture media 24 h after elicitor application. In potato cells the enzyme activity of PAL was determined 6 h after treat ment. Both cell suspension cultures were treated with 10 µg/ml glucose equivalents of crude *Pi* elicitor.

Enrichment of an elicitor activity from the culture filtrate of *Pmg* was correlated with the presence of a 42 KDa glycoprotein, as shown in Figure 2. This was purified to homogeneity and shown to elicit the full complement of coumarin phytoalexins at very low concentrations (50 ng/ml) in parsley proto-

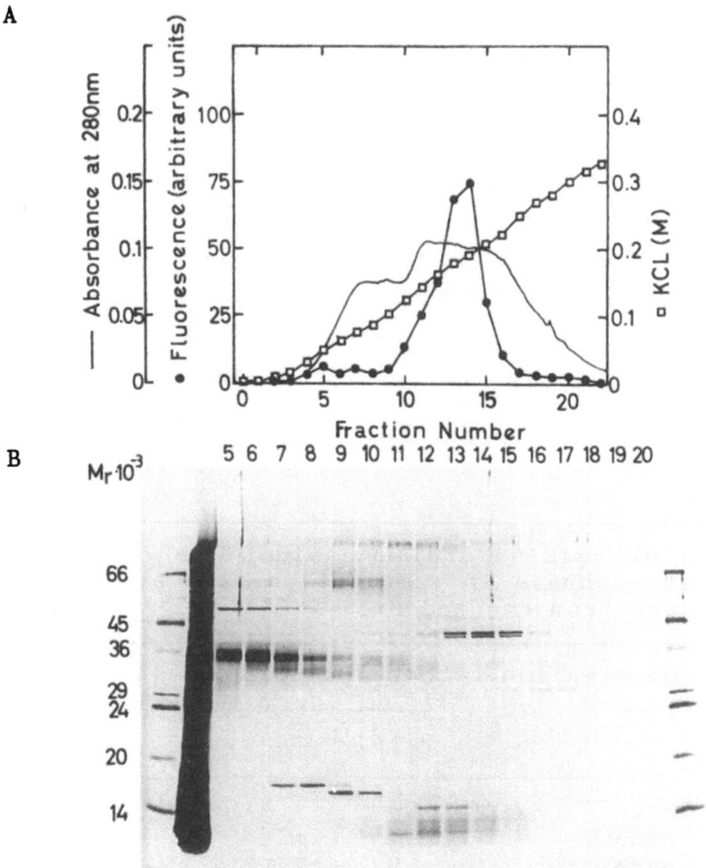

Figure 2. A) Elution of *Pmg* culture filtrate elicitor activity from a DE-cellulose column using a linear gradient of KCl in 50 mM potassium phosphate buffer, pH 6.8. Elicitor activity was determined by adding aliquots of each 8-ml fraction to $5 \cdot 10^4$ parsley protoplasts in a total volume of 0.1 ml. Furanocoumarin accumulation after 24 h incubation was measured by fluorescence at 410 nm. B) SDS-polyacrylamide gel of the crude *Pmg* culture filtrate and DE-cellulose fractions 5 - 20. The elicitor activity was resolved on the silver-stained gel as a protein doublet of approximately 42 KDa.

plasts and cells. Deglycosylation and proteinase treatments indicate that the protein portion but not the sugar is essential for activity. Although the pure protein is partially degraded by autoclaving it retains its full activity. We will attempt to isolate the intact protein and the active fragment(s) in pure form, radioactively label both and use the labelled compounds to test for saturable binding sites on the surface of parsley protoplasts.

Polyclonal antisera raised against the deglycosylated elicitor specifically recognize the protein in its glycosylated and deglycosylated form on Western blots. Using the antisera we established that the elicitor was present in higher amounts in the cell wall than in the culture filtrate, but represented a lower proportion of the total extractable cell wall protein. We are now using the antisera to test the presence of this protein in other fungi.

Table 3. Effect of calcium and potassium ions on the elicitation of defense responses in cultured parsley and potato cells by crude elicitors from *Pmg* and *Pi*.

Elicitor	Omitted ion[a]	Defense response [% of maximum][b]	
		Parsley	Potato
Pmg	none	100	100
Pi	none	44	67
None	none	9	25
None	Ca^{++}	3	25
None	K^{+}	11	31
Pmg	Ca^{++}	71	47
Pmg	K^{+}	105	150
Pi	Ca^{++}	13	40
Pi	K^{+}	45	69

[a] Cells were harvested from complete media by filtration, washed two times with medium free of Ca^{++} or K^{+} and then resuspended in the respective medium as indicated.
[b] The responses were measured as described in Table 2.

In order to identify components involved in signal transduction leading to specific gene activation in elicitor-treated parsley cells, we used a monoclonal antibody against cyclic AMP in a radioimmunoassay (Colling *et al* 1988) to determine the levels of this compound in parsley cells and protoplasts after elicitor application. The amounts of cyclic AMP remained constant in both protoplasts (3 ± 2 pmol/1·10^6 protoplasts) and cultured cells (6 ± 3 pmol/g fresh weight) within the first two hours of elicitor treatment. At this time most of the elicitor-responsive genes have reached maximum transcription rates (Somssich *et al* 1989). However, omission of Ca^{++} from the culture medium of protoplasts or cells dramatically reduced elicitor-stimulated furanocoumarin excretion (Table 3) as well as transcription rates of elicitor-responsive genes. Therefore, Ca^{++} fluxes appear to be involved in this induction process, whereas cyclic AMP does not play a significant role.

In future experiments we will use these cultured systems to screen for pathogen-derived elicitor molecules from different fungi and bacteria. The purified elicitors will then be used to identify their targets on the plant cell and study the subsequent signal chain leading to activation of the defense responses.

ACKNOWLEDGEMENTS

We thank Andrea Gierten, Magdalena Jung and Rosemarie Voigt for skilled technical assistance. This work was supported by the Max-Planck-Gesellschaft, a research grant to Dierk Scheel from the Deutsche Forschungsgemeinschaft (Sche 235/2-1) and an EMBO postdoctoral fellowship to Jane Parker.

REFERENCES

Ayers AR, Ebel J, Valent B, Albersheim P (1976) Host-pathogen interactions. X. Fractionation and biological activity of an elicitor isolated from the mycelial walls of *Phytophthora megasperma* var. *sojae*. Plant Physiol **57**:760-765

Colling C, Gilles, R, Cramer M, Nass N, Moka R, Jaenicke L (1988) Measurement of 3':5' cyclic AMP in biological samples using a specific monoclonal antibody. Second Messengers and Phosphoproteins **12**:123-133

Dangl JL, Hauffe KD, Lipphardt S, Hahlbrock K, Scheel D (1987) Parsley protoplasts retain differential responsiveness to u.v. light and fungal elicitor. EMBO J **6**:2551-2556

DiCosmo F, Misawa M (1985) Eliciting secondary metabolism in plant cell cultures. Trends in Biotechnology **3**:318-322

Ebel J (1986) Phytoalexin synthesis: the biochemical analysis of the induction process. Ann Rev Phytopathol **24**:235-264

Fritzemeier KH, Cretin C, Kombrink E, Rohwer F, Taylor J, Scheel D, Hahlbrock K (1987) Transient induction of phenylalanine ammonia-lyase and 4-coumarate:CoA ligase mRNAs in potato leaves infected with virulent or avirulent races of *Phytophthora infestans*. Plant Physiol **85**:34-41

Hahlbrock K, Scheel D (1987) Biochemical responses of plants to pathogens. In: Chet I (ed) Innovative approaches of plant disease control, Wiley & Sons, New York, p 229

Hahlbrock K, Scheel D (1989) Physiology and molecular biology of phenylpropanoid metabolism. Annu Rev Plant Physiol Plant Mol **40**:347-369

Hauffe KD, Hahlbrock K, Scheel D (1986) Elicitor-stimulated furanocoumarin biosynthesis in cultured parsley cells: S-adenosyl-L-methionine:bergaptol and S-adenosyl-L-methionine:xanthotoxol O-methyltransferases. Z Naturforsch **41c**:228-239

Jahnen W, Hahlbrock K (1988) Cellular localization of nonhost resistance reactions of parsley (*Petroselinum crispum*) to fungal infection. Planta **173**:197-204

Kombrink E, Hahlbrock K (1986) Responses of cultured parsley cells to elicitors from phytopathogenic fungi. Timing and dose dependency of elicitor-induced reactions. Plant Physiol **81**:216-221

Palavan N, Galston AW (1982) Polyamine biosynthesis and titer during various developmental stages of *Phaseolus vulgaris*. Physiol Plant **55**:438-444

Parker JE, Hahlbrock K, Scheel D (1988) Different cell-wall components from *Phytophthora megasperma* f.sp. *glycinea* elicit phytoalexin production in soybean and parsley. Planta **176**:75-82

Rohwer F, Fritzemeier KH, Scheel D, Hahlbrock K (1987) Biochemical reactions of different tissues of potato (*Solanum tuberosum*) to zoospores or elicitors from *Phytophthora infestans*. Accumulation of sesquiterpenoid phytoalexins. Planta **170**:556-561

Scheel D, Hauffe KD, Jahnen W, Hahlbrock K (1986) Stimulation of phytoalexin formation in fungus-infected plants and elicitor-treated cell cultures of parsley. In: Lugtenberg B (ed) Recognition in microbe-plant symbiotic and pathogenic interactions, Springer, Berlin, p 325

Scheel D, Dangl JL, Douglas C, Hauffe KD, Herrmann A, Hoffmann H, Lozoya E, Schulz W, Hahlbrock K (1987) Stimulation of phenylpropanoid pathways by environmental factors. In: Wettstein Dv, Chua NH (eds) Plant molecular biology, Plenum, New York, p 315

Sharp JK, Valent B, Albersheim P (1984) Purification and partial characterization of a β-glucan fragment that elicits phytoalexin accumulation in soybean. J Biol Chem **259**:11312-11320

Somssich IE, Bollmann J, Hahlbrock K, Kombrink E, Schulz W (1989) Differential early activation of defense-related genes in elicitor-treated parsley cells. Plant Mol Biol **12**:227-234

Thurston HD, Schultz O (1981) Late blight. In: Hooker WJ (ed) Compendium of potato diseases, American Phytopathol Soc, St Paul, p 40

Tietjen KG, Hunkler D, Matern U (1983) Differential response of cultured parsley cells to elicitors from two non-pathogenic strains of fungi. 1. Identification of induced products as coumarin derivatives. Eur J Biochem **131**:401-407

A FAMILY OF GENES FOR PHYTOALEXIN DETOXIFICATION IN THE PLANT PATHOGEN *NECTRIA HAEMATOCOCCA*

Hans VanEtten[1], David Matthews, Patty Matthews, Vivian Miao, Al Maloney and David Straney
Department of Plant Pathology
Cornell University
Ithaca, NY 14853
USA

INTRODUCTION

Among microorganisms that cause disease on plants, fungi are the most prevalent. Some fungi are pathogenic to only a single host species and spend their whole life cycles, except for resting structures, on that plant. Others have broad host ranges and may even be able to grow as saprophytes separate from a host plant. In either case plant pathogenic fungi must have unique traits that distinguish them from strict saprophytes and allow for their intimate association with higher plants.

One of the properties that some fungal pathogens require for this parasitic life style is a means to detoxify the toxic chemicals (phytoalexins) produced by host plants in response to their presence (VanEtten *et al* 1989). This paper deals with *Nectria haematococca* (anamorph: *Fusarium solani*), a pathogen of this type. There is a family of phytoalexin detoxifying genes in this species, some members of which are known to determine the pathogenic potential of the fungus, not only being required for pathogenicity but also controlling host range of the organism. In some cases, the phytoalexin induces expression of the genes, and the level of their expression appears to control the degree of virulence of the organism. These genes apparently have specifically evolved for pathogenicity on the plants that produce the phytoalexins. The function in nature of other members of this gene family is unknown. They may be important for the saprophytic portion of the life cycle or for

[1]Current address: Departments of Plant Pathology and Molecular and Cellular Biology, University of Arizona, Tucson, AZ 85721, USA

NATO ASI Series, Vol. H 36
Signal Molecules in Plants and Plant-Microbe Interactions
Edited by B. J. J. Lugtenberg

pathogenicity on other hosts, and/or they may be ancestors of the genes now required for pathogenicity. We will end with a brief discussion of recent findings that some of the phytoalexin detoxifying genes are located on small, meiotically unstable chromosomes.

NECTRIA HAEMATOCOCCA — A BROAD HOST RANGE ASCOMYCETE

The mating population VI (MP VI) subgroup of *N. haematococca* is able to cause disease on nine plant species and one animal (VanEtten and Kistler 1988). Members of MP VI also can exist as soil-inhabiting saprophytes, and isolates are readily obtained from diseased tissue of plants of which they are not a primary pathogen. Most of the investigations on genetics and mechanisms of pathogenicity in this fungus have involved studies of the diseases it causes on garden pea (*Pisum sativum*) and chickpea (*Cicer arietinum*). Isolates of *N. haematococca* MP VI may be pathogenic on both, neither or only one of these plants.

Although there are undoubtedly a number of traits needed to be pathogenic on pea or chickpea, maximum pathogenicity on either plant requires the ability to detoxify the specific phytoalexins characteristic of each plant (VanEtten *et al* 1989). Detoxification of the pea phytoalexin pisatin is accomplished by demethylation (Figure 1 reaction 1), while detoxification of the chickpea phytoalexins medicarpin and maackiain can occur by at least three different means (Figure 1 reactions 2, 3 and 4). Metabolism of medicarpin and maackiain by *N. haematococca* normally occurs by the same reaction (Denny and VanEtten 1982), though we routinely use only maackiain as the substrate.

The enzyme responsible for pisatin demethylation is a cytochrome P-450 monooxygenase (Matthews and VanEtten 1983). Two of the reaction types involved in maackiain and medicarpin detoxification, 6a- and 1a-hydroxylation, are known to be catalyzed by monooxygenases (Kistler and VanEtten 1981; Matthews *et al* 1987). The third reaction has not been studied in this regard but the product does result from the net addition of one atom of oxygen. Our current hypothesis is that all of these reactions are catalyzed by cytochrome P-450 monooxygenases.

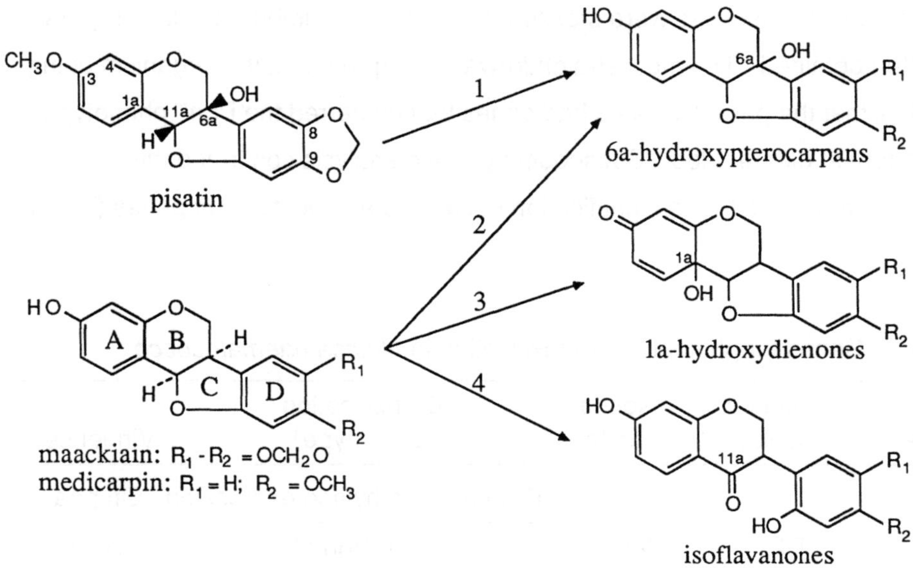

Figure 1. Detoxification of the phytoalexins pisatin, maackiain and medicarpin by *Nectria haematococca*.

GENES CONTROLLING DETOXIFICATION

A collection of 130 field isolates of *N. haematococca* MP VI obtained from soil or diseased tissue of 17 plant species has been assayed for the ability to carry out the above phytoalexin detoxification reactions (Lucy *et al* 1988). All 27 isolates obtained from pea were able to demethylate pisatin (Pda^+), while 31 of the 33 isolates obtained from chickpea or other plants that produce medicarpin or maackiain as a phytoalexin (alfalfa, red clover, sainfoin) were able to metabolize maackiain (Mak^+) (Lucy *et al* 1988). A number of additional Pda^+ and Mak^+ isolates were identified among isolates from other habitats, along with 29 Pda^- and 24 Mak^- isolates. Interestingly, 22 of the Mak^- isolates were also Pda^-, suggesting that in many Mak^+Pda^+ isolates both reactions might be catalyzed by the same enzyme, so that lack of that enzyme would normally produce the Pda^-Mak^- phenotype.

However, evidence for more than one enzyme responsible for phytoalexin detoxification was obtained in the initial genetic studies on the inheritance of pisatin detoxifying ability (Tegtmeier and VanEtten 1982; Kistler and VanEtten 1984). When different Pda$^+$ field isolates were crossed, a substantial number of Pda$^-$ progeny were often obtained. The most straightforward interpretation, this organism being haploid, is that the parent isolates had genes that conferred the Pda$^+$ phenotype (*Pda* genes) at different loci. Subsequent genetic analysis confirmed this interpretation and led to the identification of six different loci for *Pda* genes (Table 1).

Table 1. Genes for phytoalexin detoxification in *Nectria haematococca*

Gene	Source isolate	Isolate habitat	Detoxification phenotype[1]	Virulence[2]
			Pisatin demethylase expression	On pea
Pda1	T-2	Pea	High (Pda^{SH})	High
Pda2	T-219	Soil	Low (Pda^{LL})	Low
Pda3	T-2	Pea	Low (Pda^{LL})	Low
Pda4	T-23	Pea	Intermediate (Pda^{SM})	High
Pda5	Unknown	Unknown	High (Pda^{SH})	High[3]
Pda6-1	T-161	Unknown	Low (Pda^{LL})	Low
Pda6-2	T-200	Red clover	Low (Pda^{LL})	Low
None			None (Pda$^-$)	Low
			Maackiain metabolism	On chickpea
Mak1	T-161	Unknown	1a-hydroxylation	High
Mak2	T-161	Unknown	1a-hydroxylation	High
Mak3	T-200	Red clover	6a-hydroxylation	Unknown
None			None (Mak$^-$)	Low[4]

[1]Pisatin demethylase expression is characterized by the time required for metabolism of pisatin by non-induced mycelium in liquid culture (lag time for metabolism), and by the rate of demethylation after induction. Pda^{SH} indicates a *S*hort lag (2-3 hrs) and *H*igh induced rate (ca. 100 pmol/min/mg fresh wt), Pda^{SM} a *S*hort lag and *M*oderate rate, and Pda^{LL} a *L*ong lag and *L*ow rate.

[2]Virulence of the most virulent isolates possessing only the indicated gene. Highly virulent isolates produced lesions of 18 to 20 mm or more, 6 days after inoculating wounded epicotyls of pea or chickpea seedlings.

[3]The isolate that is the source of *Pda5* is highly virulent but the effect of this gene in other virulent backgrounds has not yet been critically evaluated.

[4]An exceptional Mak$^-$ field isolate has been found that is moderately virulent on chickpea; this isolate is also relatively tolerant of maackiain (Lucy *et al* 1988).

Thus far, no field isolate has been found to contain more than two *Pda* genes, and only two field isolates have been found whose *Pda* genes or alleles are at the same locus (*Pda6*).

Following a similar approach, three genes for metabolism of maackiain have also been identified (Table 1) (V. Miao, unpublished). *Mak2* and *Mak3* segregate independently of the *Pda* genes, and isolates with these genes can be Pda^-, indicating that separate enzymes are responsible for metabolism of pisatin and maackiain in these instances. However, *Mak1* is closely linked to or identical with *Pda6-1*. The possibility exists in this case that a single gene is encoding an enzyme that acts on both maackiain and pisatin.

PHENOTYPIC DIFFERENCES AMONG THE DETOXIFICATION GENES

Pisatin demethylase is an inducible enzyme, pisatin being the best inducer known (VanEtten and Barz 1981). The expression of this enzyme activity by different isolates of *N. haematococca* has been compared by measuring the time for pisatin to be metabolized by non-induced cultures (lag time for metabolism) as well as the rate of demethylation after induction with pisatin (Mackintosh *et al* 1989). Using these parameters, three quantitative phenotypes can be distinguished among the six identified genes (Table 1). Only the genes conferring the higher levels of demethylation activity (Pda^{SH} and Pda^{SM} phenotypes) are associated with virulence greater than that of Pda^- isolates. Thus the genes conferring the Pda^{LL} phenotype do not appear to give any selective advantage to the fungus on pea, and their function in nature is not known. If *Pda6-1* proves to be the same as *Mak1*, its function actually may be for pathogenicity on chickpea or other hosts of *N. haematococca* that produce medicarpin and/or maackiain as a phytoalexin.

Whether there are regulatory differences among the *Mak* genes has not yet been determined. However, it is clear that these genes code for at least two different products, as the enzyme under the control of *Mak1* or *Mak2* generates a different maackiain metabolite than that produced by *Mak3* (Table 1). Both *Mak1* and *Mak2* have been linked to increased virulence on chickpea, but *Mak3* has not yet been fully evaluated. Since highly virulent field isolates can hydroxylate maackiain by either reaction it would appear that either route of detoxification suffices for the pathogenicity of this fungus on chickpea.

The conventional genetic studies have revealed a family of phytoalexin detoxification genes in *N. haematococca,* and indicate that specific members are not only needed for pathogenicity but can determine which species of plant the organism is able to parasitize.

AN ISOLATED *PDA* GENE IDENTIFIES A FAMILY OF DETOXIFYING GENES AT THE MOLECULAR LEVEL

A *Pda* gene has been isolated from *N. haematococca* by detecting its expression in *Aspergillus nidulans* (Weltring *et al* 1988), and has been partially sequenced (A. Maloney, unpublished). This gene was obtained from isolate T-9, which has not been characterized by conventional genetics but has the Pda^{SH} phenotype. Some of the characteristics of the 3.2 kb fragment carrying this gene are shown in Figure 2. The gene encodes an approximately 1.8 kb mRNA, and RNAse protection assays as well as sequence data indicate the presence of four introns. Comparison of the sequence with other cytochrome P-450 genes revealed some homologies but the extent of homology in the region most highly conserved among cytochrome P-450 genes, the domain containing the cysteine that acts as the iron atom's fifth ligand, is not yet known: the coding sequence and the reading frame of this region is still uncertain due to the presence of an intron within or near it in the gene. Completion of the sequence analysis of the corresponding cDNA will answer this question.

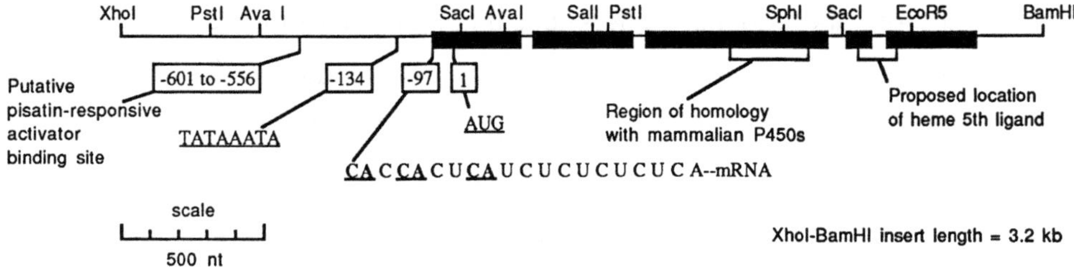

Figure 2. Molecular features of the *PdaT9* gene of *Nectria haematococca.*

The *PdaT9* gene has been used to isolate *Pda1,* and it has been demonstrated that induction of both of these genes by pisatin occurs at the level of transcription (D. Straney and A. Maloney, unpublished). A putative pisatin-responsive activator region approximately 550 bases upstream of the translation start site has been identified, based on its ability to bind a protein factor specifically enhanced in pisatin-treated mycelium (D. Straney, unpublished). This same sequence may be involved in amino-acid repression of this gene, as a factor obtained from mycelium grown in the presence of amino acids also binds to this region.

The cloned gene has also been used to compare restriction sites and hybridization intensities of homologous DNA in isolates carrying single copies of the other known *Pda* genes (A. Maloney, unpublished). Two distinct structural types of *Pda* genes can be detected. One includes the genes *Pda1*, *Pda5*, *PdaT9* (Pda^{SH}) and *Pda4* (Pda^{SM}). The other group, less homologous to *PdaT9*, consists of the genes that confer low demethylase activity and low virulence on pea (*Pda2*, *Pda3* and *Pda6*, both alleles). The differences between the two groups are distributed throughout the coding portions of the genes.

DNA from Pda^- isolates possessing *Mak3* did not hybridize to *PdaT9* DNA, indicating it is different from any of the known *Pda* genes (V. Miao, unpublished). As indicated above, *Mak1* may be identical to *Pda6-1*. Whether this is the case or the two genes are only tightly linked will most likely not be resolved until this gene is isolated. However, lack of homology with *Mak3* indicates that it, at least, represents a phytoalexin detoxification gene distinct from the *Pda* genes.

SOME *PDA* GENES ARE ON SMALL, MEIOTICALLY UNSTABLE CHROMOSOMES

Crosses involving *Pda* genes have occasionally produced segregation ratios that are inconsistent with normal Mendelian patterns (Tegtmeier and VanEtten 1982; Kistler and VanEtten 1984; Mackintosh *et al* 1989; V. Miao, unpublished). In such crosses, *Pda* genes are physically lost from the genome, as detected phenotypically by demethylase activity and recently by hybridization to *PdaT9*. Pulsed-field gel electrophoresis of *Pda6* parent isolates and progeny that had lost *Pda6* showed that this gene is on a small (<2 megabase) chromosome, and that loss of the gene is associated with large size changes in or complete disappearance of this chromosome (V. Miao, unpublished).

Analysis of isolates representing characterized *Pda* genes as well as a number of Pda^{+} field isolates indicated that *Pda* genes are often located on small (1-2 Mb) chromosomes (V. Miao, unpublished). These chromosomes resemble the "B" chromosomes previously studied in many higher organisms, in that they are dispensable for normal growth and development, meiotically unstable, variable in size among individuals of the species, and relatively small (Jones and Rees 1982). Unlike conventional B chromosomes, however, these *N. haematococca* chromosomes carry phenotypically detectable genes.

One consequence of this phenomenon will be a reconsideration of the number of *Pda* genes in this fungus. The standard allelism test, used to determine whether an unknown gene is at the same locus as a previously identified gene, is whether recombinant progeny lacking both genes can be obtained from a cross between the two. This test is now seen to be invalid for *Pda* genes in *N. haematococca*, since for example a *Pda6-1* × *Pda6-1* cross gave many Pda^{-} progeny (V. Miao, unpublished). At the very least, this fungus possesses three *Pda* genes and two *Mak* genes that are different from each other in the sense that they are phenotypically distinguishable. But whether these reside at different loci, and perhaps even what we mean by the term "locus" in this situation, are now unresolved questions.

SUMMARY

Our understanding of the *N. haematococca* genes and gene products involved in detoxifying the phytoalexins pisatin, medicarpin and maackiain is obviously incomplete. It is clear that in some cases the genes are coding for different enzymes. It seems likely that all of these enzymes are P-450 cytochromes, and there appear to be at least two distinct subfamilies: Pda and Mak. Within the subfamilies are differences such as the levels of pisatin demethylase expression in the Pda subfamily (Pda^{SH}, Pda^{SM}, Pda^{LL}) and the site of maackiain hydroxylation (1a vs. 6a) in the Mak subfamily. It is in the Pda subfamily that we have found the clearest signs of specific evolutionary adaptation of a cytochrome P-450 for parasitism on a particular host. In addition to the genetic studies indicating that pisatin demethylase is required for pathogenicity on pea, the enzyme in Pda^{SH} isolates shows a strong specificity for pisatin as its inducer and its substrate (VanEtten *et al* 1989).

The frequent co-occurrence of the Pda^{+} and Mak^{+} phenotypes in field isolates and the linkage of *Pda6-1* to *Mak1* suggest that some members of these two subfamilies could be the same. It seems likely that the *Pda* and *Mak* genes evolved from some ancestral cytochrome P-450 whose function was advantageous for the saprophytic growth of *N. haematococca.* Possibly some of these genes also conferred the ability to detoxify several phytoalexins before evolving into a more highly specific form during the coevolution of the pathogen with its different hosts. We should be able to answer some of these questions now that these genes can be isolated and characterized at the molecular level.

The recent observation that *Pda* genes can be on small dispensable chromosomes is intriguing. As far as is currently known, the instability of these chromosomes is only expressed during meiosis. Meiotic recombination of this organism in nature does not appear to be common, as the sexual cycle has only been observed in Japan on mulberry trees (Matuo and Snyder 1972). Why it would be an advantage to have genes that are required for pathogenicity be located on unstable chromosomes is not intuitively obvious. It will be of great interest to determine whether other genes for plant parasitism are located on these chromosomes and whether this plasticity is a general feature of pathogenicity genes in *N. haematococca.*

ACKNOWLEDGEMENTS

Much of the unpublished research cited was supported by Department of Energy contract DE-FGO2-86ER13540 and NSF grant DCB-85 17964.

REFERENCES

Denny TP, VanEtten HD (1982) Metabolism of the phytoalexins medicarpin and maackiain by *Fusarium solani.* Phytochemistry 21:1023-1028
Jones RN, Rees H (1982) B Chromosomes. Academic Press, London

Kistler HC, VanEtten HD (1981) Phaseollin metabolism and tolerance in *Fusarium solani* f. sp. *phaseoli*. Physiol Plant Pathol 19:257-271

Kistler HC, VanEtten HD (1984) Three non-allelic genes for pisatin demethylation in the fungus *Nectria haematococca*. J Gen Microbiol 130:2595-2603

Lucy MC, Matthews PS, VanEtten HD (1988) Metabolic detoxification of the phytoalexins maackiain and medicarpin by *Nectria haematococca* field isolates: relationship to virulence on chickpea. Physiol Mol Plant Pathol 33:187-199

Mackintosh SF, Matthews DE, VanEtten HD (1989) Two additional genes for pisatin demethylation and their relationship to the pathogenicity of *Nectria haematococca* on pea. Mol Plant-Microbe Interact (In press)

Matthews DE, VanEtten HD (1983) Detoxification of the phytoalexin pisatin by a fungal cytochrome P-450. Arch Biochem Biophys 224:494-505

Matthews DE, Weiner EJ, Matthews PS, VanEtten HD (1987) Role of oxygenases in pisatin biosynthesis and in the fungal degradation of maackiain. Plant Physiol 83:365-370

Matuo T, Snyder W (1972) Host virulence and the Hypomyces stages of *Fusarium solani* f. sp. *pisi*. Phytopathology 62:731-735

Tegtmeier KJ, VanEtten HD (1982) The role of pisatin tolerance and degradation in the virulence of *Nectria haematococca* on peas: a genetic analysis. Phytopathology 72:608-612

VanEtten HD, Barz W (1981) Expression of pisatin demethylating ability in *Nectria haematococca*. Arch Microbiol 129:56-60

VanEtten HD, Kistler HC (1988) *Nectria haematococca*, mating populations I and VI. In: Sidhu GS (ed) Advances in Plant Pathology, vol 6. Academic Press, New York, p 189

VanEtten HD, Matthews DE, Matthews PS (1989) Phytoalexin detoxification: importance for pathogenicity and practical implications. Annu Rev Phytopathol 27:143-164

Weltring K-M, Turgeon BG, Yoder OC, VanEtten HD (1988) Isolation of a phytoalexin-detoxification gene from the plant pathogenic fungus *Nectria haematococca* by detecting its expression in *Aspergillus nidulans*. Gene 68:335-344

3ucheli, P. Albersheim, and A.G. Darvill
ıydrate Research Center and
ıt of Biochemistry
orgia
ₜd.
502

rsensitive response is a commonly observed phenomenon in plant iyama *et al* 1979). A plant, when challenged by a microbe that is either not a pathogen of that plant or an incompatible race of a species that is her cultivars of the plant, commonly exhibits a hypersensitive resistance is phenotypically observed as the rapid necrosis of one or a few plant of attempted infection. This necrosis is presumed to result in processes read of the pathogen.

emical mechanism by which hypersensitive cell death is triggered has lished. Various molecules or preparations have been suggested as hypersensitive response; these include cell wall preparations from *festans* that kill potato tuber tissue (Currier 1981; Doke & Tomiyama om *P. infestans* that kill potato tissue (Currier 1981; Preisig & Kuc peptide from intercellular fluids of fungal-infected tomato that kills De Wit *et al* 1985), and a glycan that causes cell death in wheat leaves r *et al* 1986). The glycoprotein from which the glycan was obtained was ie germ tube walls of the uredospores of the fungal pathogen *Puccinia tici*. It should be noted that the origin of all of these suggested inducers ty (with the possible exception of the small peptide from fungal-infected s the pathogen rather than the host.

physiological responses in the plant have been described that are the hypersensitive response, including activation of the plasmalemma ge (Atkinson *et al* 1985), increased consumption of ATP (Tomiyama *et* ased protein synthesis (Tomiyama *et al* 1979), production of superoxide 983), increased lipoxygenase activity (Ocampo *et al* 1986), an increase ition (Beardmore *et al* 1983; Hammerschmidt *et al* 1984), increased xyproline content (Hammerschmidt *et al* 1984), and a decrease in the

NATO ASI Series, Vol. H 36
Signal Molecules in Plants and Plant-Microbe Interactions
Edited by B. J. J. Lugtenberg

plasma membrane potential (Tomiyama *et al* 1979). It is not known whether these physiological responses are a cause of or the result of plant cell death. Thus, the nature of the signal for hypersensitive cell death remains to be clearly demonstrated.

Several years ago our laboratory began investigating the possibility that fragments, derived from the cell walls of the plant itself, may possess killing activity towards plant cells. We demonstrated that wall fragments, capable of killing suspension-cultured sycamore cells, can be generated from isolated sycamore cell walls by partial acid hydrolysis (Yamazaki *et al* 1983). Cell death was demonstrated by the use of vital stains, by the inability of treated cells to grow, and by inhibition of the ability of the cells to incorporate [^{14}C]-leucine into proteins. We have demonstrated that the sycamore cell wall-derived killing fragments are capable of killing suspension-cultured maize cells, and acid-solubilized fragments of isolated maize cell walls kill both sycamore and maize cells. Furthermore, acid-solubilized fragments of rice cell walls possess killing activity when assayed against maize cells in the [^{14}C]-leucine incorporation bioassay. Thus, the killing fragments seem to be widely distributed across the plant kingdom, and fragments derived from the cell walls of divergent species of plants are able to kill the cells of that species as well as the cells of evolutionarily distant species.

The observation that plant cell wall fragments possess killing activity was extended by our demonstration that fragments released from plant cell walls, by the action of cell wall-degrading enzymes secreted by pathogens, are able to kill plant cells. This observation lends support to the idea that the killing fragments are a component of the plant resistance response and may in fact be the *in planta* signal for plant cells to react hypersensitively. Support for the concept that the plant cell wall-derived killing fragments may be released enzymically can be found in the many literature reports that pathogen-secreted cell wall-degrading enzymes kill plant cells (for a review, see Hahn *et al* (1989)). It had been suggested that this cell death occured because the enzymes weakened the cell wall to the point that the plant cells burst from osmotic pressure (Bateman & Basham 1976), but this hypothesis was refuted by a description of a pectate lyase from *Erwinia* that possessed low wall-degrading activity and yet was highly toxic to potato cells (Quantick *et al* 1983) and by the demonstration that purified fungal polygalacturonases, when assayed in the presence of an excess of polygalacturonase-inhibiting protein, retained the ability to kill potato tissue even though the macerating activity had been negated (Cervone *et al* 1989).

We have been using the ascomycetous pathogen *Magnaporthe grisea* as a source of enzymes that release killing fragments from plant cell walls. *Magnaporthe grisea* is the causal agent of rice blast, economically the most serious disease of that plant. The fungus can be readily crossed in the laboratory (Valent *et al* 1986) and has been successfully transformed using molecular genetic techniques (Parsons *et al* 1987). These characteristics make *M. grisea* an excellent pathogen for use in studies of the hypersensitive response, for proof that experimentally induced cell death is in fact representative of hypersensitive cell death will require genetic studies.

We have established that a heat-labile activity that kills suspension-cultured maize cells (assayed by [^{14}C]-leucine incorporation into protein) was secreted by *M. grisea* strain Ken 60-19 when the fungus was grown in liquid shake culture on certain carbon sources. This heat-labile killing activity was produced when the carbon source was citrus pectin, and the fungus secreted 20 to 50-fold more killing activity when the carbon source was isolated rice cell walls. No heat-labile killing activity was detected when *M. grisea* was grown on sucrose or xylan, nor was killing activity secreted when the fungus was grown on a complete medium containing yeast extract, casein hydrolysate, and sucrose. The heat-labile killing activity was concentrated and partially purified by CM-Sephadex chromatography and further purified by FPLC gel-permeation chromatography using a Superose 12 column. Gel-permeation chromatography yielded two distinct peaks of killing activity; mixing of four fractions from the two peaks of killing activity caused a synergistic level of killing activity, *i.e.*, a level of killing activity that was 2.5-fold higher than the additive level of killing activity exhibited by the four fractions independently.

Treatment of cell walls isolated from suspension-cultured maize cells with the CM-Sephadex-purified heat-labile killing activity released a steadily increasing amount of heat-stable killing activity. The release from isolated maize cell walls of heat-stable killing fragments occurred whether the heat-labile killing activity had been derived from the fungus grown on citrus pectin or on rice cell walls. The heat-stable killing activity was stable, in the presence of the fungal enzymes, for at least 50 hours. No heat-stable killing activity was solubilized when maize cell walls were treated with CM-Sephadex-chromatographed culture filtrate from the fungus grown on xylan. Treatment of maize cell walls with a fraction of either of the two peaks of killing activity from the Superose 12 gel-permeation column also caused the release of heat-stable killing fragments. Interestingly, a highly purified xylanase from *M. grisea* caused a lower yet significant level of release of killing fragments from maize cell walls. These results provided evidence the the culture filtrate of *M. grisea* must contain heat-labile killing activity in order for the filtrate to be able to solubilize, from plant cell walls, fragments that kill plant cells.

Our results agree with the observation by Hahne and Lörz (1988) that a mixture of commercially available enzymes, capable of killing suspension-cultured wheat and tobacco cells, released heat-stable components from the wheat and tobacco cells that possessed killing activity. Basham and Bateman (1975) found no release of heat-stable components that were able to kill the cells of potato tuber, cucumber fruit, or tobacco pith tissues when isolated tobacco pith tissue cell walls were treated with a pectate lyase from *Erwinia*. We were also unable to obtain wall fragments with killing activity by treating isolated maize cell walls with the pectin lyase secreted by *M. grisea*. Our results also do not confirm those of Ishii (1988), who reported that the filtrates of rice cells treated with a commercially available xylanase did not contain heat-stable components that kill rice cells, even though Ishii reported that the xylanase itself was able to kill rice cells.

We are focusing our present efforts on the purification of the active components of the *M. grisea*-secreted heat-labile (presumably enzymic) killing activity and on purification and characterization of the fungal enzyme-released heat-stable plant cell wall fragments that kill plant cells. Preliminary work with the heat-stable cell wall material suggests that it is heterogeneous in size and is probably derived from the hemicellulosic fraction of the plant cell wall. Identification and cloning of the killing enzymes secreted by *M. grisea* will enable us to use mutants of the fungus deficient in one or more of these enzymes to establish whether they are involved in hypersensitivity by determining whether the mutants have the ability to induce the hypersensitive response, and, if they do not, by transforming the mutants with a gene encoding the deficient enzyme and then determining whether the transformed fungal strains once again elicit a hypersensitive response.

This work was supported by grants from the US Department of Energy (grant number DE-FG09-85ER13425), the National Science Foundation (grant number DMB-8518488), and the Swiss National Science Foundation (post-doctoral fellowship for PB). The Complex Carbohydrate Research Center is a member of the three agency (DOE, NSF, USDA) Plant Science Center Program (grant number DE-FG09-87ER13810).

Atkinson MM, Huang J-S, Knopp JA (1985) The hypersensitive reaction of tobacco to *Pseudomonas syringae* pv. *pisi*. *Plant Physiol* 79:843-847

Basham HG, Bateman DF (1975) Killing of plant cells by pectic enzymes: The lack of direct injurious interaction between pectic enzymes or their soluble reaction products and plant cells. *Phytopathol* 65:141-153

Bateman DF, Basham HG (1976) Degradation of plant cell walls and membranes by microbial enzymes. In: Heitefuss R, Williams PH (eds) Encyclopedia of Plant Physiology, Vol. 4. Springer-Verlag, Berlin, p 316-355

Beardmore J, Ride JP, Granger JW (1983) Cellular lignification as a factor in the hypersensitive resistance of wheat to stem rust. *Physiol Plant Pathol* 22:209-220

Cervone F, De Lorenzo G, D'Ovidio R, Hahn MG, Ito Y, Darvill A, Albersheim P (1989) Phytotoxic effects and phytoalexin-elicitor activity of microbial pectic enzymes. In: Ballio A, Durbin RD, Graniti A (eds) Phytotoxins & Plant Pathogenesis, NATO ASI Series. in press

Currier WW (1981) Molecular controls in the resistance of potato to late blight. *Trends Biochem Sci* 6:191-194

De Wit PJGM, Hofman AE, Velthius GCM, Kuc JA (1985) Isolation and characterization of an elicitor of necrosis isolated from intercellular fluids of compatible interactions of *Cladosporium fulvum* (syn. *Fulvia fulva*) and tomato. *Plant Physiol* 77:642-647

Doke N (1983) Involvement of superoxide anion generation in the hypersensitive response of potato tuber tissues to infection with an incompatible race of *Phytophthora infestans* and to the hyphal wall components. *Physiol Plant Pathol* 23:345-357

Doke N, Tomiyama K (1980) Suppression of the hypersensitive response of potato tuber protoplasts to hyphal wall components by soluble glucans isolated from *Phytophthora infestans*. *Physiol Plant Pathol* 16:177-186

Hahn MG, Bucheli P, Cervone F, Doares SH, O'Neill RA, Darvill A, Albersheim P (1989) The roles of cell wall constituents in plant-pathogen interactions. In: Nester E, Kosuge T (eds) Plant-Microbe Interactions, Vol. 3. McGraw Hill, New York, p in press

Hahne G, Lörz H (1988) Release of phytotoxic factors from plant cell walls during protoplast isolation. *J Plant Physiol* 132:345-350

Hammerschmidt R, Lamport DTA, Muldoon EP (1984) Cell wall hydroxyproline enhancement and lignin deposition as an early event in the resistance of cucumber to *Cladosporium cucumerinum*. *Physiol Plant Pathol* 24:43-47

Ishii S (1988) Factors influencing protoplast viability of suspension-cultured rice cells during isolation process. *Plant Physiol* 88:26-29

Moerschbacher B, Kogel KH, Noll U, Reisener HJ (1986) An elicitor of the hypersensitive lignification response in wheat leaves isolated from the rust fungus *Puccinia graminis* f. sp. *tritici* I. Partial purification and characterization. *Z Naturforsch* 41c:830-838

Ocampo CA, Moerschbacher B, Grambow HJ (1986) Increased lipoxygenase activity is involved in the hypersensitive response of wheat leaf cells infected with avirulent rust fungi or treated with fungal elicitor. *Z Naturforsch* 41c:559-563

Parsons KA, Chumley FG, Valent B (1987) Genetic transformation of the fungal pathogen responsible for rice blast disease. *Proc Natl Acad Sci USA* 84:4161-4165

Preisig CL, Kuc JA (1988) Metabolism by potato tuber of arachidonic acid, an elicitor of hypersensitive resistance. *Physiol Mol Plant Pathol* 32:77-88

Quantick P, Cervone F, Wood RKS (1983) Isoenzymes of a polygalacturonate *trans*-eliminase produced by *Erwinia atroseptica* in potato tissue and in liquid culture. *Physiol Plant Pathol* 22:77-86

Tomiyama K, Doke N, Nozue M, Ishiguri Y (1979) The hypersensitive response of resistant plants. In: Daly JM, Uritani I (eds) Recognition and Specificity in Plant Host-Parasite Interactions. Japan Scientific Societies Press (University Park Press), Tokyo (Baltimore), p 69-84

Valent B, Crawford MS, Weaver CG, Chumley FG (1986) Genetic studies of fertility and pathogenicity in *Magnaporthe grisea (Pyricularia oryzae)*. *Iowa St J Res* 60:569-594

Yamazaki N, Fry SC, Darvill AG, Albersheim P (1983) Host-pathogen interactions XXIV. Fragments isolated from suspension-cultured sycamore cell walls inhibit the ability of the cells to incorporate [^{14}C]leucine into proteins. *Plant Physiol* 72:864-869

ne

di Biologia Vegetale
di Roma "La Sapienza"

ıstituents of plant cell wall may function as .gnals for the activation of gene expression in ɛnic fungi (Cooper and Wood 1980; Cooper 1983; al 1986; Crawford and Kolattukudy 1987). For :tin, one of the first plant cell wall components by a fungus at the onset of pathogenesis, is known :he synthesis of pectic enzymes that are not ɛn the fungus is grown in a culture medium lacking Lorenzo et al 1987; Cervone et al 1988; Dean and .989a and 1989b). Except for pectic enzymes and ıccharidases which have been studied in several ɪes et al 1972; Cooper et al 1978; Collmer and Keen functions expressed de novo by a fungus which invade a host plant are largely unknown. A study of the gene expression of phytopathogenic ɪe onset of pathogenesis is lacking, probably :he lack of suitable molecular probes.

vitro exposure of a fungus to the different ɔf the plant tissue normally encountered by the sm during pathogenesis may enable us to identify ɐ sequences corresponding to genes that are ɪring pathogenesis. It has been reported that ɔes the synthesis of endopolygalacturonase and ins in Fusarium moniliforme grown in vitro (De

NATO ASI Series, Vol. H 36
Signal Molecules in Plants and Plant-Microbe Interactions
Edited by B. J. J. Lugtenberg

Lorenzo et al 1987). Here we report the isolation and characterization of cDNA copies of mRNA species whose synthesis is induced by pectin in F. moniliforme and the identification of clones containing sequences coding for endopolygalacturonase.

RESULTS

A cDNA library was constructed in λgt10 using poly(A)-containing RNA isolated from Fusarium moniliforme that had been grown in Czapek-Dox broth containing 1% pectin as a sole carbon source. The library was screened by differential hybridization using radioactive cDNAs prepared from poly(A)-containing RNA isolated from the pectin-induced and non-induced fungus. Fourteen clones hybridized only with pectin-induced cDNA and did not hybridize at all with the non-induced cDNA. The DNAs from the fourteen recombinant clones were digested to determine the size of the inserts. The inserts were recovered and subcloned into the pGEM 7z(f)+ plasmid. Upon EcoRI digestion, electrophoresis and blotting onto nitrocellulose membranes, the recombinant plasmid DNAs were also hybridized with each of the nick-translated inserts. On the basis of cross hybridization characteristics, the fourteen clones were classified into six homology groups each of which did not hybridize to the others (tab 1). Assuming that the clones of each group represented different sized cDNA copies of the same mRNA species, from each group the clone containing the largest cDNA insert ("type" clone) was used for Northern analysis of poly(A)-containing RNA isolated from either pectin-grown and glucose-grown mycelium.

"Type" clones from groups III, IV, V and VI showed a hybridization signal to mRNA species present only in pectin-induced mycelium. Clones from groups I and II gave a strong signal to mRNA species present only in pectin-grown mycelium and a non-specific hybridization signal to the 1.9 kb species (group I) and the 3.6 kb species (group II) of rRNA. The "type" clone of group I hybridized to two distinct synthetic oligonucleotide probes corresponding to the amino acid sequence of either the N-terminal and a tryptic peptide from endopolygalacturonase.

Table 1. Characteristics of selected cDNA clones

Group	No. of members	Largest insert size (kb)	mRNA size (kb)
I	4	1.65	2.15 1.95
II	4	1.15	from 0.7 to 3.6
III	2	0.78	1.9
IV	2	0.71	1.8
V	1	1.35	2.1 1.9
VI	1	0.7	1.65

DISCUSSION

A cDNA library was constructed from poly(A)-containing RNA of the phytopathogenic fungus Fusarium moniliforme grown on pectin as a sole carbon source. By differential screening, we isolated fourteen cDNA copies of mRNAs that were expressed when the fungus was grown on pectin and not detectably expressed when the fungus was grown on glucose. Since pectin is one of the first plant cell wall components encountered by an invading fungus, the peptide translation products of these pectin-induced mRNAs are expected to be synthesized early in pathogenesis and to play an active role in infection. For example, mRNA for endopolygalacturonase is known to be induced by pectin (De Lorenzo et al 1987) and endopolygalacturonase is thought to play an important role in assisting phytopathogenic fungi in colonization of the plant tissue (Hahn et al 1989). Among the fourteen clones isolated, clones putatively harboring sequences coding for endopolygalacturonase have been identified by using two synthetic oligonucleotide probes corresponding to the partial aminoacid sequence of the enzyme. At present, we have no information about the functions coded by the other clones. Some of them are likely to correspond to mRNAs encoding proteins that are synthesized at the onset of pathogenesis. These fungus-derived pathogenesis proteins should deserve the same interest shown for the plant-derived pathogenesis-related proteins and corresponding genes that are induced in the plant during pathogenesis.

Gene activation occurs in both partners of a plant-fungus interaction. Nevertheless, while gene expression has been studied in some detail in plants, the number of studies on fungal gene expression during pathogenesis is still very limited (Hahn et al 1989). We are confident that the pectin-induced cDNA clones we have isolated provide a path to the study of gene expression of _F. moniliforme_ during the early stages of interaction with host plants. The _Fusarium_ pectin-induced cDNA clones may also be used for isolating fungal pathogenesis probes by heterologous hybridization in those systems (obligate parasites) where an alternative strategy is rather complicated.

The cDNA clones we have isolated are copied from mRNA species that are induced by pectin. The molecular signals for the induction of our clones are not known yet. It is known that autoclaved pectin produces various pectic fragments with different degree of polymerization (York et al 1985). It is likely that one or several of these fragments are the messages for the induction of the fungal mRNAs identified in this study. The possibility also exists that different fungal mRNAs are induced by different pectic fragments and that the same fragments (oligogalacturonides with DP from 10 to 13) which function as "endogenous" elicitors (Hahn et al 1981; Cervone et al 1989) for phytoalexin synthesis in plants also induce the synthesis of some mRNAs in fungi. Experiments are now in progress to identify the possible translation products corresponding to the isolated cDNA clones and to study their temporal expression following treatments with oligogalacturonides of various DP. Experiments are also underway to study the expression of these clones when the fungus is inoculated to a compatible and/or an incompatible plant host.

ACKNOWLEDGEMENTS

This work was supported by a grant MPI (60%).

REFERENCES

Cervone F, Castoria R, Spanu P, Bonfante-Fasolo P (1988) Pectinolytic activity in some ericoid mycorrhizal fungi. Trans Brit Mycol Soc 91: 537-539

Cervone F, De Lorenzo G, Salvi G, Camardella L (1986) Molecular evolution of fungal polygalacturonase. In: Bailey J (ed) Biology and Molecular Biology of Plant-Pathogen Interactions. NATO ASI Series, Vol. H1, Springer-Verlag, Berlin Heidelberg, pp 385-392

Cervone F, Hahn MG, De Lorenzo G, Darvill AG, Albersheim P (1989) Host-Pathogen Interactions XXXIII. A plant protein converts a fungal pathogenesis factor into an elicitor of plant defense responses. Pl Physiol 90: 542-548

Collmer A, Keen NT (1986) The role of pectic enzymes in plant pathogenesis. Ann Rev Phytopathol 24: 383-409

Cooper RM (1983) The mechanisms and significance of enzymic degradation of host cell walls by parasites. In: Callow JA (ed) Biochemical Plant Pathology. John Wiley and Sons, New York, pp.101-135

Cooper RM, Rankin B, Wood RKS (1978) Cell wall degrading enzymes of vascular wilt fungi II. Properties and modes of action of polysaccharidases of Verticillium alboatrum and Fusarium oxysporum f. sp. lycopersici. Physiol Pl Pathol 13: 101-134

Cooper RM, Wood RKS (1980) Cell wall degrading enzymes of vascular wilt fungi III. Possible involvement of endo-pectin lyase in Verticillium wilt of tomato. Physiol Pl Pathol 16: 285-300

Crawford MS, Kolattukudy PE (1987) Pectate lyase from Fusarium solani f. sp. pisi: purification, characterization, in vitro translation of the mRNA, and involvement in pathogenicity. Arch Biochem Biophys 258: 196-205

Dean RA, Timberlake WE (1989a) Production of cell wall-degrading enzymes by Aspergillus nidulans: a model system for fungal pathogenesis of plants. The Plant Cell 1: 265-273

Dean RA, Timberlake WE (1989b) Regulation of the Aspergillus nidulans pectate lyase gene (pelA). The Plant Cell 1: 275-284

De Lorenzo G, Salvi G, Degra' L, D'Ovidio R Cervone F (1987) Induction of extracellular polygalacturonase and its mRNA in the phytopathogenic fungus Fusarium moniliforme. J Gen Microbiol. 133: 3365-3373

Hahn MG, Bucheli P, Cervone F, Doares SH, O'Neill RA, Darvill A Albersheim P (to be published) The roles of cell wall constituents in plant-pathogen interactions. In: Nester E and Kosuge T (eds) Plant-Microbe Interactions. Vol.3, McGraw Hill, New York

Hahn MG, Darvill AG, Albersheim P (1981) Host-Pathogen Interactions XIX. The endogenous elicitor, a fragment of a plant cell wall polysaccharide that elicits phytoalexin accumulation in soybean. Pl Physiol 68: 1161-1169

Jones TM, Anderson AJ, Albersheim P (1972) Host-Pathogen Interactions IV. Studies on the polysaccharide-degrading enzymes secreted by Fusarium oxysporum f. sp. lycopersici. Physiol Pl Pathol 2: 153-166

York WS, Darvill AG, McNeil M, Stevenson TT, Albersheim P (1985) Isolation and characterization of plant cell walls and plant cell wall components. Meth Enzymol 118: 3-40

LOCAL AND SYSTEMIC GENE ACTIVATION IN FUNGUS-INFECTED POTATO LEAVES

Klaus Hahlbrock, Nikolaos Arabatzis, Michael Becker-Andre, Hans-Jürgen Joos, Erich Kombrink, Martin Schröder, Günter Strittmatter, Janet Taylor

Max-Planck-Institut für Züchtungsforschung
Abteilung Biochemie
D - 5000 Köln 30
F.R.G.

NATO ASI Series, Vol. H 36
Signal Molecules in Plants and Plant-Microbe Interactions
Edited by B. J. J. Lugtenberg

LOCAL AND SYSTEMIC GENE ACTIVATION IN FUNGUS-INFECTED POTATO LEAVES

The response of potato (***Solanum tuberosum***) leaf tissue to infection by the late-blight fungus ***Phytophthora infestans*** (***Pi***) is very rapid. Inoculation of either the lower or the upper side of a potato leaf with ***Pi*** zoospores under proper environmental conditions is followed by germination of the fungus within minutes and penetration of hyphae through stomata or subsidiary cells, or occasionally between epidermal cells (Cuypers and Hahlbrock 1988). The earliest readily detectable responses of immediately affected plant cells are callose deposition and the accumulation of phenolic compounds in the cell wall. Both effects are clearly visible about 2-3 h postinoculation. The accumulation of cell wall-associated phenolics at this early stage is an indication of hypersensitive cell death and is detected most easily by autofluorescence of the cells under UV light (Fig. 1).

With few exceptions (Cuypers and Hahlbrock 1988, Cuypers et al 1988, Rohwer et al 1987), we have confined our studies on potato-***Pi*** interactions to mature leaves, the primary infection site in the field. A major difference between compatible and incompatible interactions of the potato cultivar Datura (carrying ***Pi***-resistance gene R1) and ***Pi*** races 1 and 4, respectively, was observed in the mesophyll (Cuypers and Hahlbrock 1988, Hahlbrock et al 1988). Hypersensitive cell death occurred in the mesophyll almost invariably in the incompatible interaction, but only rarely in the compatible interaction. The final result was a more or less unimpaired penetration of ***Pi*** race 1 through the leaf and sporulation within about 24 h, whereas race 4 was confined in small necrotic regions where it died eventually.

A differential response was also seen at the level of phenylalanine ammonia-lyase (PAL) and 4-coumarate:CoA ligase (4CL) mRNA induction (Fritzemeier et al 1987, Hahlbrock et al 1988). These mRNAs encode two closely related enzymes of general phenylpropanoid metabolism and are coordinately regulated under all conditions tested (Fitzemeier et al 1987, Hahlbrock and Scheel 1989). The mRNAs are induced in all infected plant tissues (Hahlbrock and Scheel 1989) and are among the most rapidly accumulating transcripts in *Pi*-infected potato leaves (Cuypers et al 1988, Fritzemeier et al 1987, Hahlbrock et al 1988). Recently, the induction of mRNA encoding a "pathogenesis-related protein" (PR1) was found to follow the same time course as that previously observed for PAL and 4CL mRNAs (unpublished result).

The structures (Fig. 2) and genomic organization of genes encoding PAL, 4CL and PR1 have recently been established as a prerequisite for more detailed studies of their modes of expression. In potato, PAL is encoded by an unusually large

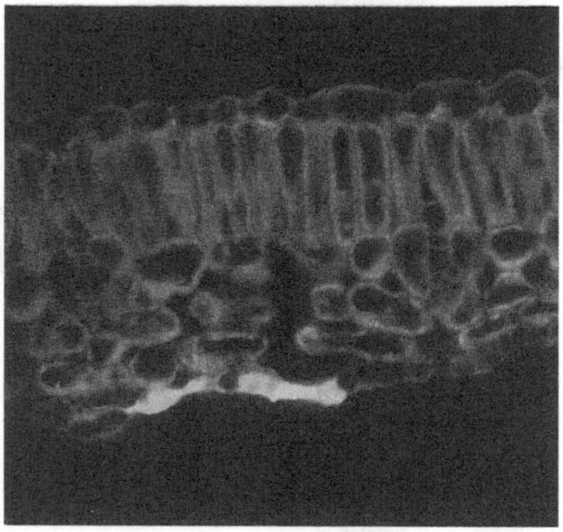

Fig. 1. Autofluorescence of immediately affected plant cells 3 h postinoculation of a potato leaf with *Pi* zoospores. Inoculation was on the lower epidermis, embedding in L.R. gold.

family of about 30-40 genes per haploid genome. In contrast, 4CL is encoded by two, and PR1 by a family of approximately 10, very similar genes.

Another major question concerns the spacial distribution of transient mRNA accumulation in relation to the fungal penetration site. This question was approached by *in situ* RNA hybridization using labelled antisense RNAs as hybridization probes. A typical example is shown in Figure 3 for PR1. At 24 h postinoculation, a time point when PAL, 4CL and PR1 mRNA levels were high in compatible as well as incompatible interactions (Fritzemeier et al 1987 and unpublished results), strong hybridization signals have been observed for PAL and PR1 mRNAs around the central necrotic spot. Although a detailed analysis has not been carried out, preliminary results seem to indicate that the response seen at this level is qualitatively similar in both types of interaction, but is somewhat delayed and more diffuse in the compatible as compared with the incompatible

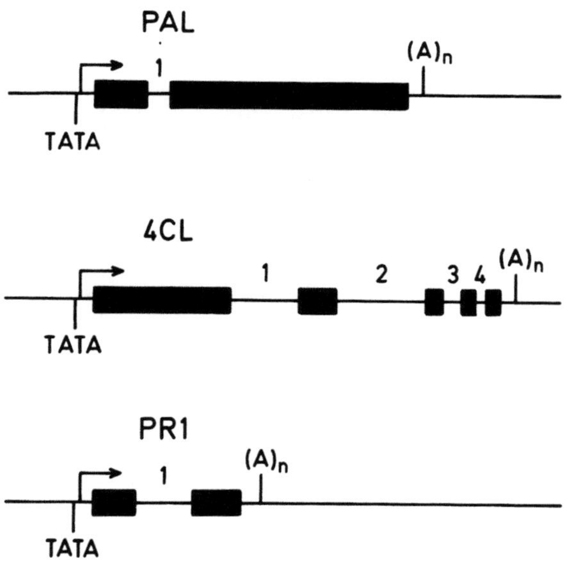

Fig. 2. Schematic representation of basic structural features of individual members from three selected gene families encoding PAL, 4CL and PR1 in potato. See text for copy number estimations.

interaction. In both cases, however, the accumulation of PAL and PR1 mRNAs is more or less strictly localized around the site of fungal penetration.

This is in contrast to the accumulation patterns of several other PR proteins which were recently identified as glucanase and chitinase isoenzymes (Kombrink et al 1988). These proteins accumulate systemically throughout the infected leaf with very similar time courses over several days in compatible and incompatible interactions. Typical results on the tissue localization are shown in Figure 4 for the β-1,3-glucanases.

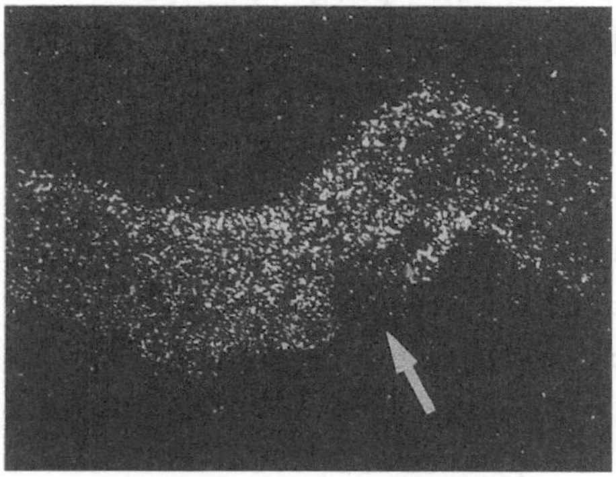

Fig. 3. Autoradiography of a cross-section from a *Pi*-infected potato leaf (incompatible interaction 24 h postinoculation) hybridized with ^{35}S-labelled PR1 antisense RNA. Methods were essentially as described by Cuypers et al (1988). Arrow points at center of infection site.

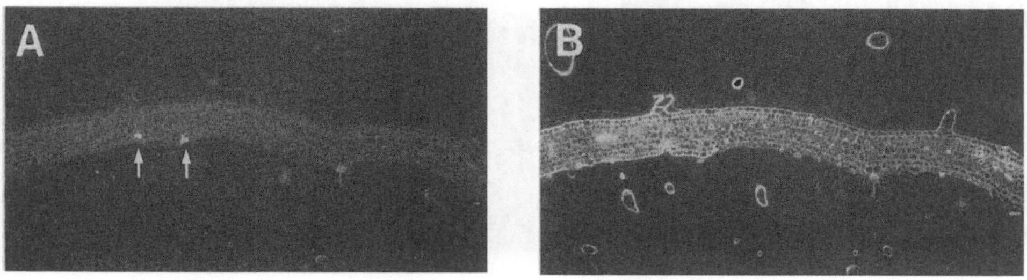

Fig. 4. Serial sections of a potato leaf 48 h postinoculation with *Pi* zoospores (incompatible interaction), showing (A) autofluorescence of infection sites (arrows) and (B) immunohistochemical localization of β-1,3-glucanase using fluorescein-labelled second antibodies.

Using a polyclonal antiserum raised against one of the purified glucanases, which detects all isoforms, we demonstrated the systemic, partially extracellular occurrence of the enzyme protein in all parts of an infected leaf.

In situ hybridization with antisense RNA indicates that β-1,3-glucanase accumulation is preceded by systemic activation of the corresponding gene(s) (Fig. 5). The RNA probe used for these initial studies was generated from tobacco β-1,3-glucanase cDNA (Mohnen et al 1985). Analysis of the genomic structure and organization of the various potato β-1,3-glucanase genes is in progress (L. Beerhues and E. Kombrink, unpublished results) and suggests a large nucleotide sequence similarity between the corresponding cDNAs from potato and tobacco.

In summary, the results obtained so far can be schematically depicted and interpreted as follows (Fig. 6).

In the incompatible interaction, all potato leaf cells establishing contact with the penetrating fungus undergo hypersensitive death. The fungus is thus confined to small necrotic lesions and is prevented from further penetration of the tissue and consequently from sporulation. The surrounding tissue is triggered to activate numerous genes encoding several defence-related pathways. This response involves local as well as systemic gene activation. Local gene activation appears to

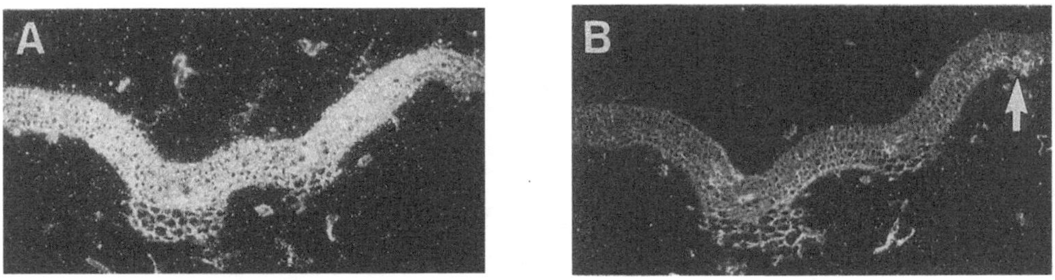

Fig. 5. Autoradiography of serial sections of a potato leaf 24 h postinoculation with *Pi* zoospores (incompatible interaction) following hybridization *in situ* with ^{35}S-labelled (A) antisense and, as a control, (B) sense RNA specific for β-1,3-glucanase mRNA. Probes were generated from tobacco β-1,3-glucanase cDNA (Mohnen et al 1985).

be transient, whereas late stages in the timing of systemic effects have not been studied. It remains open whether these two types of gene activation are strictly distinct or are extremes of various, as yet undetected intermediate types. Also uncertain at this stage is the ultimate fate of the cells involved in transient, local gene activation. One possibility, which would be consistant with the final size of the remaining necrotic spots, is the death of all or most of these cells

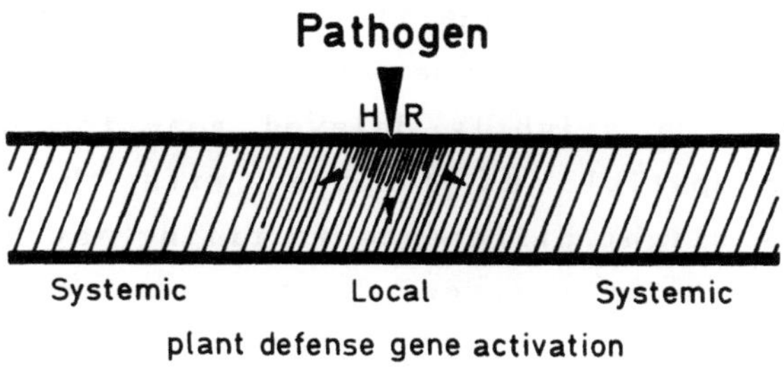

	Relative timing
HR=Hypersensitive response Cell death Autofluorescence (phenolics) Callose deposition etc.	very rapid
Local gene activation Phenylpropanoid pathways Pathogenesis-related (PR) proteins Phytoalexins etc.	rapid/ intermediate
Systemic gene activation 1.3-ß-glucanases Chitinases Other PR proteins etc.	intermediate/ slow

Fig. 6. Scheme depicting various defence reactions of fungus-infected plant leaves in incompatible interactions. Note that this idealized scheme applies to infections through either the upper or the lower epidermis.

subsequent to the period of transient gene activation. An early, sharp peak of PAL and 4CL mRNA accumulation (Fritzemeier et al 1987) suggests that a more detailed analysis may reveal rapid, local transcript accumulation in association with the first instance of hypersensitive cell death.

The compatible interaction differs in this system from the incompatible interaction by an apparent suppression of hypersensitive cell death in the mesophyll, resulting in relatively unimpaired fungal penetration of the leaf and efficient sporulation. Although no qualitative difference has been observed at the mRNA and enzyme activity levels, the overall result seems to be a slightly delayed reaction of the plant tissue in the compatible as compared with the incompatible interaction. This slight difference in timing may be sufficient to allow the fungus to outgrow the plant's attempted lesion limitation.

Thus, although many detailed questions remain open, the emerging general picture indicates rather complex signalling mechanisms leading to the observed differential temporal and spacial distribution of the various components of the plant's defense response. It is noteworthy in this connection that essentially similar patterns were seen in the nonhost incompatible interaction of parsley leaves (***Petroselinum crispum***) with ***Phytophthora megasperma*** f.sp. ***glycinea***, a ***Pi***-related fungal pathogen of soybean (E. Schmelzer, S. Krüger-Lebus and K. Hahlbrock, manuscript submitted), suggesting extensive similarity of the signal-reaction chains in the two systems. Comparative studies in this direction are in progress (*e.g.*, Scheel et al, this volume).

ACKNOWLEDGEMENTS

Previously unpublished work reported here was supported in part by the Ministry of Research and Technology of the FRG (BMFT "Genzentrum Köln", Project 1.02). We thank Dr. F. Meins, Basel, for providing the tobacco β-1,3-glucanase cDNA clone.

REFERENCES

Cuypers B, Hahlbrock K (1988) Immunohistochemical studies of compatible and incompatible interactions of potato leaves with *Phytophthora infestans* and of the nonhost response to *Phytophthora megasperma*. Can. J. Bot. **66**:700-705

Cuypers B, Schmelzer E, Hahlbrock K (1988) *In situ* localization of rapidly accumulated phenylalanine ammonia-lyase mRNA around penetration sites of *Phytophthora infestans* in potato leaves. Mol. Plant Microbe Interactions **4**:157-160

Fritzemeier KH, Cretin D, Kombrink E, Rohwer F, Taylor J, Scheel D, Hahlbrock K (1987) Transient induction of phenylalanine ammonia-lyase and 4-coumarate:CoA ligase mRNAs in potato leaves infected with virulent or avirulent races of *Phytophthora infestans*. Plant Physiol. **85**:34-41

Hahlbrock K, Cretin C, Cuypers B, Fritzemeier KH, Hauffe KD, Jahnen W, Kombrink E, Rohwer F, Scheel D, Schmelzer E, Schröder M, Taylor J (1988) Tissue specificity and dynamics of disease resistance responses in plants. In: Plant Molecular Biology, NATO ASI Series (D von Wettstein, ed.), Plenum Publishing Corporation, pp. 399-406

Hahlbrock K, Scheel D (1989) Physiology and molecular biology of phenylpropanoid metabolism. Annu. Rev. Plant Physiol. Mol. Biol. **40**, in press

Kombrink E, Schröder M, Hahlbrock K (1988) Several 'pathogenesis-related' proteins in potato are 1,3-β-glucanases and chitinases. Proc. Natl. Acad. Sci. USA **85**:782-786

Mohnen D, Shinshi H, Felix G, Meins F (1985) Hormonal regulation of β-1,3-glucanase messenger RNA levels in cultured tobacco tissues. EMBO J. **4**:1631-1635

Rohwer F, Fritzemeier KH, Scheel D, Hahlbrock K (1987) Biochemical reactions of different tissues and cultivars of potato (*Solanum tuberosum*) to zoospores or elicitors from *Phytophthora infestans*. Accumulation of sesquiterpenoid phytoalexins. Planta **170**:556-561

CONTROL OF THE HYPERSENSITIVE REACTION OF LETTUCE TO THE DOWNY MILDEW FUNGUS *BREMIA LACTUCAE*

Mansfield, J.W., Woods, A.M., Fagg, J. & Bennett, M.

Department of Biochemistry & Biological Sciences, Wye College (University of London), Ashford, Kent TN25 5AH, U.K.

The term hypersensitive reaction (HR) has been defined as the rapid and localized death of challenged plant cells associated with the restriction of microbial colonization. Plant cell death and restriction of microbial growth are essential features of all hypersensitive reactions but biochemical events associated with the HR may vary. Thus, in some plants, phytoalexins may accumulate whereas in others, different antimicrobial systems (lignification and chitinase synthesis for example) may be activated (Mansfield 1986; Hahlbrock & Scheel 1987).

It is becoming increasingly apparent that damage to the plant's plasma membrane is an early event in the onset of the HR (Pavlovkin et al 1986; Koga et al 1988; Ádám et al 1989). When membrane damage becomes irreversible, the responding cell should be considered to be dead because tonoplast dysfunction and loss of compartmentation soon follow. Irreversible membrane damage (IMD) can be detected at the cellular level by the failure of cells to plasmolyse. Plasmolysis provides a more critical test of membrane function than the use of certain vital stains such as Evans blue, neutral red and fluorescein diacetate (Stadelman 1966; Gahan 1984; Koga et al 1988). For example, working with lettuce cells in suspension culture, we have found that certain cells which failed to plasmolyse still showed some accumulation of fluorescein. In the few interactions in which it has been studied in detail, the occurrence of IMD has been found to precede other responses such as phytoalexin accumulation (Sato et al 1971; Doke & Tomiyama 1975; Doke et al 1976) and the deposition of silica and autofluorescent phenolics (Koga et al 1988). Studies of infection development *in planta* therefore implicate IMD as a very early response, perhaps chronologically close to a race-specific recognition event. These findings contrast with many studies on the responses of isolated protoplasts and cultured cells to non-specific

NATO ASI Series, Vol. H 36
Signal Molecules in Plants and Plant-Microbe Interactions
Edited by B. J. J. Lugtenberg

elicitors. In these *in vitro* systems IMD apparently does not occur but various processes are activated. Recent examples of this phenomenon concern the effect of reduced glutathione as a very effective elicitor of the transcription of genes controlling phenylpropanoid biosynthesis (Dron et al 1988) and the demonstration that parsley protoplasts retain differential responsiveness to u.v. light and a fungal elicitor (Dangl et al 1987). It seems likely that the transduction pathways identified using *in vitro* systems are not those effective during the initial interaction between the avirulent microbe and resistant plant but that they may operate in cells surrounding those undergoing IMD. The death of the initially challenged cell (often this is the *only* cell to die in an incompatible interaction) may serve to amplify responses in surrounding cells. Primary and secondary signalling may be envisaged to be linked via IMD as outlined in Fig. 1.

Primary recognition
↓
Irreversible membrane damage
↓
Release of secondary signals
↓
Secondary recognition in surrounding cells
↓
Activated, amplified response

Fig. 1. Recognition, signal transduction and activated defence during infection.

We have chosen the gene for gene interaction between lettuce and the downy mildew fungus *Bremia lactucae* as a model to test various aspects of this hypothesis. Here, we describe experiments on the control of IMD and the HR occurring during incompatible interactions.

IRREVERSIBLE MEMBRANE DAMAGE AND THE HR IN LETTUCE

B. lactucae is an obligate parasite and is biotrophic, absorbing nutrients from living cells (Ingram et al 1976). During invasion of susceptible cultivars the fungus forms primary and secondary vesicles and intracellular hyphae within the first penetrated epidermal cell, and then produces intercellular hyphae and haustoria. In susceptible cultivars the first penetrated cell remains alive at least until hyphae grow from the secondary vesicle into the intercellular space; cells containing haustoria remain alive for at least 7 days when the fungus has begun to sporulate.

In Lactuca sativa, 16 different factors for resistance to B. lactucae are now recognised and 13 have been characterized as single dominant alleles and mapped to four linkage groups (Farrara et al 1987). Resistance is expressed by the HR and the time of IMD in penetrated cells varies depending on the resistance gene being expressed. Thus, we found that in cultivars expressing Dm5/8, IMD occurs during expansion of the fungal primary vesicle, within 1.5h of penetration whereas in expression of Dm1 or Dm3, IMD is not observed until 2-2.5h later, during expansion of the secondary vesicle. Data for the interaction between cvs. Valmaine (Dm5/8) and Dandie (Dm3) and the isolate VO/11 which is known to be heterozygous at A5/8 and A3 are included in Fig. 2 (avirulence is dominant in the diploid fungus, Norwood & Crute, 1984). Certain other gene-for-gene interactions, i.e. Dm6/A6 and Dm7/A7, affect the later stages of infection development (Woods et al 1988).

IS PROTEIN SYNTHESIS REQUIRED FOR IRREVERSIBLE MEMBRANE DAMAGE?

The speed of occurrence of IMD, particularly during the expression of Dm5/8 suggested that the early response was caused by the direct effect of a fungal product on the plant cell membrane. Thus, referring to Fig. 1, IMD would be caused and the HR initiated without any activation of protein synthesis. In order to test this hypothesis we examined the effects of heat-shock and inhibitors of protein synthesis on the timing of IMD. Inhibitors of translation (blasticidin S, BcS) or mRNA synthesis (actinomycin D and cordycepin) were applied to cut stem ends of cvs. Valmaine (Dm5/8) and Dandie (Dm3), cotyledons were then detached and petioles inoculated with conidiosporangia of isolate VO/11. The occurrence of IMD in penetrated cells was observed. Treatment with BcS led to a delay in onset of IMD in both cultivars (Fig. 2). In cotyledons of cv. Valmaine pre-treated with water alone, plasmolysis was observed in less than 40% of cells which contained a primary vesicle $>15\mu m$ in diameter. After treatment with BcS (5, 12.5, 25 or $50\mu g\ ml^{-1}$) IMD was delayed in cv. Valmaine until secondary vesicles had expanded to between 10 and $25\mu m$. In cotyledons of cv. Dandie pre-treated with water, the Dm3 resistance gene was expressed by the occurrence of IMD when the fungus had produced a secondary vesicle between 10 and 25um in diameter. Following treatment with BcS, IMD usually did not occur until secondary vesicles were 25um in diameter and often was not observed until the formation of intracellular hyphae. In contrast to the effects caused by BcS, the inhibitors affecting mRNA synthesis caused little if any delay in the onset of IMD.

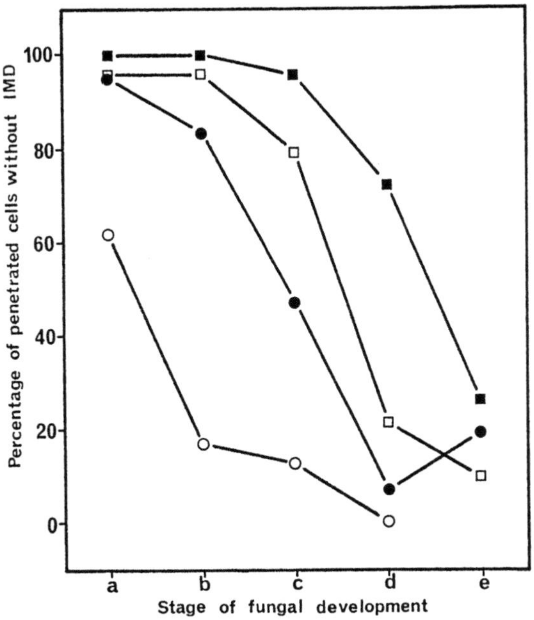

Fig. 2. Effect of BcS (50μg ml^{-1}) on the stage of fungal development at which membrane damage occurred in cvs. Valmaine (Dm5/8) and Dandie (Dm3) following inoculation with B. lactucae isolate VO/11 (A1-8). The percentage of penetrated cells in which IMD had not occurred was determined for cells containing infection structures at the following stages of development, a) primary vesicle ≤15μm in diameter, b) primary vesicle >15μm in diameter, c) secondary vesicle ≤10μm in diameter, d) secondary vesicle >10μm but ≤25μm in diameter and e) secondary vesicle >25μm in diameter with or without developing intracellular hyphae. As controls, BcS was omitted and cotyledons treated with sterile distilled water alone. Results were compiled from two experiments; ● and ○, cv. Valmaine; ■ and □ cv. Dandie; closed and open symbols, treatment with BcS or water respectively.

The heat-shock treatment selected involved placing detached cotyledons on the base of a Petri dish which was floated for 45s in a covered water bath regulated at 55°C. Treated cotyledons were allowed to cool to room temperature for at least 0.5h before inoculation. The effects of heat-shock were examined using cultivars and isolates which allowed the expression of Dm1, Dm3 and Dm5/8 to be studied in different genetic backgrounds. In all genotypes heat-shock immediately before inoculation caused IMD to occur at a later stage of fungal development. With some cultivar/isolate combinations, particularly those involving Dm5/8 and A5/8, the plant's response following heat shock was essentially the same as that observed during compatible interactions.

In Dm1:A1 and Dm3:A3 interactions, the effect of heat-shock on IMD and the subsequent HR was lost if treated tissues were allowed to recover for 17h before inoculation. By contrast, in Dm5/8:A5/8 interactions an interval of up to 3 days between heat-shock and inoculation did not reduce suppression of IMD and *B. lactucae* was able to sporulate on heat-treated tissues. Without heat-shock, growth of avirulent isolates on cultivars with Dm5/8 is typically restricted to the production of a short intra- or intercellular hypha (Woods et al 1988). The gene-specific effect of heat-shock was clearly demonstrated in cv. Diana which has Dm1, 3, 7 and 5/8. Ability to undergo the HR controlled by Dm1 and Dm3 was restored within 17h of heat-shock whereas the more rapid expression of Dm5/8 within the same tissue continued to be suppressed.

Heat-shock is known to affect protein synthesis by inhibiting the transcription and translation of mRNAs other than those newly synthesized to code for heat-shock proteins (Nagao et al 1986). Thus the suppression of IMD by heat-shock and BcS indicates that *de novo* protein synthesis by lettuce cells is probably essential for IMD and subsequent necrobiosis. This contrasts with *Phytophthora infestans*/potato interactions in which protein synthesis is not required for IMD providing inoculated tissues have acquired the ability to react hypersensitively (Doke & Tomiyama 1975). Both actinomycin D and cordycepin have been reported to inhibit mRNA synthesis in lettuce (Halmer 1989) but they did not delay the onset of IMD. Clearly, results obtained with inhibitors must be treated with caution, but assuming that they reached and were active in penetrated cells it would appear that IMD is regulated by activated translation (Woods et al 1989) and not by the direct effect of a toxic fungal metabolite. Membrane damage during the HR is thought to be caused by the highly reactive superoxide anion ($O_2^{\cdot -}$) and derived superoxides ($\cdot OH$, H_2O_2 or $^1O_2$). Superoxide generation may be initiated by the action of newly synthesized lipoxygenase (Keppler & Novacky 1987) or an O_2^- generating NADPH oxidase (Doke 1985). Inhibition of synthesis of these enzymes may explain the suppression of IMD by BcS and heat-shock.

SIGNIFICANCE OF THE GENE SPECIFIC EFFECT OF HEAT-SHOCK

The temporary effect of heat-shock on protein synthesis may cause a temporary delay in IMD as observed in Dm1:A1 and Dm3:A3 interactions. A temporary inhibition of protein synthesis cannot, however, explain the more permanent gene-specific suppression of resistance observed in Dm5/8:A5/8 interactions. The effect of heat-shock on resistance expressed by the Dm5/8 gene did not diminish until 2-3 days after treatment (Woods et al 1989). This result suggests that a structural component with a low turnover rate is affected rather than, for example, an inducible protein whose rapid synthesis would be restored within several hours of heat-shock. In studies of protein turnover, the half lives of plasma membrane proteins have been found to vary from 2-100h (Morre et al 1979). Our hypothesis is, therefore, that heat-shock modifies a receptor protein in the host plasma membrane which is specific to the Dm5/8 gene and has a low rate of turnover.

Experiments have established that the essential properties of the Dm5/8 gene-specific receptor protein are that it should be i) inactivated by the minimum heat-shock required to prevent the HR; ii) removed from the PM and replaced slowly, reaching functional levels at the time of recovery of the ability to undergo the HR and iii) be found only in cultivars with the Dm5/8 gene. By analysis of turnover of membrane proteins in different cultivars it should be possible to identify proteins fulfilling these criteria and therefore isolate the putative receptor which may be the product of the Dm5/8 gene.

ACTIVATED DEFENCE

As *B. lactucae* is an obligately biotrophic fungus, restriction of fungal growth during the HR may be by starvation due to failure to take up nutrients from host cells with a damaged plasma membrane. There is a dearth of information on the biochemistry of the responses of lettuce to infection. Lettuce tissues are rich in polyphenols (e.g. chlorogenic and isochlorogenic acids; Sharples 1964). Histochemical analyses have demonstrated the accumulation of phenolics and lignin in cells undergoing the HR to B. lactucae. These processes, associated with resistance in many other plants (Ride 1983; Ampomah & Friend 1988) may also contribute to production of an antifungal environment in lettuce. Fluorescence microscopy has revealed the localized accumulation of wall bound phenolics around penetration sites at the time of, or in some cases slightly preceding, IMD during incompatible interactions. Lignification of the walls and within the cytoplasm of the penetrated cell (as indicated by staining with phloroglucinol and persistence of some fluorescence after treatment with 0.5M NaOH for 2h at 80°C) was not detected until several hours after IMD. Regulation of the lignification response should be amenable to molecular dissection at the cellular level using reporter gene fusions (Dron et al 1988) and in situ hybridisation (Somssich et al 1988; Lamb et al 1989).

CONCLUDING REMARKS

Several aspects of Fig. 1 have now been tested by experimentation using lettuce downy mildew disease. Clearly more work is needed focussing on responses occurring in and around the penetrated cell but it is already possible to propose a model (Fig. 3) more directly applicable to interactions between Bremia and lettuce.

As outlined in Fig. 3, there is now evidence indicating candidates for molecules associated with recognition (receptor protein), control of IMD (now confirmed as an active response) and markers available for dissection of one amplified response (lignification).

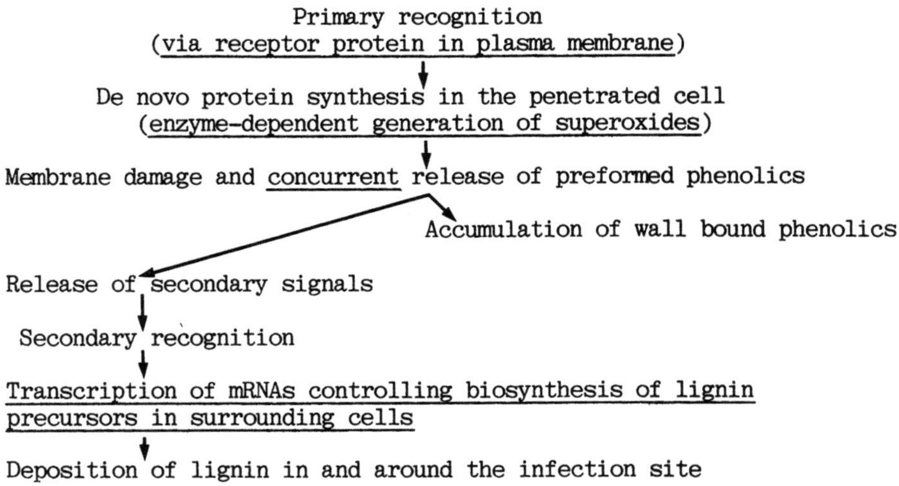

Fig. 3. Recognition, signal transduction and activated defence during the HR in lettuce to Bremia lactucae (hypothetical steps are underlined).

The well defined biology of the lettuce/Bremia infection provides a sound basis for analysis of the plant's responses during attempted colonization. Unfortunately, we are unable to report significant progress in our search for race and cultivar specific elicitors of the HR. The fungal "cognon" side of the interaction remains poorly understood (Mansfield et al 1988).

REFERENCES

Ádám A, Farkas T, Somlyai G, Hevesi M & Király Z (1989) Consequence of O_2^- generation during bacterially induced hypersensitive reaction in tobacco: deterioration of membrane lipids. Physiol & Molec Plant Pathol **34**: 13-26

Ampomah YA & Friend J (1988) Insoluble phenolic compounds and resistance of potato tuber disc to Phytophthora and Phoma. Phytochemistry **27**: 2533-2541

Dangl JL, Hanfle KD, Lipphardt S, Hahlbrock K & Scheel D (1987) Parsley protoplasts retain differential responsiveness to u.v. light and fungal elicitor. EMBO J: 2551-2556

Doke N (1985) NADPH-dependent O_2^- generation in membrane fractions isolated from wounded potato tubers inoculated with Phytophthora infestans Physiol Plant Pathol **27**: 311-322

Doke N, Nakae Y & Tomiyama K (1976) Effect of Blasticidin S on the production of rishitin in potato tuber tissue infected by an incompatible race of Phytophthora infestans. Phytopath Z **87**: 337-344

Doke N & Tomiyama K (1975) Effect of Blasticidin S on hypersensitive death of potato leaf petiole cells caused by infection with an incompatible race of Phytophthora infestans. Physiol Plant Pathol **6**: 169-175

Dron M, Clouse SD, Lawton MA, Dixon RA & Lamb CJ (1988) Glutathione and fungal elicitor regulation of a plant defense gene promoter in electroporated protoplasts. Proc Natl Acad Sci USA **85**: 6738-6742

Farrara BF, Ilott TW & Michelmore RW (1987) Genetic analysis of factors for resistance to downy mildew (Bremia lactucae) in species of lettuce (Lactuca sativa and L. serriola). Plant Pathology **36**: 499-514

Gahan PB (1984) Cell senescence and death in plants. In: Bowen ID & Lockshin RA (eds) Cell Death in Biology and Pathology. Chapman and Hall. London/New York p. 145-169

Hahlbrock K & Scheel D (1987) Biochemical responses of plants to pathogens. In: Chet I (ed) Innovative approaches to plant disease control. Wiley-Interscience, New York. p. 229-254

Halmer P (1989) De novo synthesis of mannanase by the endosperm of Lactuca sativa. Phytochemistry **28**: 371-378

Ingram DS, Sargent JA & Tommerup IC (1976) Structural aspects of infection by biotrophic fungi. In: Friend J & Threlfall DR (eds) Biochemical aspects of plant-parasite relationships. Academic Press, London. p. 43-78

Keppler LD & Novacky A (1987) The initiation of membrane lipid peroxidation during bacteria-induced hypersensitive reaction. Physiol & Molec Plant Pathol **30**: 233-345

Koga H, Zeyen RJ, Bushnell WR & Ahlstrand GC (1988) Hypersensitive cell death, autofluorescence and insoluble silicon accumulation in barley leaf epidermal cells under attack by Erysiphe graminis f.sp. hordei. Physiol & Molec Plant Pathol **32**: 395-411

Lamb CJ, Lawton MA, Dron M & Dixon RA (1989) Signals and transduction mechanisms for activation of plant defenses against microbial attack. Cell **56**: 215-224

Mansfield JW (1986) Recognition, elicitors and the hypersensitive reaction. In: Lugtenberg B (ed) Recognition in Microbe-Plant Symbiotic and Pathogenic Interactions. Springer-Verlag, Berlin. p. 434-437

Mansfield JW, Woods AM, Street PFS & Rowell PM (1988) Recognition processes in lettuce downy mildew disease. In: Chapman GP, Ainsworth CC & Chatham CJ (eds) Eukaryote Cell Recognition. Cambridge University Press, Cambridge. p. 241-256

Morré DJ, Kartenbeck J & Franke WW (1979) Membrane flow and interconversions among endomembranes. Biochim et Biophys Acta (MR) **559**: 71-152

Nagao RT, Kimper JA, Vierling E & Key JL (1986) The heat-shock response: a comparative analysis. In: Miflin BJ (ed) Oxford Surveys of Plant Molecular and Cell Biology Volume 3. Oxford University Press, Oxford. p. 384-438

Norwood JM & Crute IR (1984) The genetic control and expression of specificity in Bremia lactucae. Plant Path **33**: 385-400

Pavlovkin J, Novacky A, Ullrich-Eberius CI (1986) Membrane potential changes during bacteria-induced hypersensitive reaction. Physiol Plant Pathol **28**: 125-135

Ride JP (1983) Cell walls and other structural barriers in defence. In: Callow JA (ed) Biochemical Plant Pathology. Wiley & Sons, Chichester, p. 215-236

Sato N, Kitazawa K & Tomiyama K (1971) The role of rishitin in localizing the invading hyphae of Phytophthora infestans in infection sites at the cut surface of potato tubers. Physiol Plant Pathol **1**: 289-295

Sharples GC (1964) Polyphenol content of head lettuce. Proc Am Soc Hort Sci **84**: 356-363

Somssich IE, Schmeizer E, Kawalleck P and Hahlbrock K (1986) Rapid activation by fungal elicitor of genes encoding pathogenesis-related proteins in cultured parsley cells. Proc Natl Acad Sci USA **83**: 2427-2430

Stadelmann EJ (1966) Evaluation of turgidity, plasmolysis and deplasmolysis of plant cells. In: Prescott DM (ed) Methods of Cell Physiology. Academic Press, New York. p. 143-216

Woods AM, Fagg J & Mansfield JW (1988) Fungal development and irreversible membrane damage in cells of Lactuca sativa undergoing the hypersensitive reaction to the downy mildew fungus Bremia lactucae. Physiol & Molec Plant Pathol **32**: 483-498

Woods AM, Fagg J & Mansfield JW (1989) Effects of heat-shock and inhibitors of protein synthesis on irreversible membrane damage occurring during the hypersensitive reaction of Lactuca sativa L. to Bremia lactucae Regel. Physiol & Molec Plant Pathol **34** (in press)

BIOCHEMICAL AND MOLECULAR ASPECTS OF THE PHENYLAMIDE FUNGICIDE-RECEPTOR INTERACTION IN PLANT PATHOGENIC *PHYTOPHTHORA* SPP.

Leen C. Davidse and Grardy C.M. van den Berg-Velthuis
Wageningen Agricultural University, Department of Phytopathology,
Binnenhaven 9, 6709 PD Wageningen
The Netherlands

INTRODUCTION

Among plant pathogenic *Phytophthora* spp., *Phytophthora infestans* is worldwide known because it causes late blight of potatoes, one of the most devastating diseases of agricultural crops. Its control in most potato-growing regions depends upon the application of fungicides. In the Netherlands only, costs of late blight control are between 75 and 100 million Dutch guilders annually and amount to 10 % of production costs.

Successes of breeding for late blight resistance in potato until now are meagre, because the fungus readily overcomes major single genes for resistance and a high level of durable resistance is difficult to achieve because of its complex inheritance. The fact that old varieties that are highly susceptible to the disease are still widely grown (e.g. in the Netherlands the variety Bintje, introduced in 1910; in the USA Russett Burbank, introduced before 1890; and in the UK King Edward, introduced in 1902) illustrates on one hand the difficulties of potato breeders to come up with agronomically acceptable varieties with a high level of late blight resistance and on the other hand the success of chemical disease control. Especially the dithiocarbamates and triphenyltin compounds have been and still are the fungicides potato growers rely on for late blight control.

The increasing concern about the deterioration of the environment by the use of these compounds that have to be used in a preventive way at rates up to 4 kg per ha dictates other ways of late blight control. In the immediate future integrated crop management programmes have to be developed aiming at i) growing of varieties with a high level of durable resistance and ii) availability of highly active fungicides (50 - 150 g per ha) for curative use to back up host resistance if required. It is obvious that fundamental research is needed to achieve this. Modern ways of breeding for resistance will benefit from understanding the relation between the host and pathogen

NATO ASI Series, Vol. H 36
Signal Molecules in Plants and Plant-Microbe Interactions
Edited by B. J. J. Lugtenberg

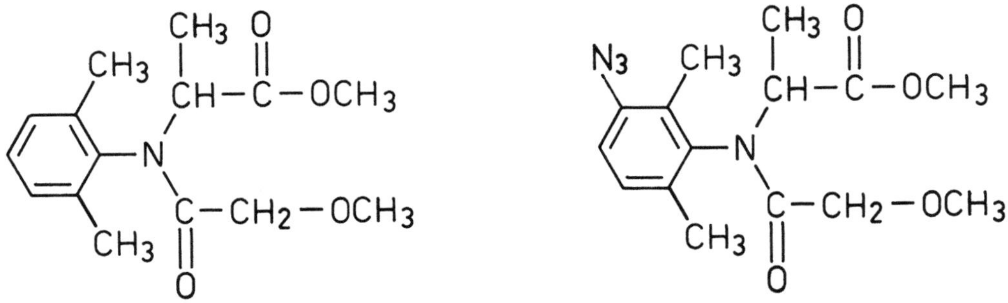

Fig.1 Structures of metalaxyl (left) and azidometalaxyl (right)

at the molecular level and fungicide design requires knowledge of potential target sites in the fungus.

In this contribution progress in the research on the molecular interaction between phenylamide fungicides and their target site in the fungal cell will be discussed. Phenylamides of which metalaxyl (Fig. 1) is the best-known compound, are highly active against late blight (100 - 150 g per ha) but development of resistance in the fungal population forbid their use as single active ingredient products in curative treatments. Knowledge of molecular interactions involved in binding of phenylamides at the receptor site and of structural changes of the receptor that result in resistance is essential to redesign phenylamide structures that might be less prone to development of resistance or to design compounds that selectively inhibit strains with resistance to the existing phenylamides. Practical experiences on combating pathogens that have become resistant to benzimidazole fungicides with the *N*-phenyl carbamates, MDPC and diethofencarb, which selectively act against benzimidazole-resistant strains of *Botrytis cinerea* (Kato 1988) indicate that this approach can be successful. Understanding this phenomenon at the molecular level, however, is necessary to evaluate the full potential of this approach.

The highly specific mechanism of action of the phenylamides on Oomycete fungi makes them also excellent tools to study the physiology of these organisms many details of which are still unknown. Methods used to characterize and identify the fungicide receptor in the fungal cell are similar to those employed in research on molecular signals involved in host-pathogen interactions. We believe that studies like this will contribute to a better understanding of the physiology of *P. infestans* which may ultimately lead to the development of highly effective and environmental safe ways to control late blight.

MECHANISM OF ACTION OF PHENYLAMIDES

The mechanism of action of the phenylamides has been most intensively studied using metalaxyl as a representative member of this group. Metalaxyl specifically inhibits incorporation of [5,6-3H]uridine into RNA by living mycelia of *P. infestans* and *Phytophthora megasperma* f.sp. *medicaginis* (*Pmm*), the causal organism of lucerne root rot (Davidse et al 1988). Incorporation of [methyl-^{3}H]thymidine into DNA and [^{14}C]phenylalanine into protein is affected to a lesser extent. Since metalaxyl at low concentrations does not affect the uptake of precursors or the conversion of nucleosides into nucleotides, inhibition of RNA synthesis must be responsible for the observed effects on uridine incorporation. At metalaxyl concentrations that are fully inhibitory to growth, a complete inhibition of uridine incorporation does not occur. In *Pmm* uridine incorporation is reduced to circa 20 % of the control value; in *P. infestans* it is reduced to circa 60 %. Apparently only part of the cellular RNA synthesis is sensitive to phenylamides, indicating a specific inhibition of one of the RNA polymerases. Uridine incorporation by phenylamide resistant strains of both *Phytophthora* spp. is not affected by metalaxyl.

In vitro studies with crude cell free extracts or partially purified RNA polymerases failed to demonstrate any inhibitory effect on RNA synthesis (Davidse et al 1983). Endogenous RNA polymerase activity of isolated nuclei, however, appeared to be sensitive (Table 1). α-Amanitin, a specific inhibitor of RNA polymerase II was also inhibitory in this assay (Table 2). The effects of metalaxyl and α-amanitin are additive, indicating interference of metalaxyl with an RNA polymerase activity different from the α-

TABLE 1. Effect of metalaxyl on endogenous RNA polymerase activity of nuclei isolated from phenylamide-sensitive (S) and phenylamide-resistant (R) strains of *Pmm* and *P. infestans*[a]

Concentration of metalaxyl in assay mixture (μg.ml-1)	Polymerase activity as % of control in nuclei from:			
	*Pmm*S	*Pmm*R	*P.inf*S	*P.inf*R
0.1	68	100	68	96
1	56	99	60	95
10	51	100	57	88

[a]For methods and origin of strains, see Davidse et al (1988)

TABLE 2. Effects of metalaxyl and α-amanitin on endogenous RNA polymerase activity of nuclei isolated from a phenylamide-sensitive strain of *Pmm*

Inhibitor(s) added	Concentration ($\mu g.ml^{-1}$)	Activity as % of control
metalaxyl	10.0	64
α-amanitin	0.1	76
α-amanitin	5.0	72
α-amanitin	100.0	71
α-amanitin	250.0	78
metalaxyl + α-amanitin	10.0 + 5.0	38
metalaxyl + α-amanitin	10.0 + 100.0	32
metalaxyl + α-amanitin	10.0 + 250.0	42

amanitin sensitive polymerase II. Analysis of radiolabeled RNA formed in vivo during incubation with metalaxyl showed that synthesis of poly(A)denylated RNA that represents most of the m(essenger)RNA was less sensitive than that of total RNA, the majority of which is r(ibosomal)RNA. Fractionation of in vivo labeled total RNA by agarose gel electrophoresis showed that in the presence of metalaxyl both high molecular weight rRNA species are not labeled. These results indicate that metalaxyl selectively inhibits rRNA synthesis by interference with the activity of the RNA polymerase I-template complex. In resistance strains the target site has been changed resulting in insensitivity.

CHARACTERIZATION OF THE PHENYLAMIDE-RECEPTOR IN THE RNA POLYMERASE I-TEMPLATE COMPLEX

To substantiate the idea that a target site change is responsible for phenylamide resistance, binding studies were performed with [^{3}H]metalaxyl and cell-free mycelial extracts of sensitive and resistance strains. Binding was determined with a Sephadex G25 gel filtration assay. Specific binding of metalaxyl occurred only in extracts of sensitive strains (Davidse 1989). As determined from Scatchard plots K_d-values were 0.17 and 0.07 μM and binding capacities were 5 and 3 pmole per mg crude cellular protein for *Pmm* and *P.inf*, respectively. Other phenylamides competively inhibited [^{3}H]metalxyl binding indicating that the mechanism of action of the different phenylamides is similar. Although binding activity could be easily demonstrated in crude cell free mycelial extracts of phenylamide-sensitive *Phytophthora* strains, binding activity decreased considerably when the extracts were subjected to Heparin Sepharose affinity chromatography, a first step in the purification of RNA polymerases (Hammond and Holland 1983). Purification of the phenylamide receptor, therefore, cannot be followed using the binding assay. An alternative approach would be to covalently label the receptor with a photoaffinity probe. Purification of the labeled complex can then be followed either by a radioassay or an immunoassay, which respectively require a radiolabeled photoaffinity probe or antiserum that selectively recognizes covalently bound phenylamide residues.

A suitable photoaffinity probe would be 3-azidometalaxyl (Figure 1), since substituents like amino, nitro and chloro at the 3-and 5-position do not cause a drastic change in biological activity of the compound (Hubele et al 1983). This compound, therefore, was synthezised and its biological properties were determined. Azidometalaxyl indeed inhibited mycelial growth of the phenylamide-sensitive strain of *P. megasperma* f. sp. *medicaginis* (EC_{50}: 4 μg/ml), whereas the compound was almost inactive against growth of the phenylamide-resistant strain. Like metalaxyl, azidometalaxyl inhibited endogenous RNA polymerase activity from isolated nuclei, although at least a ten-fold higher concentration was required to achieve a similar level of inhibition (Table 3). Azidometalaxyl bound to the phenylamide receptor, as indicated by its ability to compete with [^{3}H]metalaxyl. Hence azidometalaxyl is a suitable photoaffinity probe for the receptor.

Photoaffinity labeling experiments up till now failed to demonstrate specifically labeled protein bands in a Western blot using rabbit antibodies raised again a formaldehyde-conjugated 3-aminometalaxyl-bovine serum albumin complex and goat antirabbit-IgG conjugated alkaline phosphatase as second antibody to detect phenylamide bound residues. Similar labeling patterns were observed using mycelium of *Pmm*S and *Pmm*R as source of the

TABLE 3. Effect of metalaxyl and azidometalaxyl on endogenous RNA polymerase activity of nuclei isolated from a phenylamide-sensitive strain of *Pmm*

Inhibitor	Polymerase activity as % of control in the presence of the inhibitor at a concentration (μg ml^{-1}) of:		
	0.1	1.0	10.0
metalaxyl	63	53	52
azidometalaxyl	92	67	57

crude protein extract indicating aspecific labeling. Preincubation of the extracts with or without metalaxyl prior to incubation with azidometalaxyl and photolabelling and including dithiothreitol as a suppressor of aspecific labeling in the extracts, also did not result in differentially labeled protein bands. Fractionation of labeled extracts that were either preincubated with or without metalaxyl on Heparin Sepharose to enrich the nucleic acid bound RNA polymerases also did not result in a differential labeling pattern of the enriched fraction.

CONCLUDING REMARKS

Identification of the phenylamide receptor in the RNA polymerase I-template complex proves to be difficult. Photoaffinity labeling, a traditional biochemical approach to characterize receptors and especially those that are labile during purification, has not been successful until now. A molecular biological approach by cloning the phenylamide resistance gene cannot be followed yet, because a tranformation system for *Phytophthora* fungi is still lacking and the gene cannot be selected for in established transformation systems because the fungi used in these systems are naturally resistant to phenylamides.

REFERENCES

Davidse LC (1989) Biochemical basis of resistance to phenylamide fungicides. In: Green MB, Moberg WK, LeBaron HM (eds) Fundamental and practical approaches to combating resistance. ACS Books, Washington, in press

Davidse LC, Hofman AE, Velthuis GCM (1983) Specific interference of metalaxyl with endogenous RNA polymerase activity in isolated nuclei from *Phytophthora megasperma* f.sp. *medicaginis*. Exp. Mycol. 7:344-361

Davidse LC, Gerritsma OCM, Ideler J, Pie K, Velthuis GCM (1988) Antifungal modes of action of metalaxyl, cyprofuram, benalaxyl and oxadixyl in phenylamide-sensitive and phenylamide-resistant strains of *Phytophthora megasperma* f.sp. *medicaginis* and *Phytophthora infestans*. Crop Protection 7:347-355

Hammond CI, Holland MJ (1983) Purification of yeast RNA polymerases using Heparin agarose affinity chromatography. J. Biol. Chem. 258:3230-3241

Hubele A, Kunz W, Eckhardt W, Sturm E (1983) The fungicidal activity of acylalanines. In: Doyle P, Fujita T (eds) Pesticide chemistry, human welfare and the environment, vol 1: Synthesis and structure-activity relationships. Pergamon Press, Oxford, p 233

Kato T (1988) Negative cross resistance activity of MDPC and diethofencarb against benzimidazole-resistant fungi. In: Delp CJ (ed) Fungicide resistance in North America. APS Press, St Paul, p 40

MECHANISMS IN SYSTEMIC INDUCED DISEASE RESISTANCE

Andres Binder, Gianni Baer, Christina Hofmann and Kati Kovấts
Maag-Roche Ltd.
CH-8157 Dielsdorf
Switzerland

INTRODUCTION

Systemic induced disease resistance in whole plants has been observed with a number of plant pathogen systems (for a review see Kuc 1987). The best resistance effects against diseases caused by viruses, bacteria and fungi have been achieved by preinfection with the same or different microorganisms. To explain these effects, systemic signals (chemical and/or electrical) transported over long distances in the plant have to be postulated. Until now evidence for such signals originate mainly from observations of enhanced resistance against pathogens on leaves distant from the induction site.

This paper describes changes in biochemical-physiological events, resulting from long distance systemic signals in cucumber and tomato, induced with a preinfection: (i) Systemic induction of enzymes represent a measure of the induced state of the plant. (ii) Local cell reactions upon challenge infection visualize the practical consequences of the systemic induced state - a measure of enhanced defense. All these events are compared with the observed induced resistance of the plants against the pathogens.

MATERIALS AND METHODS

Plant pathogen systems

Young cucumber plants were induced on the first leaf by infection with a spore suspension of Colletotrichum lagenarium. After 7-12 days the second leaf was either used for enzyme measurements or for challenge infection, applied with the same pathogen as the inducer infection. This challenged

NATO ASI Series, Vol. H 36
Signal Molecules in Plants and Plant-Microbe Interactions
Edited by B.J.J. Lugtenberg

leaf was either probed for defense mechanisms or taxed after 7 days for induced resistance against the pathogen (0% = full infection of the control plants, 100% = no disease symptoms). Similarly, tomato plants were induced on the first 4 leaves by infection with a zoospore suspension of Phytophthora infestans. Enzyme experiments, challenge infection, induced defense experiments and taxation for induced resistance were done with the fifth and sixth leaf as described above for cucumbers.

Enzyme activities

Leaf material was homogenized at pH 2.8 using a pestle and mortar. The extract was centrifuged twice and the supernatant assayed for the following enzyme activities, immediately or after storage at -20°C:
Chitinase according to Mauch et al. (1984) using tritiated chitin (activities in equiv. (mg) Streptomyces griseus chitinase per g fresh weight). Glucanase also according to Mauch et al (1984) using reduced laminarin (activities in umol Glc equiv. per g fresh weight per 30 min). Peroxidase according to Goldberg et al (1983) using guaiacol as substrate (activities in OD units at 450 nm per g fresh weight per min). Protease according to Vera and Conejero (1988) using azocasein as substrate (activities in equiv.(ug) Streptomyces griseus pronase per g fresh weight per 30 min). Protein concentration was determined according to Bradford (1976) using BSA as standard. Results are from one experiment with 3 replicates (5 plants per rep). Each experiment was repeated at least 2 times.

Plant defense reactions and fungal growth

Infection sites of challenge infections were stamped out, fixed in chloroform-ethanol-trichloracetic acid and stored in 50% ethanol. The discs were examined for plant defense reactions and fungal structures using the following methods:
Lignin was stained (i) with phloroglucinol-HCl according to Conti et al (1986), (ii) with toluidine blue-O according to O'Brien et al (1965) and (iii) by the yellow autofluorescence according to Beardmore et al (1983). Callose was measured after staining with the fluorescent anilin blue according to Eschrich and Currier (1964). Fungal structures were visualized by staining with the fluorescent diethanol (Uvitex 2B) according to Kuck et al (1981). Observations, image analysis and correlations of plant reactions and fungal growth were done with a Zeiss microscope equipped with epifluorescence. All results are means of 3 independent experiments.

RESULTS AND DISCUSSIONS

Induced state: enhanced enzyme activities as systemic signal perception

In cucumber plants, the systemic enhancement of chitinase, glucanase and peroxidase was significant, whereas protease activity was not induced (Fig. 1). On the other hand, at the site of induction (local), all four activities were induced (results not shown). In tomato, a similar systemic enhancement of chitinase and glucanase was observed, yet in this case protease was also induced systemically (Fig. 2). Without obvious reasons, peroxidase activity in tomato turned out to be very variable in different experiments. As in cucumbers, all four activities were enhanced locally in tomato (data not shown). The observed systemic induced resistance against the pathogens amounted to 70 - 90% in both species.

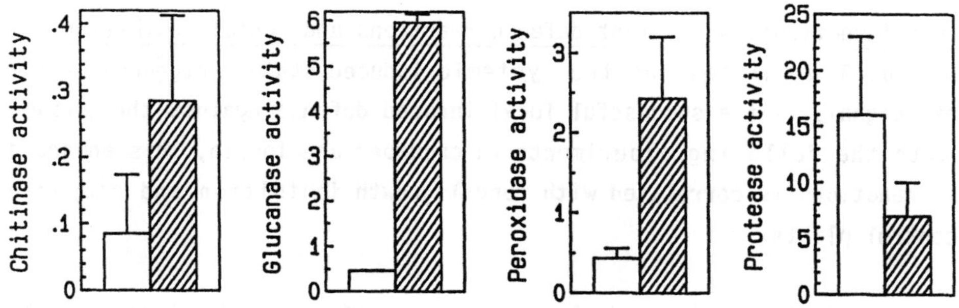

Fig. 1. Systemic induction of enzyme activities in cucumber. Experimental details, including activity units are given in Materials and Methods. □: control, ▨: induced. Induced disease resistance after challenge infection with <u>C:lagenarium</u> was <u>77%</u>.

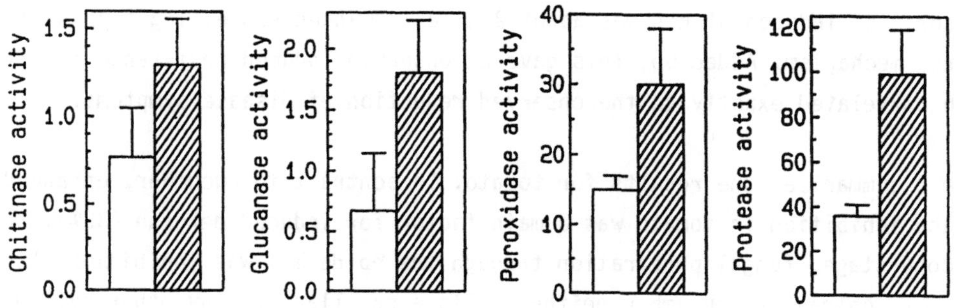

Fig. 2. Systemic induction of enzyme activities in tomato. Experimental details, including activity units are given in Materials and Methods. □: control, ▨: induced. Induced disease resistance after challenge infection with <u>P.infestans</u> was <u>85%</u>.

From these results the following conclusions can be drawn: (i) Systemic increase of selected enzyme activities such as chitinase and glucanase coincides with systemic induction of disease resistance to the pathogens in cucumber as well as in tomato. (ii) Other systemic induced activities are more specific for certain plants, such as protease in tomato. (iii) The pattern of induced enzyme activities may be different in local versus systemic leaves (e.g. protease in cucumber). (iv) The variability of the peroxidase activity shows that this is a questionable marker for the systemic induced resistance state, since the activity is probably also enhanced by other, uncontrollable parameters (e.g. stress).

Generally it can be postulated that the systemic signal could pre-activate the regulation of potential defense genes, measurable as enhanced activities of lytic enzymes.

Challenge infection: local plant defense reactions and fungal inhibition

The real biological test of the systemic induced state, triggered by the systemic signal, is the successful local induced defense against the pathogen. With the following experiments in cucumber and tomato, this enhanced defense reactions is correlated with fungal growth inhibition and compared with control plants.

In cucumbers (Table 1), induced plants showed a 13% enhanced defense on the leaf surface i.e. appressoria formation of the fungus was slightly more inhibited in induced plants. Yet, by far the most prominent enhanced defense mechanism was the inhibition of penetration hyphae through the epidermis. Correlated with this effect, fluorescent stained callose papillae under appressoria of induced plants was about 2 times enhanced suggesting a possible defense mechanism. Added up, this gave an observed induced resistance of 95% which correlated exactly to the observed reduction of disease symptoms.

Table 2 summarizes the results for tomato. In contrast to cucumber, enhanced surface inhibition in tomato was a main factor for induced defense (32%). In a second stage, fungal penetration through the epidermis was inhibited 44%. Yet, in contrast to cucumber neither callose papillae nor any other cytological phenomena could be observed. Hence, another unknown defense mechanisms has to be postulated. In a third stage fungal growth was also inhibited in the mesophyll (46%). Correlated with this effect, hypersensitive reaction in mesophyll cells of induced plants (autofluorescence of lignin) was about 3 times enhanced suggesting a possible defense mechanism. In cucumber as well as in tomato all the described defense mechanisms could indeed be observed in systemic induced as well as in control plants, yet in the latter these activities were obviously insufficient to prevent disease symptoms. A detailed publication will appear soon (Kovats, to be published).

Table 1. Cucumber - C.lagenarium: Local events after challenge infection in control and in systemic induced plants. The results shown are from 1 infection drop (500 spores). More details are given in Materials and Methods

Stage of plant defense	Fungal development of 500 original spores control	induced	Plant resistance enhancement in % of control*
Surface: Inhibition of appressoria formation, after 14 hrs.	321	279	13
Epidermis: Inhibition of fungal penetration, after 72 hrs.	70	3	95

*) reduction of fungal development compared to control

Table 2. Tomato - P.infestans: Local events after challenge infection in control and in systemic induced plants. The results shown are from 1 infection drop (300 spores). More details are given in Materials and Methods

Stage of plant defense	Fungal development of 300 original spores control	induced	Plant resistance enhancement in % of control*
Surface: Inhibition of spore germination, after 14 hrs.	63	43	32
Epidermis: Inhibition of fungal penetration, after 14-48 hrs.	11	4	62 (44)
Mesophyll: Inhibition of hyphal growth, after 48 hrs.	9	2	79 (46)

*) reduction of fungal development compared to control (in parathesis: resistance per stage)

From these experiments the following points can be postulated:
(i) Upon pathogen infection, plants in the state of systemic induced resistance (triggered by systemic signals) show the same defense reactions as the control plants, but stronger and/or faster, i.e. the induced state could represent a more efficient regulated state of already existing defense genes. (ii) Successful defense consists of several mechanisms (toxins, callose, lignin like materials etc.) at various levels (surface, epidermis, mesophyll etc.) whereby the crucial steps are different from species to species. Yet, one has to bear in mind, that only a few of the possible induced defense mechanisms could be investigated here. (iii) It is sensible to draw such conclusions but the ultimate proof whether these events are indeed primary causes for the observed induced disease resistance or only secondary symptoms of fungal infection in the plant, cannot be given with these experiments. Hence, further work - also using molecular biological techniques - has to be carried out and is currently under way.

REFERENCES

Beardmore J, Ride JP, Granger, JW (1983) Cellular lignification as a factor in the HR of wheat to stem rust. Physiol Plant Pathology 22:209-220

Bradford M (1976) A rapid and sensitive method for the quantitation of protein. Anal Biochem 72:248-254

Conti GC, Bassi M, Maffi D, Bocci AM (1986) Host parasite relationship in a susceptible and a resistant rose cultivar inoculated with Spherotheca pannosa. J Phytopathology 117:312-320

Eschrich W, Currier HB (1964) Identification of callose by its diachrome and fluorochrome reactions. Stain Technology 39:303-307

Goldberg R, Cateson AM, Czaninski Y (1983) Some properties of syringaldazine oxidase, a peroxidase specifically involved in the lignification processes. Z Pflanzenphysiol 110:267-279

Kovats K (to be published) Comparison of systemic induced resistance of tomato and cucumber. Dissertation, University of Zurich

Kuc J (1987) Plant immunization and its applicability for disease control. In: Chet I (ed) Innovative approaches to plant disease control. Wiley & Sons, New York, p 255-274

Kuck KH, Tiburzy R, Hänssler G, Reisener HJ (1981) Vizualization of rust haustoria in wheat leaves by using fluorochromes. Physiol Plant Pathology 19:439-441

Mauch F, Hadwiger LE, Boller T (1984) Ethylene: Symptom, not signal for the induction of chitinase and b-1,3-glucanase in pea pods by pathogens and elicitors. Plant Physiol 76:607-611

Vera P, Conejero V (1988) Pathogen-related proteins in tomato. Plant Physiol 87:58-63

APOPLASTIC PROTEINS INVOLVED IN COMMUNICATION BETWEEN TOMATO AND THE FUNGAL PATHOGEN CLADOSPORIUM FULVUM

Pierre J.G.M. De Wit, Guido F.J.M. van den Ackerveken, Matthieu H.A.J. Joosten and Jan A.L. van Kan
Wageningen Agricultural University
Department of Phytopathology
P.O.Box 8025 3700 EE Wageningen
The Netherlands

INTRODUCTION

<u>Cladosporium</u> <u>fulvum</u> is a biotrophic fungal parasite which colonizes the intercellular spaces between the mesophyll cells of tomato leaves and is confined to the apoplast during the main part of the disease cycle (De Wit 1977; Lazarovits and Higgins 1976). The apoplastic fluid of infected leaves contains proteins which are constitutively produced by the plant, plant proteins specifically induced by the fungus and fungal proteins constitutively produced or specifically induced by the plant. Several of these induced plant proteins belong to the class of pathogenesis related (PR) proteins, some of which appear to have 1,3-β-glucanase and chitinase activity. The latter proteins have antifungal activity. One class of fungal proteins may be important for obtaining or establishing basic compatibility, while another class of proteins plays a role in race-or cultivar-specific resistance. These latter proteins are also known as race-specific elicitors.

ISOLATION,PURIFICATION AND CHARACTERIZATION OF PROTEINS OCCURRING IN THE APOPLAST OF C. FULVUM INFECTED TOMATO

Apoplastic fluids from tomato leaves can easily be obtained. Leaves are infiltrated with water <u>in</u> <u>vacuo</u>. For this purpose a beaker containing <u>C. fulvum</u>-infected tomato leaves was filled with water (a perforated ceramic lid was placed in the beaker to prevent the leaves from floating) and placed in a desiccator connected to a suction pump. After 15 min under reduced pressure, at room temperature, air was readmitted which caused the intercellular spaces to be filled with water. The infiltrated leaves were surface-dried, placed into a centrifuge tube specifically designed for this purpose (De Wit and Spikman 1982) and the apoplastic fluid was obtained by

NATO ASI Series, Vol. H 36
Signal Molecules in Plants and Plant-Microbe Interactions
Edited by B. J. J. Lugtenberg

centrifugation for 15 min at 3000g. The proteins occurring in the apoplastic fluids were purified and analysed by a combination of column chromatography (sieving, cationic, anionic, chromatofocusing), high pressure liquid chromatography (HPLC), fast protein liquid chromatography (FPLC), low pH non-denaturing polyacrylamide gel electrophoresis (PAGE), high pH non-denaturing PAGE and SDS-PAGE.

For the compatible-specific proteins no bioassay is yet available. Race-specific elicitor activity was determined by assaying the necrosis-inducing activity of apoplastic fluids. Apoplastic fluids and fractions originating from different compatible interactions between C. fulvum and tomato were injected into the intercellular space of different cultivars of tomato carrying different genes for resistance to the fungus (De Wit and Spikman 1982). The PR proteins were assayed for 1,3-β-glucanase and chitinase activity as described by Joosten and De Wit (1989).

FUNGAL PROTEINS SPECIFIC FOR THE COMPATIBLE INTERACTION

In apoplastic fluids of C. fulvum-tomato interactions a number of proteins accumulates after infection which are specific for the compatible interaction (De Wit et al 1986; Joosten and De Wit 1988). The first member of this class of proteins has been purified to homogeneity by cation-exchange chromatography on CM-Sephadex C25, followed by chromatofocusing. The molecular weight of this protein was estimated at 14 kD. Polyclonal antibodies raised against this protein were used to study its origin and the time course of its accumulation. The protein could be detected in apoplastic fluids by Western blotting. Blots containing proteins from C. fulvum grown in vitro, healthy plants, compatible and incompatible interactions between C. fulvum and tomato, were probed with antibodies raised against the 14 kD protein. These blots showed that the protein was specific for compatible interactions between C. fulvum and tomato. It could neither be detected in compatible Phytophthora infestans-tomato interactions nor in the mycelium of different races of C. fulvum grown in vitro, indicating that this protein is specifically induced when the fungus is growing in the tomato plant. No serological relationship with other fungal or plant proteins could be detected on Western blots. In time course experiments it was found that this compatible interaction-specific protein could be detected 6 days after inoculation of tomato plants with a virulent race of C. fulvum. From these experiments it still could not be concluded whether the protein is of plant or fungal origin, as it could only be detected in the interaction. Preliminary immunocytological studies, however, indicated that the protein is most probably of fungal origin. The biological function of this protein is not known yet. It might play a role in obtaining or maintaining basic compatibility.

The sequence of the N-terminal 36 amino acids of the protein has been determined. A partly degenerated synthetic DNA probe derived from the amino acid sequence of the N-terminus was used to screen genomic DNA of plant and fungus to check the origin of the encoding gene. Until now this synthetic DNA probe has not detected any positive signal on Southern blots with DNA of the fungus or the host. Also on Northern blots containing poly(A) RNA from the plant or fungus no clear positive signal could be detected. However, on Northern blots containing poly(A) RNA from compatible interactions between C. fulvum and tomato a mRNA of approximately 700 nucleotides could be detected. This mRNA was found only in poly(A) RNA of compatible but not in poly(A) RNA of incompatible interactions. Apparently the gene is only expressed in compatible interactions. The mRNA encoding the 14 kD protein could be detected between 4 and 6 days after inoculation with virulent races of C. fulvum. Primer extension studies with the synthetic oligonucleotide as a primer revealed an extension product of ca. 250 base pairs leaving a coding capacity for 30-40 amino acids in front of the N-terminus of the mature protein. Experiments are in progress to clone the encoding gene in order to confirm its origin and to study its regulation and function. Currently a cDNA library derived from poly(A) RNA from a compatible interaction is being screened for the presence of the cDNA for this protein. The cDNA library has been prepared in the expression vector lgt11 to enable us to screen with antiserum raised against the 14 kD protein or with the synthetic oligonucleotide. The 14 kD protein is not the only protein which is specific for compatible C. fulvum-tomato interactions. There are a number of other proteins between 14 and 20 kD which are specific for compatible interactions (see Table 1). These are now being purified by FPLC. Once purified these proteins will be used to raise antibodies and their amino acid sequence will be determined prior to cloning.

In the interaction between C. fulvum and tomato there is also evidence for the existence of suppressors (Peever and Higgins 1989). They counteract the activity of glycoproteinaceous non-specific elicitors (De Wit and Kodde 1981). Intercellular fluids of all compatible interactions contained suppressor activity. When co-injected with non-specific elicitors necrosis induction and callose deposition, but not electrolyte leakage were sup pressed. This suppressor, however, is most probably of host origin as apoplastic fluids from uninoculated plants also contained suppressor activity. The suppressors showed no race-cultivar specificity and their activity might be due to enzymatic breakdown of the non-specific elicitors. These suppressors would thus have a very general activity.

RACE-SPECIFIC PROTEINS AS ELICITORS OF DEFENSE RESPONSES

The intercellular fluids isolated from different compatible C. fulvum-tomato interactions contain race-specific proteinaceous elicitors (De Wit and Spikman 1982). One of these elicitors, the putative product of avirulence gene A9 which specifically interacts with a product of resistance gene Cf9 has been purified and its amino acid sequence has been determined (Scholtens-Toma and De Wit 1988). It was isolated from apoplastic fluids of compatible interactions with races carrying the avirulence gene A9. The elicitor was not detectable in compatible interactions involving race 2.4.5.9, which does not carry avirulence gene A9. Recently we have found that new Dutch, French and Polish isolates of C. fulvum carrying virulence gene a9 (race 2.5.9, 2.4.9.11 and 2.4.5.9.11, respectively) also did not produce the necrosis-inducing peptide indicating that presence of this protein and the presence of avirulence gene A9 are very strongly associated (Scholtens-Toma et al 1989). The peptide was not detected in culture filtrates of the fungus grown in vitro. The peptide could also not be detected in incompatible interactions, presumably because it is only present in very low amounts and it may bind to receptor sites as soon as it is produced. The peptide contains 27 amino acids, 6 of which are cysteine residues. The peptide must be processed from a precursor protein, but it is not known when and where this occurs during the infection cycle. The purified peptide causes severe necrosis when injected into the leaves of tomato cultivars carrying the Cf9 resistance gene. The necrosis-inducing activity of the elicitor was destroyed after reduction with β-mercaptoethanol or dithiothreitol. Surprizingly the chemically synthesized peptide was not biologically active. Several attempts to oxidize the chemically synthesized peptide under different conditions in order to obtain an active molecule have failed thusfar.

A partly degenerated synthetic oligonucleotide probe derived from the amino acid sequence has been used to screen Southern blots containing DNA from the fungus and the plant. On these Southern blots no clear positive signals were obtained with the degenerated synthetic oligonucleotide. Hybridization was observed on Northern blots containing poly(A) RNA from compatible interactions involving races carrying avirulence gene A9. These gave clear positive signals when probed with the synthetic oligonucleotide. No signal was obtained from blots containing poly(A) RNA from compatible interactions involving races lacking avirulence gene A9. Also no positive signals were obtained on Northern blots containing poly(A) RNA from healthy tomato leaves nor from the fungus grown in vitro. These results clearly indicate that the encoding gene is only expressed in vivo when the fungus colonizes the tomato leaves. It also supports our data obtained at the protein level, where the necrosis-inducing elicitor was only detected in vivo but not in vitro. In a time course experiment the mRNA encoding the necrosis-inducing elicitor could be detected in poly(A) RNA from a compatible interaction at 6 days after inoculation. The size of the mRNA was assessed to be around 700 base pairs. Primer extension studies with the synthetic oligonucleotide revealed that the extension product was around 280 bases long, leaving a coding capacity for approximately 75 codons upstream of the N-terminus of the mature peptide. Presently we are screening a cDNA library of poly(A) RNA isolated from compatible interactions involving races carrying avirulence gene A9. From this library we expect to isolate a clone which can be used to probe Southern blots of DNA from the fungus and the host, which will enable us to identify the origin of the encoding gene.

FUNGUS-MEDIATED ACCUMULATION OF PATHOGENESIS RELATED(PR) PROTEINS

The incompatible interaction between C. fulvum and tomato is characterized by a typical hypersensitive response (HR) observed as the formation cell wall appositions and necrosis. The accumulation of PR proteins is one of the early indicators of incompatibility. In tomato a number of different PR proteins accumulates after inoculation with C. fulvum (De Wit and Van der Meer 1986; De Wit et al 1986; Joosten and De Wit 1989). Two of the predominant PR proteins, a 1,3-β-glucanase (35 kD, pI ca.6.1) and a chitinase (26 kD, pI ca. 6.4), were purified by ion exchange chromatography and chromatofocusing. By using antibodies raised against the two enzymes it was found that apoplastic fluids and leaf homogenates of the tomato-C. fulvum interaction contain one additional 1,3-β-glucanase (33 kD, pI > 8) and three additional chitinases (27 kD; 30 kD, pI>8; 32 kD, pI>8). The acidic 1,3-β-glucanase and chitinase are predominantly present in the apoplast, while the basic forms of both enzymes predominantly occur intracellularly. The accumulation of both enzymes occurs earlier in the incompatible interaction than in the compatible one (Joosten and De Wit 1988).

Another class of PR proteins, the P14-like proteins, also accumulates differentially in the compatible and incompatible interactions (De Wit and Van der Meer 1986). Antibodies raised against P14 detected at least one other P14-like protein with a different iso-electric point (Joosten and De Wit unpublished). These two P14-like PR proteins are very homologous to the PR 1a, PR 1b and PR 1c of tobacco (Nassuth and Sänger 1986) as they cross-reacted with the P14 antibodies and vice versa. Most of the apoplastic PR proteins in tomato leaves which are separated by low pH non-denaturing PAGE (De Wit et al 1986) have now been purified and partly identified. Antibodies raised against these proteins will be used for immunocytological localization of these proteins *in planta*.

The PR proteins of tomato with 1,3-β-glucanase and chitinase activity are presently being tested for antifungal activity *in vitro*. Protein preparations from apoplastic fluids and leaf homogenates of infected tomato were tested against a number of different plantpathogenic fungi grown *in vitro*. Preliminary results indicate that the proteins occurring in total leaf homogenates of infected plants have higher antifungal activities than the proteins present in intercellular fluids.

SUMMARY

All the proteins isolated, purified and (partly) characterized from apoplastic fluids of *C.* fulvum-infected tomato leaves are summarized in Table 1. Some of the non-race-specific fungal proteins are most probably involved in eshtablishing basic compatibility or in maintaining the biotrophic phase. Race-specific proteins are believed to play a role in the induction of HR in cultivars carrying the different genes for resistance. Until now however we have only been able to isolate these proteins from compatible *C. fulvum*-tomato interactions (where the fungus successfully colonizes the tomato leaves, which do not respond with HR to the race-specific elicitor).
The nine PR proteins present in the apoplast of *C. fulvum* infected tomato leaves are highly related to those occurring in TMV-infected tobacco (Fritig et al, this volume).

Table 1. Proteins occurring in apoplastic fluids from C. fulvum infected tomato leaves.

CLASS OF PROTEINS	MW	PROPERTY/FUNCTION	REFERENCE
Compatible interaction-specific proteins	14 kD	specific for compatible C. fulvum-tomato interactions (non-race-specific)	Joosten and De Wit 1989
	15 kD	,,	Unpublished
	17 kD	,,	Unpublished
	19 kD	,,	Unpublished
Race-specific proteins	3 kD	induces necrosis in tomato cultivars carrying resistance gene Cf9	Scholtens-Toma and De Wit 1988
	5-7 kD	induces chlorosis in tomato cultivars carrying resistance gene Cf4	Unpublished
	4-6 kD	induces chlorosis in tomato cultivars carrying resistance gene Cf5	Unpublished
Pathogenesis related (PR) proteins	14 kD (two)	no known enzyme activity; related to PR 1a, PR 1b, and PR 1c of tobacco	De Wit and Van Der Meer 1986; De Wit et al 1986; Unpublished
	14 kD	no serological relationship with the previous two	Unpublished
	26 kD	acidic chitinase	Joosten and De Wit 1989
	27 kD	chitinase	Unpublished
	30 kD	basic chitinase	Unpublished
	32 kD	basic chitinase	Unpublished
	33 kD	basic glucanase	Joosten and De Wit 1989
	35 kD	acidic glucanase	,,

REFERENCES

De Wit PJGM (1977) A light and scanning-electron miroscopic study of infection of tomato plants by virulent and avirulent races of Cladosporium fulvum. Neth J Plant Path 83: 109-122

De Wit PJGM and Kodde E (1981) Further characterization and cultivar specificity of glycoprotein elicitors from culture filtrates and cell walls of Cladosporium fulvum (syn. Fulvia fulva). Physiol Plant Pathol 18: 298-314

De Wit PJGM and Spikman G (1982) Evidence for the occurrence of race and cultivar-specific elicitors of necrosis in intercellular fluids of compatible interactions of Cladosporium fulvum and tomato. Physiol Plant Pathol 21: 1-11

De Wit PJGM, Buurlage MB and Hammond KE (1986) The occurrence of host, pathogen and interaction-specific proteins in the apoplast of Cladosporium fulvum (syn. Fulvia fulva) infected tomato leaves. Physiol Mol Plant Pathol 29: 159-172

De Wit PJGM and Van der Meer FE (1986) Accumulation of the pathogenesis-related tomato leaf protein P14 as an early indicator of incompatibility in the interaction between Cladosporium fulvum (syn. Fulvia fulva) and tomato. Physiol Mol Plant Pathol 28: 203-214

Fritig B, Rouster J, Kauffmann S, Stinzi A, Geoffroy P, Kopp M and Legrand M (1989) Virus-induced glycanhydrolases and effects of oligosaccharide signals of plant-virus interactions. In: Molecular signals in microbe-plant symbiotic and pathogenic systems (B Lugtenberg ed) Springer Berlin Heidelberg New York In press

Joosten MHAJ and De Wit PJGM (1988) Isolation, purification and preliminary characterization of a protein specific for compatible Cladosporium fulvum (syn. Fulvia fulva) -tomato interactions. Physiol Mol Plant Pathol 33: 142-253

Joosten MHAJ and De Wit PJGM (1989) Identification of several pathogenesis-related proteins in tomato leaves inoculated with Cladosporium fulvum (syn.Fulvia fulva) as 1,3-β-glucanases and chitinases. Plant Physiol 59: 945-951

Lazarovits G and Higgins VJ (1976) Ultrastructure of susceptible, resistant and immune reactions of tomato to races of Cladosporium fulvum. Can J Bot 54: 255-249

Nassuth A and Sänger HL (1986) Immunological relationship between "pathogenesis-related" leaf proteins from tomato, tobacco and cowpea. Vir Res 4: 229-242

Peever TL and Higgins VJ (1989) Suppression of the activity of non-specific elicitor from Cladosporium fulvum by intercellular fluids from tomato leaves. Physiol Mol Plant Pathol In press

Scholtens-Toma IMJ and De Wit PJGM (1988) Purification and primary structure of a necrosis inducing peptide from apoplastic fluids of tomato infected with Cladosporium fulvum (syn. Fulvia fulva). Physiol Mol Plant Pathol 33: 59-67

Scholtens-Toma IMJ, De Wit GJM and De Wit PJGM (1989) Characterization of apoplastic fluids isolated from tomato lines inoculated with new races of Cladosporium fulvum. Neth J Plant Pathol 95:161-168

logie,
stanz
public of Germany

ON

ngi have a complex life cycle (see Littlefield 1981) e different spore forms. The basidiospores, aecio- uredospores develop highly specialized infection in order to penetrate into the leaf. The germ tube aryotic uredospore follows the array of wax crystals icle in a typical way and grows towards a stoma (Fig the tip of the germ tube contacts the lip of the sto- ing (see Hoch and Staples 1987) appressorium forma- s. From the lower surface of the appressorium, an in- g develops and penetrates into the substomatal cavi- leaf where the substomatal vesicle with the infection elops. After close contact with the parenchymatous , a haustorial mother cell develops which, in turn, ates a haustorium in the host cell (Staples et al 1985).

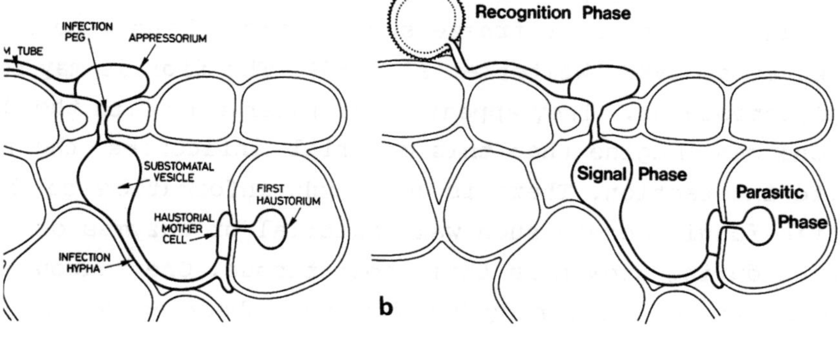

ction structures of the dikaryotic uredospore of most rust fungi.
norphological characterization
functional characterization (Mendgen 1988).

NATO ASI Series, Vol. H 36
Signal Molecules in Plants and Plant-Microbe Interactions
Edited by B. J. J. Lugtenberg

HOST RECOGNITION

This sequence of infection structures is the product of more or less specific interactions between the host and the parasite. In the recognition phase (Mendgen et al 1988), the germ tube recognizes (Fig 1b) the surface of the cuticle and the stomatal lip. This surface signal can be replaced by nearly any plastic surface (e.g. polyethylene) scratched with sand paper. During development of the substomatal vesicle and the infection hypha, more specific signals are exchanged between host and parasite. During this signal phase, the rust fungus suppresses defence reactions of the host plant such as the activity of phenylalanine ammonia-lyase (Yamamoto et al 1977; Moerschbacher et al 1988). Mainly the host plant promotes further fungal development and after differentiation of the haustorial mother cell, induces the first haustorium in the host cell (Mendgen 1982). Now, the parasitic phase of the rust fungus begins. Very specific interactions between host and parasite induce a flow of metabolites from host to parasite. This complex balance between the two partners can be disturbed eventually by the products of the host plant's resistance genes.

During the signal phase and the parasitic phase, the hyphae of the rust fungus exhibit special features. The surface of these hyphae contains ß $1 \rightarrow 3$ and ß $1 \rightarrow 6$ glucans (Mendgen et al 1985). In the electron microscope, brush like fibrils on the intercellular hyphae can be seen after freeze substitution of infected tissue (Mendgen et al 1988; Welter et al 1988). The fibrils may represent glycoproteins. They appear to be interwoven with the host wall. One can imagine that this material influences the host parasite interaction. There is not much information on how parasitic fungi secrete such wall material or enzymes or other molecules during growth in their host tissue. One reason might be that a typical Golgi body has not been observed yet in rust fungi.

SECRETION IN ANIMALS AND PLANTS

In animals, a well developed Golgi apparatus is responsible for secretion (Farquhar 1985; Burgess and Kelly 1987). This means a functional continuum of membranous cell compounds consisting of the nuclear envelope, endoplasmic reticulum, Golgi apparatus and a number of quite different vesicle types of very different size. Some of these vesicles finally merge with the plasma membrane. The Golgi apparatus itself consists of regions of stacked cisternae (dictyosomes) without ribosomes. The cisternae may occur singly, as flattened cavities or very often consist of a central plate-like region continous with a peripheral system of tubules and vesicles. The dictyosomes seem to be polarized. One pole is associated with the nuclear envelope or with the endoplasmic reticulum. It is called the proximal pole or the forming face. The opposite pole or the maturing face of the Golgi stacks seems to be involved with the formation of secretory vesicles. Here, an elaborate trans Golgi network is observed very often.

In plants, it is also assumed that a large part of the intracellular transport of macromolecules takes place via dictyosomes (Robinson and Kristen 1982). The Golgi apparatus of plants can be very similar to the structures observed in animals (Schnepf 1969; Sievers and Schnepf 1981). However, a biochemical and a structural analysis has indicated significant differences (Morré 1977; Jones and Robinson 1989). Because most plant cells secrete much smaller amounts of proteins, vesicle movement between ER and Golgi apparatus is considerably reduced. Also, the process of vesicle-mediated secretion and membrane recycling seems different from what has been observed in animal cells. Turgor pressure forces acting on the plasma membrane seem to induce unique types of plasma membrane configurations such as flattened vesicles with slit shaped membrane fusion sites and horse shoe-shaped membrane infoldings (Staehelin and Chapman 1987). There is a very close and characteristic association of ER cisternae and the plasmalemma. This may indicate a molecular recycling of membrane lipids from the plasma membrane to the ER (Craig and Staehelin 1988).

SECRETION IN FUNGI

In fungi, growth takes place mainly at the hyphal tip. This tip is also responsible for secretion of different types of molecules. Glucans and chitin are obviously synthesized at the plasma membrane (see Wessels 1987). Glycoproteins, e.g. elicitors (Kogel et al 1988) should follow the normal way of synthesis, processing, sorting and delivery.

Some "lower" fungi, such as *Pythium*, *Chytridium*, *Saprolegnia*, and *Phytophthora*, have a typical Golgi apparatus (Grove et al 1968, Taylor and Fuller 1981, Heath et al 1985, Förster and Mendgen 1987). The dictyosomes appear to pinch off vesicles that accumulate at special areas of the fungal protoplast or at the hyphal tip. The vesicles consist of different populations of size and their contents react with stains for carbohydrates. With immunological techniques, it was demonstrated that these vesicles, depending on their size, may have different contents (Gubler and Hardham 1988). Pectinmethylesterase was localized in the dictyosome, in three different types of vesicles and over the plasma membrane (Förster and Mendgen 1987). This labeling pattern suggests the "normal" way of protein glycosylation and secretion as described for animals.

In other fungi, the situation is different. *Saccharomyces cerevisiae* has a small, but still typical dictyosome (Baba and Osumi 1987). In *Schizosaccharomyces pombe*, only single tubular structures with vesicles or single fenestrated cisternae were detected after ultrarapid freezing and freeze substitution (K. Tanaka, pers. comm.). Also in other Ascomycetes and Basidiomycetes, such very small structures reminiscent of single Golgi cisternae have been detected: smooth cisternae (Howard and Aist 1979), centers of presumed Golgi activity (Hoch and Howard 198o), modified dictyosomes (Dargent et al 1982), Golgi-like bodies (Dahmen and Hobot 1986), Golgi equivalents (Roberson and Fuller 1988), Golgi cisternae (Newhouse et al 1983), or Golgi bodies (McLaughlin 1974; Mims et al 1988).

In the dicaryotic stage of the rust fungus *Uromyces*, the appearance of the Golgi equivalent seems to change during its development. In the germ tubes, small, variously shaped segments of inflated cisternae were detected (Hoch and Staples 1983). In infection hyphae, the Golgi equivalent consists of numerous small tubular-vesicular structures (Scheffold, unpublished result) which were called tubular-vesicular complex type 2 (TVC 2) (Welter et al 1988). TVC 2 is similar to the structures observed in germ tubes, but larger in size and the cisternae are more elaborated. Subsequently, in the parasitic phase (in intercelluar hyphae and haustoria), TVC 2 becomes less frequent (Knauf et al 1989). Now, a different, quite prominent structure, called tubular-vesicular complex type 1 (TVC 1), becomes obvious in intercellular hyphae and haustoria (Fig. 2). It consists of more or less big complexes of cisternae. In developing spores, TVC 1 covers most of the spore's cytoplasm. In areas where the tubules approach the plasmalemma, sometimes vesicles fuse with the plasma membrane suggesting secretion events. These events in the rust fungi (Knauf and Mendgen 1988) are different from secretion of vesicles in plants (Craig and Staehelin 1988).

Do these tubular-vesicular structures have the secretory functions of a Golgi apparatus? Are proteins glycosylated there and are the vesicles designated for secretion? In order to study these processes in the rust fungi, the following strategies are being pursued:

1. Monoclonal antibodies are raised against the surface of infection hyphae and haustorial mother cells. We are looking now whether these antibodies bind to glycoprotein common to tubular-vesicular complexes and to glycoproteins secreted on the hyphal surface.

2. Antibodies against pectinesterase from *P. infestans* also recognize such an enzyme in the rusts. Since the enzyme is glycosylated, we look for enzymes responsible for this. A GDP-Man:Dol-P mannosyltransferase and a galactosyltransferase have been detected (Welter, Fink and Frittrang, unpublished result). The first enzyme is localized in the ER Tanner and Lehle 1987), the latter may be localized in the

smooth cisternae of the dictyosome. With antibodies against these enzymes, we hope to characterize the Golgi equivalent in the rust fungi.

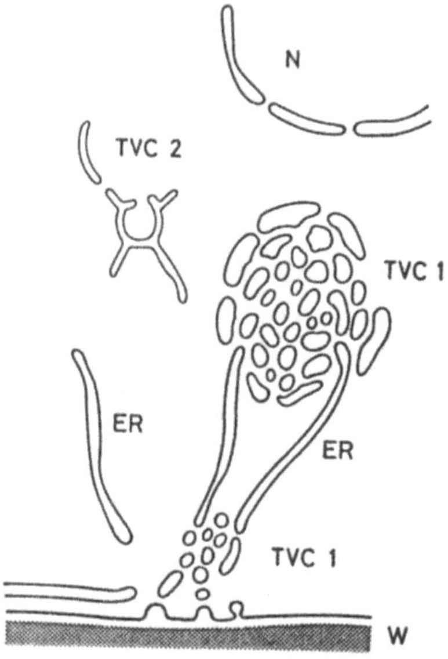

Fig. 2
Secretion in the parasitic phase of rust fungi
TVC 1= tubular-vesicular complex type 1
TVC 2= tubular-vesicular complex type 2
ER= endoplasmic reticulum
N= nucleus W= wall

ACKNOWLEDGEMENTS

W. R. Bushnell, W. Fink and K. Welter helped with the manuscript and the Deutsche Forschungsgemeinschaft with a grant (Me 523/12).

REFERENCES

Baba M, Osumi M (1987) Transmission and scanning electron microscopic examination of intracellular organelles in freeze-substituted *Kloeckera* and *Saccharomyces cerevisiae* yeast cells. J Electron Microsc Techn 5: 249-261

Burgess TL, Kelly RB (1987) Constitutive and regulated secretion of proteins. Ann Rev Cell Biol 3: 243-293

Craig ST, Staehelin LA (1988) High pressure freezing of intact plant tissues. Evaluation and characterization of novel features of the endoplasmic reticulum and associated membrane systems. Eur J Cell Biol 46: 8o-93

Dahmen H, Hobot JA (1986) Ultrastructural analysis of *Erysiphe graminis* haustoria and subcuticular stroma of *Venturia inaequalis* using cryosubstitution. Protoplasma 131: 92-1o2

Dargent R, Touzé-Soulet JM, Rami J, Montant C (1982) Cytochemical characterization of the Golgi apparatus in some filamentous fungi. Exp Mycol 6: 1o1-114

Farquhar MG (1985) Progress in unraveling pathways of Golgi traffic. Ann Rev Cell Biol 1: 447-488

Förster H, Mendgen K (1987) Immunocytochemical localization of pectinesterases in hyphae of *Phytophthora infestans*. Can J Bot 65: 26o7-2613

Grove SN, Bracker CE, Morré DJ (1968) Cytomembrane differentiation in the endoplasmic reticulum-Golgi apparatus-vesicle complex. Science 161: 171-173

Gubler F, Hardham AR (1988) Secretion of adhesive material during encystment of *Phytophthora cinnamomi* zoospores, characterized by immunogold labelling with monoclonal antibodies to components of pheripheral vesicles. J Cell Science 9o: 225-235

Heath IB, Rethoret K, Arsenault AL, Ottensmeyer FP (1985) Improved preservation of the form and contents of wall vesicles and the Golgi apparatus in freeze substituted hyphae of *Saprolegnia*. Protoplasma 128: 81-93

Hoch HC, Howard RJ (198o) Ultrastructure of freeze-substituted hyphae of the basidiomycete *Laetisaria arvalis*. Protoplasma 1o3: 281-297

Hoch HC, Staples RC (1983) Ultrastructural organization of the non-differentiated uredospore germling of *Uromyces phaseoli* var. *typica*. Mycologia 75: 795-824

Hoch HC, Staples RC (1987) Structural and chemical changes among the rust fungi during appressorium development. Ann Rev Phytopathol 25: 231-247

Howard RJ, Aist JR (1979) Hyphal tip cell ultrastructure of the fungus *Fusarium*: Improved preservation by freeze-substitution. J Ultrastructure Res 66: 224-234

Jones RL, Robinson DG (1989) Protein secretion in plants. New Phytol 111: 567-597

Knauf MG, Mendgen K (1988) Secretion systems and membrane-associated structures in rust fungi after high pressure freezing and freeze-fracturing. Biol Cell 64: 363-37o

Knauf MG, Welter K, Müller M, Mendgen K (1988) The haustorial host-parasite interface in rust-infected bean leaves after high pressure freezing. Physiol Mol Plant Pathol 34: 519-53o

Kogel G, Beissmann B, Reisener HJ, Kogel KH (1988) A single glycoprotein from *Puccinia graminis* f.sp. *tritici* cell walls elicitis the hypersensitive lignification response in wheat. Physiol Mol Plant Pathol 33: 173-185

Littlefield LJ (1981) Biology of the plant rusts. Iowa State University Press, Ames

Mendgen K (1982) Differential recognition of the outer and inner walls of epidermal cells by a rust fungus. Naturwiss 69: 5o2-5o3

Mendgen K, Lange M, Bretschneider K (1985) Quantitative estimation of the surface carbohydrates on the infection structures of rust fungi with enzymes and lectins. Arch Microbiol 14o: 3o7-311

Mendgen K, Schneider A, Sterk M, Fink W (1988) The differentiation of infection structures as a result of recognition events between some biotrophic parasites and their hosts. J Phytopathol 123: 259-272

McLaughlin DJ (1974) Ultrastructural localization of carbohydrate in the hymenium and subhymenium of *Coprinus*. Evidence for the function of the Golgi apparatus. Protoplasma 82: 341-364

Mims CW, Richardson EA, Timberlake WE (1988) Ultrastructural analysis of conidiophore development in the fungus *Aspergillus nidulans* using freeze-substitution. Protoplasma 144: 132-141

Mims CW, Roberson RW, Richardson EA (1988) Ultrastructure of freeze-substituted and chemically fixed basidiospores of *Gymnosporangium juniperi-virginianae*. Mycologia 8o: 356-364

Moerschbacher BM, Noll UM, Flott BE, Reisener HJ (1988) Lignin biosynthetic enzymes in stem rust infected resistant and susceptible near-isogenic wheat lines. Physiol Mol Plant Pathol 33: 33-46

Morré DJ (1977) Membrane differentiation and the control of secretion: A comparison of plant and animal Golgi apparatus. In: Brinkley BR and Porter KR (eds). International Cell Biology pp 293-3o3, New York, Rockefeller University Press

Newhouse JR, Hoch HC, MacDonald WL (1983) The ultrastructure of *Endothia parasitica*. Comparison of a virulent with a hypovirulent isolate. Can J Bot 61: 389-399

Roberson RW, Fuller MS (1988) Ultrastructural aspects of the hyphal tip of *Sclerotium rolfsii* preserved by freeze substitution. Protoplasma 146: 143-149

Robinson DG, Kristen U (1982) Membrane flow via the Golgi apparatus of higher plant cells. Int Rev Cytol 77: 89-127

Schnepf E (1969) Sekretion und Exkretion bei Pflanzen. Protoplasmatologia VIII, Springer Verlag

Sievers A, Schnepf E (1981) Morphogenesis and polarity of tubular cells with tip growth. In: Kiermayer O (ed). Cell Biol. Monographs, Vol 8: Cytomorphogenesis in plants. Springer Verlag, pp 265-299

Staehelin LA, Chapman RL (1987) Secretion and membrane recycling in plant cells: novel intermediary structures visualized in ultrarapidly frozen sycamore and carrot suspension-culture cells. Planta 171: 43-57

Staples RC, Hoch HC, Epstein L, Lacetti L, Hassouna S (1985) Recognition of host morphology by rust fungi. Can J Plant Pathol 7: 314-322

Tanner W, Lehle L (1987) Protein glycosylation in yeast. Biochimica et Biophysica Acta 9o6: 91-99

Taylor JW, Fuller MS (1981) The Golgi apparatus, zoosporogenesis and development of the zoospore discharge apparatus of *Chytridium confervae*. Exp Mycol 5: 35-59

Welter K, Müller M, Mendgen K (1988) The hyphae of *Uromyces appendiculatus* within the leaf tissue after high pressure freezing and freeze substitution. Protoplasma 147: 91-99

Wessels JGH (1987) Cell wall synthesis in apical hyphal growth. Int Rev Cytol 1o4: 37-79

Yamamoto H, Hokin H, Tani T, Kadota G (1977) Phenylalanine ammonia-lyase in relation to the crown rust resistance of oat leaves. Phytopath Z 9o: 2o3-211

;, Franciska Waelkens, Mark Vanstockem, Kris Michiels, Els Van
, Ann Proost, Pieternel Van Rhijn, Ann Devos, August Van Gool and
leyden
of Leuven, F.A. Janssens Laboratory of Genetics
Croylaan 42, B-3030 Heverlee

ON

a of the genus *Azospirillum* are diazotrophs associated with the of grasses (for review see Döbereiner and Pedrosa 1987). The genetic analysis of *Azospirillum* plant interaction is hampered by of plant root differentiation at the site of interaction. The .tion of bacterial genes possibly involved in *Azospirillum* *ses* has therefore been approached either by molecular .ion with cloned genes from other rhizosphere bacteria (Fogher *et* ;; Waelkens *et al* 1987) or by genetic complementation of zed mutants of *Rhizobium meliloti* with gene libraries of *nse* (Michiels *et al* 1988). By genetic complementation we isolated two *A. brasilense* Sp7 DNA fragments that correct *:i* *exoB* and *exoC* mutants respectively. These fragments are n the p90 plasmid DNA of *A. brasilense* Sp7 (Michiels *et al* :n this paper we report a more extensive inventory of *A. brasilense* are involved in exopolysaccharide synthesis. Furthermore we ;ome experiments to study the nature and regulation of these genes, y we focus on the genetic map of the p90 plasmid of *A. brasilense* '.

.TION OF *A.BRASILENSE* GENES INVOLVED IN EXOPOLYSACCHARIDE SYNTHESIS

: loci for the synthesis of exopolysaccharide in *Azospirillum* : strain Sp7 (ATCC 29145) were identified in two ways : random Tn5 s or genetic complementation of *exo* mutants of *R. meliloti* strain

NATO ASI Series, Vol. H 36
Signal Molecules in Plants and Plant-Microbe Interactions
Edited by B. J. J. Lugtenberg

Table 1. Correction of *R. meliloti exo* mutants with *A. brasilense* DNA

R. meliloti exo mutants[a] genotype	phenotype	*R. meliloti* transconjugants plasmid[b]	phenotype
exoA	Cal^-Fix^-	p90, pCal801-90	Cal^+Fix^+
exoB	Cal^-Fix^-	p90, pCal102	Cal^+Fix^+
exoC	Cal^-Fix^-	pCal111, pCal112	Cal^+Fix^-
		p90, pCal101-90	Cal^+Fix^+
exoF	Cal^-Fix^-	p90	Cal^+Fix^+
exoG	$Cal^dFix^{\pm}$	pCal301, pCal303	Cal^+Fix^+
exoH	Cal^+Fix^- (haloless)	pCal901-90	Cal^+Fix^+
exoM	Cal^-Fix^-	pCal402-90,	Cal^+Fix^+
		pCal102, pCal111	Cal^+Fix^+
exoN	Cal^dFix^+	pCal601, pCal603	Cal^+Fix^+
exoP	Cal^-Fix^-	pCal501, pCal502,	Cal^+Fix^-
		pCal503, pCal504-90	Cal^+Fix^+
exoK	Cal^dFix^+	pCal201, pCal202,	Cal^+Fix^+
		pCal111	Cal^+Fix^+

[a] Described by and obtained from G. Walker (MIT)
[b] p90 refers to the 90 MDa plasmid of *A. brasilense* Sp7 with a Tn*5*-Mob insertion near the *exoC* correcting locus; pCal000 refers to recombinant cosmids of a clone library made of total *A. brasilense* Sp7 DNA; pCal000-90 refers to recombinant cosmids made with *A. brasilense* Sp7 p90 plasmid DNA.

Table 2. *A. brasilense* Sp7 mutants affected in exopolysaccharide synthesis

Mutants[a]	Phenotype	Chromosomal or plasmid bound
AB7030Tn*5*-1	Cal^{++}	chr
AB7030Tn*5*-11, 12	Cal^d	chr
AB7030Tn*5*-21, 22, 23	Cal^-	chr
AB7001(exo*C*::Tn*5*)	Cal^+	p90
AB7002(exo*B*::Tn*5*)	Cal^+	p90

[a] Mutants were obtained either through random Tn*5* (or derivative) mutagenesis or by reversed genetics with mutant alleles obtained by Tn*5* mutagenesis of pCal clones (see Table 1).

Rm1021 with *Azospirillum* DNA. *A. brasilense* Sp7 colonies on Calcofluor containing plates exhibit a blue-purple fluorescence under UV light (Michiels *et al* 1989b). This property can be used for the isolation of Tn5 induced exopolysaccharide deficient (EPS$^-$) mutants, as described for *R. meliloti* Rm1021 by Leigh *et al* (1985). For genetic complementation we used either a clone library of *A. brasilense* total DNA or p90 plasmid DNA in the broad host range vector pLAFR1, or a Tn5-Mob derivative of the p90 plasmid. Mobilization to *R. meliloti exo* mutants was carried out either on plate (for the cosmid clones) followed by selection for complementation of the Calcofluor dark phenotype, either *in planta* (alfalfa, for the cosmid clones and the p90 plasmid), followed by the selection for complementation of the Fix$^-$ phentoype. Results are summarized in Tables 1 and 2. It can be noticed from the results in Table 1 that some *R. meliloti exo* mutants (e.g. *exoC*) are either corrected for the Calcofluor dark phenotype or for the dark Calcofluor phenotype and the Fix$^-$ phenotype, depending on the plasmid used. The Fix$^-$ phenotype of the *R. meliloti exoC* (pCa1112) transconjugant can be understood when looking at the NMR spectrum of the isolated EPS. The NMR spectrum of this EPS is quite different from that of *R. meliloti* wild type EPS both in the carbohydrate profile and in the absence of acyl substituents (results not shown). In cases where complementation was observed by mobilization of the p90 plasmid from *A. brasilense* Sp7 to the *R. meliloti exo* mutants (triparental matings *in planta*), bacteria were reisolated from the nitrogen fixing nodules. Surprisingly, none of the reisolates contained *A. brasilense* DNA, suggesting that complementation has occurred through transient expression of the non-replicating p90 plasmid DNA in *R. meliloti*. When inoculated on alfalfa, these reisolates induce Fix$^-$ nodules.

The EPS produced by *A. brasilense* Sp7 cannot be precipitated by cetrimide. Tn5 induced Calcofluor dark mutants of *A. brasilense* (see Table 2) however, produce an exopolysaccharide that can be precipitated with cetrimide. This would suggest that *A. brasilense* produces at least two different types of exopolysaccharide.

PHENOTYPE OF *A. BRASILENSE* STRAINS WITH MUTATIONS IN THE *EXO* CORRECTING LOCI

Mutations in the *exoB* and *exoC* correcting loci do not cause a Calcofluor dark phenotype in *A. brasilense* Sp7. These loci are therefore not involved in the synthesis of the neutral Calcofluor binding exopolysaccharide of *A. brasilense*. EPS samples of AB7001, AB7002 and *A. brasilense* wild-type were size-fractionated by FPLC. AB7001 (*exoC*::Tn*5*) and AB7002 (*exoB*::Tn*5*) produce EPS that have a lower molecular weight than the wild-type EPS (Michiels *et al* 1988). AB7001 has been analyzed by two-dimensional (2-D) protein analysis (De Mot and Vanderleyden, Can. J. Microbiol., in press). The mutant shows the presence of a truncated protein as compared to the wild-type. Recently, we have isolated a Tn*5* induced motility mutant (Van Rhijn *et al* , Appl. Environ. Microbiol., in press). This mutant appears to have the Tn*5* insertion in the *exoC* correcting locus. In this mutant, the protein that is truncated in AB7001, is completely absent. Like in AB7001, no alteration of calcofluor fluorescence is observed. This indicates that in *A. brasilense* like in *R. meliloti*, *exoC* mutations have pleiotropic effects.

ANALYSIS OF *LACZ* FUSIONS OF THE *EXOC* CORRECTING LOCUS OF *A. BRASILENSE* SP7

Plasmids pCL157 and pCL172 are derivatives of pCa1111 (*exoC* correcting cosmid, see Table 1) that contain *lacZ* in the *exoC* correcting locus either in sense (pCL172) or in antisense orientation (pCL157).

These plasmids were transferred to a number of *A. brasilense* and *R. meliloti* strains for which β-galactosidase activity was measured (see Table 3). *R. meliloti exoR* and *exoS* mutants overproduce succinoglycan (Doherty *et al* 1988). *A. brasilense* mutants AB7030 Tn*5*-1 and AB7030 Tn*5*-102 show extra bright fluorescence. In all these strains the expression of the *exo*C correcting locus appears to be reduced compared to the parent strain. Interestingly, the β-galactosidase activity encoded by plasmid pCL157 is markedly increased in both the *exoR* and *exoS* mutants. This would suggest the presence of another gene on pCa1111, in opposite orientation of the *exoC* correcting gene, that is subject to the same regulation as the *R. meliloti exo* genes on pSym2.

A. BRASILENSE GENES LOCATED ON THE 90 MEGADALTON (P90) PLASMID OF *A. BRASILENSE SP7*

Previous studies have suggested that most *A. brasilense* and some *A. lipoferum* strains contain a highly conserved plasmid, measuring approximately 90 MDa, which likely represents a plasmid family (Michiels *et*

Table 3. Expression of *A. brasilense exoC* gene in *Azospirillum* and *Rhizobium* backgrounds

3.A. (time = 470 min)

Strain[a]	A_{420}/A_{600}	A_{420}/A_{600}[b]
7030	0.011	0
7030(pCL157)	0.196	0.185
7030(pCL172)	5.812	5.801
7030Tn*5*-1	0.007	0
7030Tn*5*-1(pCL157)	0.023	0.016
7030Tn*5*-1(pCL172)	2.263	2.256
7030Tn*5*-102	0.032	0
7030Tn*5*-102(pCL157)	0.019	0
7030Tn*5*-102(pCL172)	1.966	1.934

3.B. (time = 495 min)

Strain[a]	A_{420}/A_{600}	A_{420}/A_{600}[b]
1021	0.080	0
1021(pCL157)	0.149	0.069
1021(pCL172)	1.432	1.352
7095	0.031	0
7095(pCL157)	0.509	0.478
7095(pCL172)	0.697	0.666
7096	0.011	0
7096(pCL157)	0.538	0.527
7096(pCL172)	0.512	0.501

[a] *R. meliloti* strains 7095 (*exoR*) and 7096 (*exoS*) are described by Doherty *et al* (1988) and were obtained from G. Walker

[b] $A_{420}/A_{600} = A_{420}/A_{600}$ (tested strain) - A_{420}/A_{600} (parent strain)

al 1989b). From Table 1, it can be concluded that the *exoA*, *exoB*, *exoC*, *exoF*, *exoH*, *exoM* and *exoP* correcting loci are located on the p90 plasmid. Recently, we have identified a chemotaxis mutant of *A. brasilense* Sp7, as the result of a Tn*5* insertion in the p90 plasmid. Interestingly we have obtained preliminary evidence that the p90 plasmid contains also a *nodC* correcting locus. In a triparental mating with a *R. meliloti nodC* mutant (obtained from J. Dénarié), *A. brasilense* (p90::Tn*5*-Mob) and *E. coli* (RP4-4) *in planta*, we obtained nitrogen fixing nodules on alfalfa.

ACKNOWLEDGEMENTS

We like to thank Dr. J. Dénarié, Dr. S. Stachel and Dr. G. Walker for providing strains and plasmids. This research was supported by funds from the F.G.W.O. (contract 3.0093.89) and the Onderzoeksraad K.U. Leuven (OT/88/20).

REFERENCES

Döbereiner J and Pedrosa FO (1987) Nitrogen-fixing bacteria in non-leguminous crop plants. Science Tech Publishers, Springer Verlag

Doherty D, Leigh JA, Glazebrook J and Walker GC (1988) *Rhizobium meliloti* mutants that overproduce acidic Calcofluor-binding exopolysaccharide. J Bacteriol *170*: 4249-4256

Fogher C, Dusha I, Barbot P and Elmerich C (1985) Heterologous hybridization of *Azospirillum* DNA to *Rhizobium nod* and *fix* genes. FEMS Microbiol Lett *30*: 245-250

Leigh JA, Signer ER and Walker GC (1985) Exopolysaccharide-deficient mutants of *Rhizobium meliloti* that form ineffective nodules. Proc Natl Acad Sci USA *82*: 6231-6235

Michiels K, Vanderleyden J, Van Gool A and Signer ER (1988) Isolation and characterization of *Azospirillum brasilense* loci that correct *Rhizobium meliloti exoB* and *exoC* mutants. J Bacteriol *170*: 5401-5404

Michiels K, Vanderleyden J, Van Gool A and Signer ER (1989*a*) Isolation and properties of *Azospirillum lipoferum* and *Azospirillum brasilense* surface polysaccharide mutants. *In* F. Skinner *et al.* (eds), Nitrogen fixation with non-legumes, Kluwer Academic Publishers, the Netherlands, pp. 189-195

Michiels K, De Troch P, Onyeochia I, Van Gool A, Elmerich C and Vanderleyden J (1989*b*) Plasmid localization and mapping of two *Azospirillum brasilense* loci that affect polysaccharide synthesis. Plasmid *21*, 142-146

Waelkens F, Maris M, Verreth C, Vanderleyden J and Van Gool A (1987) *Azospirillum* DNA shows homology with *Agrobacterium* chromosomal virulence genes. FEMS Microbiol Lett *43*: 241-246

BACTERIAL GENES INVOLVED IN THE COMMUNICATION BETWEEN SOYBEAN AND ITS ROOT NODULE SYMBIONT, *BRADYRHIZOBIUM JAPONICUM*

Michael Göttfert, Philipp Grob, Silvia Rossbach, Hans-Martin Fischer, Beat Thöny, Denise Anthamatten, Ines Kullik and Hauke Hennecke
Mikrobiologisches Institut, ETH-Zentrum, Schmelzbergstr. 7, CH-8092 Zürich, Switzerland

INTRODUCTION

During the course of the establishment of a legume root nodule, an exchange of signals between the rhizobial microsymbiont and the host plant takes place at all developmental stages. This extends from the early induction of *nod* genes to the late exchange of metabolites in the mature, nitrogen fixing symbiosis (for review see Long 1989). In this article we report on three examples of *Bradyrhizobium japonicum* genes involved in the communication with soybean (*Glycine max* L.Merr.) and other hosts, one being concerned with nodulation, the other with bacteroid development, and the third with symbiotic nitrogen fixation.

TWO NEW NODULATION GENES INVOLVED IN HOST SPECIFICITY: *NODV* AND *NODW*

The *nodYABC* genes of *B. japonicum* (and possibly *nodI* and *J* as well) are induced by isoflavones such as daidzein and genistein (Kosslak et al 1987; Göttfert et al 1988; Banfalvi et al 1988). The presence of a well-conserved *nod* box consensus sequence in the promoter region of this operon makes it likely that the mechanism of its transcriptional activation is similar to that known in fast-growing *Rhizobium* species (Hong et al 1987; Fisher et al 1988). Additional *nod* box sequences have been identified in the *B. japonicum* genome one of which is associated with a gene that plays an important role specifically in the nodulation of mungbean (*Vigna radiata*) but not of soybean and other hosts (Göttfert et al 1989). Two other *B. japonicum* DNA regions determining host specificity have been found which, however, are not connected with *nod* boxes and thus are believed to be regulated differently: one is located adjacent to (downstream of) the common *nod* genes and determines host specificity for Siratro (Nieuwkoop et al 1987; see also the article by G. Stacey and coworkers in this book); the other region (previously termed "*nod*-1") is not closely linked to any other known *nod* gene and is es–

NATO ASI Series, Vol. H 36
Signal Molecules in Plants and Plant-Microbe Interactions
Edited by B. J. J. Lugtenberg

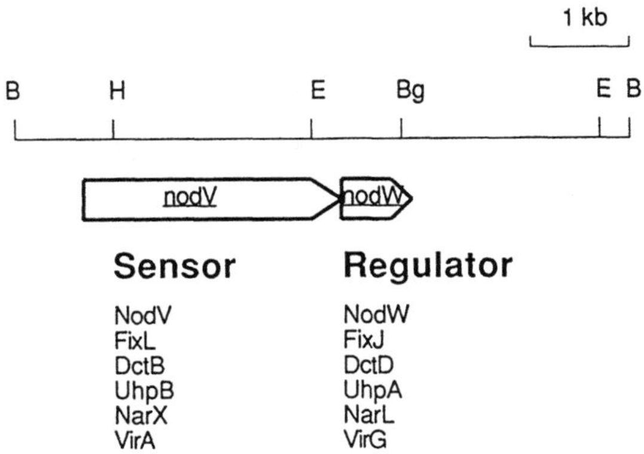

Fig. 1. Map of the *nodV* and *nodW* genes and examples of homologous sensor and regulator proteins of the family of two-component regulatory systems. References: FixL/FixJ (David et al 1988); DctB/DctD (Ronson et al 1987b); UhpB/UhpA (Weston and Kadner 1988); NarX/NarL (Stewart et al 1989); VirA/VirG (Winans et al 1986; Melchers et al 1987).

sential for the nodulation of mungbean, Siratro and cowpea (*Vigna unguiculata*) whereas it is less important for the nodulation of soybean (Hahn and Hennecke 1988).

The DNA of the "*nod*-1" region was sequenced and found to contain two open reading frames which we named *nodV* and *nodW* (Fig. 1). The stop codon of *nodV* overlapped the start codon of *nodW* which suggested that the two genes were coupled transcriptionally and translationally. The NodV and NodW proteins were homologous to members of the protein family of two-component regulatory systems (see Fig. 1, and Ronson et al 1987a for review) in which one component acts as a sensor whereas the other functions as a gene regulator. The conserved domains in the members of this family are about 200 amino acids of the C-terminal regions of the sensor proteins and at least 120 amino acids of the N-terminal regions of the regulator proteins. In addition, most of the sensor proteins have hydrophobic domains at their N-termini which suggests that they are membrane-associated. In the proposed model for signal transduction (Nixon et al 1986) the sensor protein responds to an environmental signal by sensing a stimulus near its N-terminal region and transducing this signal to its own C-terminal domain. The activated C-terminal domain may then interact with and modify the conserved N-terminal domain of the regulator protein. The so modified regulator is subsequently able to activate the expression of a specific gene or set of genes. In the case of the NodV and NodW proteins the most interesting (but still unanswered) questions now are: (i) what is the nature of the signal acting on NodV, and (ii) what kind of gene(s) is(are) activated by NodW? Bearing in mind that the *nodVW* genes play a role in host specificity, the NodV protein may respond to an exclusive host plant factor, and/or the products of the NodW-activatable genes confer host-specific functions.

A NOVEL GENE INVOLVED IN BACTEROID DEVELOPMENT

We had previously described a *B. japonicum* Tn*5* mutant (strain 3160) which elicited very small, white pseudonodules on soybean and three other legume hosts (Regensburger et al 1986; Rossbach et al 1989). Although strain 3160 was able to induce massive cell division in the inner-cortical root tissue, we could not obtain convincing evidence for the presence of infection threads and bacteroids. The affected mutant DNA and the corresponding wild-type DNA regions were cloned and the essential gene was delimited to approximately 1.5 kb by further mutagenesis and complementation experiments. Preliminary DNA sequence data suggested that this DNA carried one single gene.

Mutant 3160 did not succeed in bacteroid propagation even in the presence of other coinoculated *B. japonicum* strains, whereas it was able to restore full nodulation capability of a coinoculated *nodC* mutant (Table 1).

Ton Bisselings group (Wageningen, NL) had shown that the absence of endosymbiotic 3160 bacteria in 3160-induced pseudonodules was strictly correlated with a selective decrease in expression of the soybean early nodulin gene *ENOD55* whereas *ENOD2* and *ENOD13* were expressed normally (Rossbach et al 1989). Obviously, strain 3160 is blocked in such an early stage of bacteroid and root nodule development that this condition does not suffice to produce enough of a signal required to fully turn on the *ENOD55* gene. This does not necessarily mean that the gene affected by the 3160 mutation is directly responsible for the production of that signal; rather, its function may only be a necessary prerequisite in order for the invading bacterium to reach a stage in which that signal can be generated.

THE NIFA PROTEIN AND THE OXYGEN SIGNAL

The nitrogen fixation regulatory protein NifA is active at low O_2 concentration, a condition that prevails endosymbiotically, whereas it is irreversibly inactivated at high O_2 concentration (Fischer and Hennecke 1987; Kullik et al 1989). Anaerobic NifA activity - directly or indirectly - depends on the presence of a divalent metal ion, possibly Fe^{2+}, which lends support to a hypothesis in which the oxidation state of a NifA-bound metal or a metal-containing cofactor is influenced by the ambient oxygen concentration (Fig. 2; Fischer et al 1988). Two cysteines in the interdomain linker located between the central domain and the DNA binding domain and two other cysteines near the carboxy-terminal end of the central domain have been shown to be essential for NifA activity (Fischer et al 1988) and may thus be involved both in metal binding and oxygen respon-

Table 1. Results from coinoculation experiments on soybean[a]

B. japonicum strains used for incoculation	Relevant characteristics	Nodulation phenotype	Number of bacteria isolated from nodules[b]		
			wt	3160	168
wt	Rif^R	+	10^8		
3160	Rif^R Km^R Bdv^-	+/-		5	
168	Spc^R Km^R *nodC*$^-$	-			n.a.
Coinoculation mixture (1:1)					
3160 + wt		+	10^8	2	
3160 + 168		+		10	10^8

[a]Abbreviations and symbols: wt = wild type; Rif^R, Spc^R, Km^R = resistances for rifampicin, spectinomycin and kanamycin; Bdv = bacteroid development; + = normal nodulation, - = no nodules; +/- = abberrant nodulation as it is characteristic for strain 3160; n.a. = not applicable.
[b]The bacterial strains were identified by their antibiotic resistance markers.

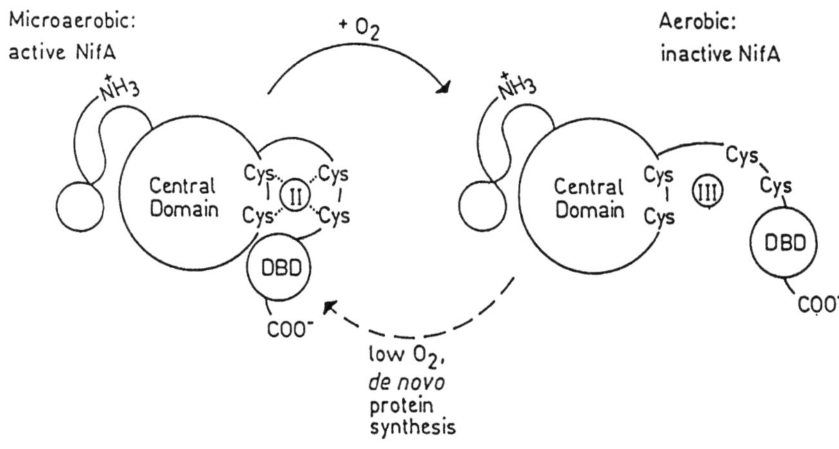

Fig. 2. Hypothetical model for the inactivation of NifA by oxygen. For further details see text. Abbreviations: DBD, DNA binding domain; Cys, cysteine; II/III, reduced and oxidized metal cofactor.

siveness (Fig. 2). In a heterologous *Escherichia coli* system, after shift of cells from aerobiosis to anaerobiosis, the oxygen-inactivated NifA could not be restored to active NifA without prior *de novo* protein synthesis: either new NifA synthesis was required, or the new synthesis of an enzyme was necessary to convert inactive to active NifA (Kullik et al 1989; Fig. 2).

The *nifA* structural gene is organized in the *fixRnifA* operon (Thöny et al 1987). The expression of this operon is partly regulated by oxygen, too (Thöny et al 1989). Aerobic expression of the *fixRnifA* operon was shown to depend on a *cis*-acting, promoter-upstream DNA sequence located between the -24/-12 promoter and position -86 relative to the transcription start site. An adenine at position -66 was essential for maximal expression. A chromosomal deletion of the upstream activator sequence (UAS) led to a symbiotically defective phenotype which was typical for *nifA* mutants. *B. japonicum* crude extracts contained a protein that bound to the UAS. Using a chromosomally integrated *fixR-lacZ* fusion, the level of expression of the *fixRnifA* operon was found to be five-fold higher under reduced oxygen tension as compared to aerobiosis. This increase was due to autoactivation by the NifA protein and was partly independent of the UAS. Based on these data we propose a model on the regulation of nitrogen fixation genes in *B. japonicum* involving a dual positive control of the *fixRnifA* operon. At high oxygen concentrations the operon is expressed at moderate level, subject to activation by a *trans*-acting factor binding to the UAS. Under such conditions the *nifA* gene product is inactive. At very low oxygen concentration - a condition favorable to NifA activity - the NifA protein is the *trans*-acting factor which (i) enhances the level of *fixRnifA* expression (hence, its own synthesis) and (ii) activates other *nif* and *fix* genes. In conclusion, the oxygen-dependent cascade control of *nif* and *fix* gene expression appears to be adapted nicely to the transition of B. *japonicum* from an aerobic soil bacterium to a microaerobic root nodule endosymbiont, and a low oxygen concentration is the primary signal that turns on the nitrogen fixation genes.

ACKNOWLEDGEMENTS

The research reported here was supported by grants from the Swiss National Science Foundation and the Federal Institute of Technology Zürich. We thank H. Paul for typing the manuscript.

REFERENCES

Banfalvi Z, Nieuwkoop A, Schell M, Besl L and Stacey G (1988) Regulation of *nod* gene expression in *Bradyrhizobium japonicum*. Mol Gen Genet 214: 420-424

David M, Daveron M-L, Batut J, Dedieu A, Domergue O, Ghaï J, Hertig C, Boistard P and Kahn D (1988) Cascade regulation of *nif* gene expression in *Rhizobium meliloti*. Cell 54: 671-683

Fischer HM and Hennecke H (1987) Direct response of *Bradyrhizobium japonicum nifA*-mediated *nif* gene regulation to cellular oxygen status. Mol Gen Genet 209: 621-626

Fischer HM, Bruderer T and Hennecke H (1988) Essential and non-essential domains in the *Bradyrhizobium japonicum* NifA protein: identification of indispensable cysteine residues potentially involved in redox reactivity and/or metal binding. Nucl Acids Res 16: 2207-2224

Fisher RF, Egelhoff TT, Mulligan JT and Long SR (1988) Specific binding of proteins from *Rhizobium meliloti* cell-free extracts containing NodD to DNA sequences upstream of inducible nodulation genes. Genes Dev 2: 282-293

Göttfert M, Weber J and Hennecke H (1988) Induction of a *nodA-lacZ* fusion in *Bradyrhizobium japonicum* by an isoflavone. J Plant Physiol 132: 394-397

Göttfert M, Lamb JW, Gasser R, Semenza J and Hennecke H (1989) Mutational analysis of the *Bradyrhizobium japonicum* common *nod* genes and further *nod* box-linked genomic DNA regions. Mol Gen Genet 215: 407-417

Hahn M and Hennecke H (1988) Cloning and mapping of a novel nodulation region from *Bradyrhizobium japonicum* by genetic complementation of a deletion mutant. Appl Environ Microbiol 54: 55-61

Hong G-F, Burn JE and Johnston AW (1987) Evidence that DNA involved in the expression of nodulation (*nod*) genes in *Rhizobium* binds to product of the regulatory gene *nodD*. Nucl Acids Res 15: 9677-9690

Kosslak RM, Bookland R, Barkei J, Paaren HE and Appelbaum ER (1987) Induction of *Bradyrhizobium japonicum* common *nod* genes by isoflavones isolated from *Glycine max*. Proc Natl Acad Sci USA 84: 74258-7432

Kullik I, Hennecke H and HM Fischer (1989) Inhibition of *Bradyrhizobium japonicum nifA*-dependent *nif* gene activation by oxygen occurs at the NifA protein level and is irreversible. Arch Microbiol 151: 191-197

Long SR (1989) Rhizobium-legume nodulation: life together in the underground. Cell 56: 203-214

Melchers LS, Thompson DV, Idler KB, Neuteboom STC, de Maagd RA, Schilperpoort RA and Hooykaas PJJ (1987) Molecular characterization of the virulence gene *virA* of the *Agrobacterium tumefaciens* octopine Ti plasmid. Plant Mol Biol 11: 227-237

Nieuwkoop AJ, Banfalvi Z, Deshmane N, Gerhold D, Schell MG, Sirotkin KM and Stacey G (1987) A locus encoding host range is linked to the common nodulation genes of *Bradyrhizobium japonicum*. J Bacteriol 169: 2631-2638

Nixon BT, Ronson CW and Ausubel FM (1986) Two-component regulatory systems responsive to environmental stimuli share strongly conserved domains with the nitrogen assimilation regulatory genes *ntrB* and *ntrC*. Proc Natl Acad Sci USA 83: 7850-7854

Regensburger B, Meyer L, Filser M, Weber J, Studer D, Lamb JW, Fischer HM, Hahn H and Hennecke H (1986) *Bradyrhizobium japonicum* mutants defective in root-nodule bacteroid development and nitrogen fixation. Arch Microbiol 144: 355-366

Ronson CW, Nixon BT and Ausubel FM (1987a) Conserved domains in bacterial regulatory proteins that respond to environmental stimuli. Cell 49: 579-581

Ronson CW, Astwood PM, Nixon BT and Ausubel FM (1987b) Deduced products of C4-dicarboxylate transport regulatory genes of *Rhizobium leguminosarum* are homologous to nitrogen regulatory gene products. Nucl Acids Res 15: 7921-7934

Rossbach S, Gloudemans T, Bisseling T, Studer D, Kaluza B, Ebeling S and Hennecke H (1989) Genetic and physiological characterization of a *Bradyrhizobium japonicum* mutant defective in early bacteroid development. Mol Plant-Microbe Interact, in press

Stewart V, Parales J and Merkel SM (1989) Structure of genes *narL* and *narX* of the *nar* (nitrate reductase) locus in *E.coli* K-12. J Bacteriol 171: 2229-2234

Thöny B, Fischer HM, Anthamatten D, Bruderer T and Hennecke H (1987) The symbiotic nitrogen fixation regulatory operon (*fixRnifA*) of *Bradyrhizobium japonicum* is expressed aerobically and is subject to a novel, *nifA*-independent type of activation. Nucl Acids Res 15: 8479-8499

Thöny B, Anthamatten D and Hennecke H (1989) Dual control of the symbiotic nitrogen fixation regulatory operon *fixRnifA* of *Bradyrhizobium japonicum*: Analysis of *cis*- and *trans*-acting elements. J Bacteriol 171, in press

Weston LA and Kadner RJ (1988) Role of *uhp* genes in expression of the *Escherichia coli* sugar-phosphate transport system. J Bacteriol 170: 3375-3383

Winans SC, Ebert PR, Stachel SE, Gordon MP and Nester EW (1986) A gene essential for *Agrobacterium virulence* is homologous to a family of positive regulatory loci. Proc Natl Acad Sci USA 83: 8278-8282

SIGNAL EXCHANGE BETWEEN *R. trifolii* AND CLOVERS

B.G. Rolfe, C.L. Sargent, J.J. Weinman, M.A. Djordjevic, J. McIver, [1]J.W. Redmond, [1]M. Batley, [1]D.C. Yuan and [1]M.W. Sutherland
Plant Microbe Interaction Group, RSBS,
Australian National University, Canberra and
[1]School of Chemistry, Macquarie University, NSW, Australia

Introduction

Recently, flavonoid molecules of specific structure have been identified as the plant signals released from roots which are responsible for the induction of bacterial nodulation (*nod*) genes (Firmin *et al* 1986; Peters *et al* 1986; Redmond *et al* 1986). Once the *nod* genes have been induced, rhizobia have been shown to elaborate and secrete signal molecules which are responsible for two obvious early nodulation events: root hair curling, resulting in aberrant, distorted, root hair growth, and the induction of cell division in the root cortical tissue (Bauer *et al* 1985; Bhuvaneswari and Solheim 1985; Faucher *et al* 1988). These responses are apparent 24 h after infection. The bacterial signal molecules responsible for these effects have not been identified. However, several studies show that heat stable, low molecular weight soluble factors can affect root morphology and root hair growth.

Signal exchange between invading rhizobia and plant cells also results in the control by the plant of the number of root nodules. After induction of nodulation in the root, a systemic response is initiated which temporarily arrests the induction of nodules in the near vicinity of a pre-existing nodule (Bhuvaneswari *et al* 1980; Pierce and Bauer 1983; Kosslak and Bohlool 1984). With the exception of the induction of several nodulin genes, other biochemical responses to the infection of rhizobia remain unknown.

Because of the availability of well characterized mutants of *Rhizobium leguminosarum* biovar *trifolii*, we have concentrated upon signal exchanges occurring early in the infection of *Trifolium repens* (white clover) and *T. subterraneum* (subterranean clover). These studies have (a) resulted in the development of a plate bioassay system for the characterisation of bacterial filtrates to identify *Rhizobium*-derived molecules capable of modifying plant growth (b) shown a differential response of white clover to the infection of nodulating "compatible" and "incompatible" *Rhizobium* strains and (c) resulted in an assay to measure the extent of the Induced Plant Systemic Inhibition of Nodulation (IPSIN).

NATO ASI Series, Vol. H 36
Signal Molecules in Plants and Plant-Microbe Interactions
Edited by B. J. J. Lugtenberg

Characterization of *Rhizobium* culture filtrates with white clover seedlings

A plate assay system was developed to test bacterial filtrates for their root hair distortion (Had) and curling (Hac) activities (Figure 1). This is a more rapid and less labour intensive assay than our previous system (Canter-Cremers *et al* 1986), and is ideal for testing the large numbers of fractions generated during chemical analysis. We have established that the Had and Hac activities are present in cell-free culture supernatants only when *nod* genes are present in the bacteria and DHF inducer has been added to cultures. Sterile supernatant cultures of flavonoid-induced ANU843 cells have been fractionated by hollow-fibre filtration. All activity is retained by a 0.1μm filter. The hair curling component is heat-stable and not extracted into organic solvent at acidic, neutral or alkaline pH. Exopolysaccharide (EPS) is present as a major component of this fraction, but purified EPS samples have no hair curling activity in the assay. Solvent precipitation of crude exopolysaccharide in the presence of added salt leaves active supernatant fractions. Chromatographic analysis of these fractions is in progress.

Induction of stimulatory flavonoid compounds in white clover in response to inoculation with *Rhizobium*

We have used a bioassay for the presence of *nod* gene inducing compounds to establish that white clover responds to *Rhizobium* inoculation by altering its normal production of flavonoid compounds. We have been consistently able to demonstrate a significant increase in the presence of flavonoid compounds able to induce the nodulation genes of *R.leguminosarum* bv. *trifolii* strain ANU843 as a result of *Rhizobium* inoculation. Figure 2 shows the effect of inoculating a "compatible" (ANU843) and an "incompatible" (*R.leguminosarum* bv. *viciae* strain RL300) strain onto white clover. Strain RL300 is capable of nodulating pea, vicia and Lens plants but is incapable of nodulating white clovers. A clear increase in the induction of *nod* gene inducing compounds in these plants with these *Rhizobium* inoculations over the mock inoculated controls is evident. We have observed fluctuations in the total levels of inducing compounds produced between experimental series, indicating that environmental conditions (eg light, temperature) are likely to be factors influencing the plant response. Despite this, statistically significant trends exist within experimental series, which indicate considerable

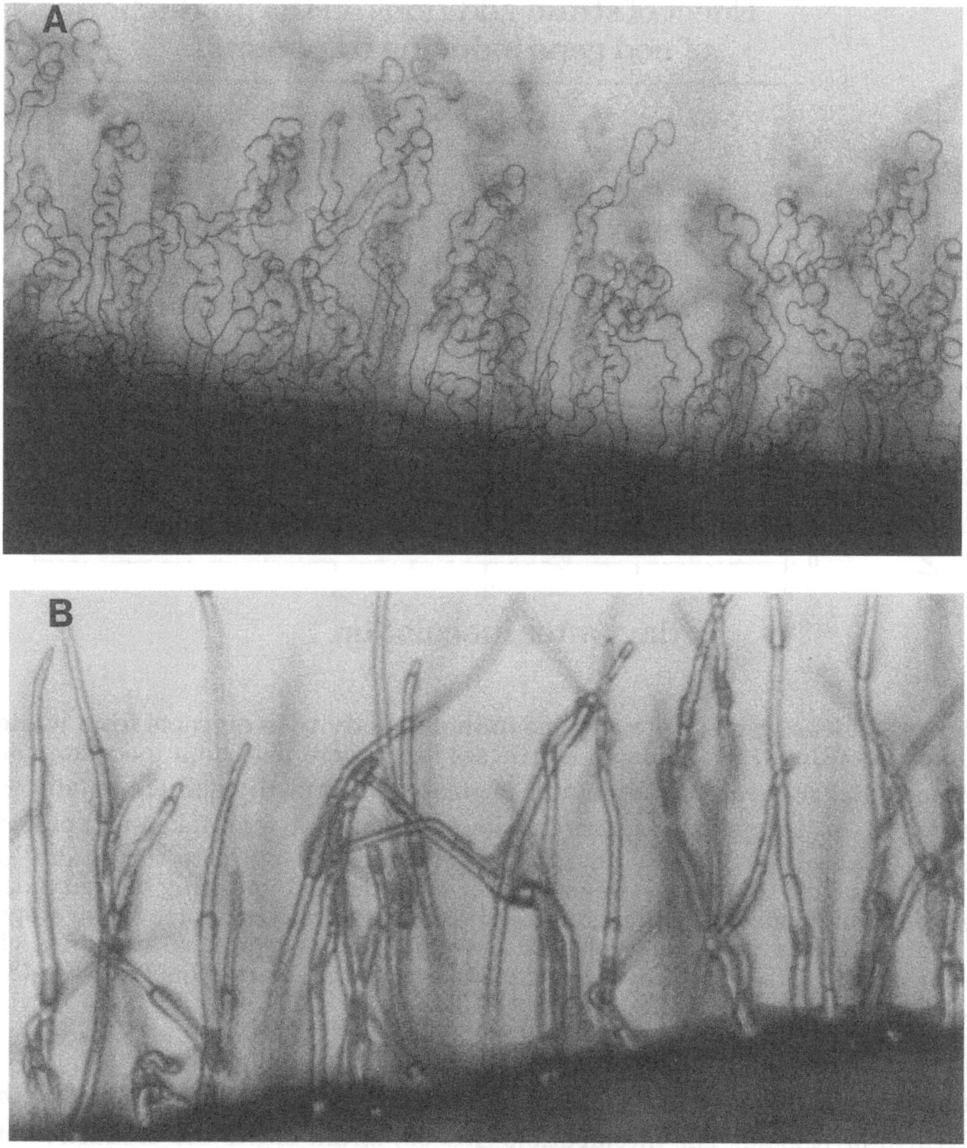

Figure 1. Root hair curling by cell free culture extracts. All *Rhizobium* strains were grown in B3 minimal medium (Dazzo *et al* 1982) containing the *nod* gene inducer DHF at 1mM. Aliquots of 100ml of sterile supernatants were spread over the surface of water-agar plates. The roots of seedlings were inserted into slits made in the agar surface to allow their growth into the agar. This procedure ensured that the roots were evenly covered with inoculated agar and that the response of the developing root hairs to added substances was reproducible. Plates were incubated under a light bank at room temperature for two days. Figure 1A and 1B show respectively the response of the root hairs of white clover plants to the presence of culture filtrates from ANU843 and ANU845 (pSym⁻) cells exposed to DHF.

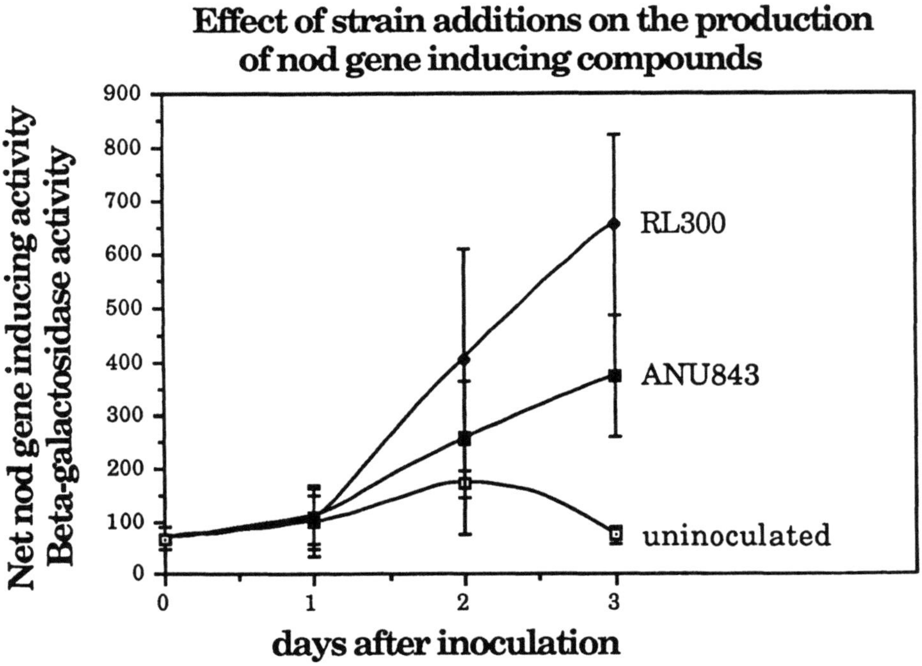

Figure 2. Measurement of *nod* gene inducing activity in extracts from white clover seedlings. Germinated sterile seedlings were placed on inoculated or uninoculated F-medium plates and incubated in growth cabinets at 450µE $sec^{-1}m^{-2}$. At daily intervals, forty plants per treatment were taken and placed in 50% ethanol for 2 hours. Plants were removed and extracts stored at $4^{0}C$ before assaying (using between 2 and 4 replicates) inducing activity as described previously (Djordjevic *et al* 1987). Results shown plot the average and standard deviation of pooled data obtained from two separate series of experiments. RL300, *R.leguminosarum* bv. *viciae* strain RL300; ANU843, *R.leguminosarum* bv. *trifolii* strain ANU843.

differences in the degree of the plant response observed when inoculated with ANU843 or RL300. The incompatible (RL300) strain elicits the production of elevated levels of *nod* gene inducing substances when compared to that induced by the compatible strain (ANU843). The introduction by conjugation of a 8.2 kb *Bam*HI fragment from ANU843 (containing the *nodFERL* and *nodMN* operons) into the RL300 strain brings the level of *nod* gene inducing compounds triggered by this strain back down to the levels seen with ANU843 (Weinman *et al* 1988). We are currently examining the signals which modulate this plant response, through the use of specific genetic constructions, culture extracts, and isolated *Rhizobium* surface components.

Physiological responses to *Rhizobium* infection

Two major physiological factors act to control root nodule induction on most legume roots. First, root cells destined to become root hairs have been shown to possess a high, but transient, susceptibility to infection by rhizobia. These cells occur just behind the root tip in an area where root hair cells have not emerged, and retain their susceptibility to infection for time periods as short as six hours (Bhuvaneswari *et al* 1980). Secondly, once infection has been initiated, systemic plant signals are induced which act to diminish infectivity elsewhere on the root (Pierce and Bauer 1983). The net effect of these two phenomena is most clearly seen in IPSIN (Induced Plant Systemic Inhibition of Nodulation) assays. Using IPSIN split root assays, Sargent *et al* (1987) showed that the simultaneous inoculation of both rootlets of a subterranean clover plant with the wild type *R. leguminosarum* bv. *trifolii* strain ANU843 led to the generation of equal numbers of nodules on both rootlets (Fig. 3B). In contrast, ANU843 mutants affected in *nodE* (region III) or *nodL* (region IV) could be inhibited from nodulating a rootlet by the inoculation of ANU843 on the opposing rootlet (see Fig. 3B)

This phenomenon is dependent upon the temperature and lighting conditions used in IPSIN assays using split root subterranean clover plants. Under lower light and temperature regimes than described by Sargent *et al* (1987), (a) strain ANU843 is not able to inhibit the nodulation of a *nodL* mutant (strain ANU251) inoculated onto the opposing rootlet, and (b) the *nodL* mutant is capable of inhibiting a *nodE* mutant from nodulating the opposing rootlet. Using the higher light and temperature conditions previously described by Sargent *et al* (1987) the precise opposite result occurred. We conclude, therefore, that the mutation in *nodL* has more severe effects upon nodulation efficiency under higher light and temperature regimes while the mutation of *nodE* has more severe effects at lower temperatures and lighting intensities. A *nodL* mutant cannot be distinguished from a wild-type strain under lower light and temperature regimes using IPSIN assays.

A range of *R. leguminosarum* bv. *trifolii* constructs with differing abilities to induce nodules were used in IPSIN split root assays to determine how nodulation-competent a strain had to be before it was able to establish an IPSIN response. Only strains which retained the ability to induce fully functional nodules induced a detectable IPSIN response. Neither root hair curling ability (Hac^+), nor the ability to infect and induce nodule foci (Hac^+ Inf^+ Noi^+) alone, was sufficient to establish on IPSIN response (Fig. 3A).

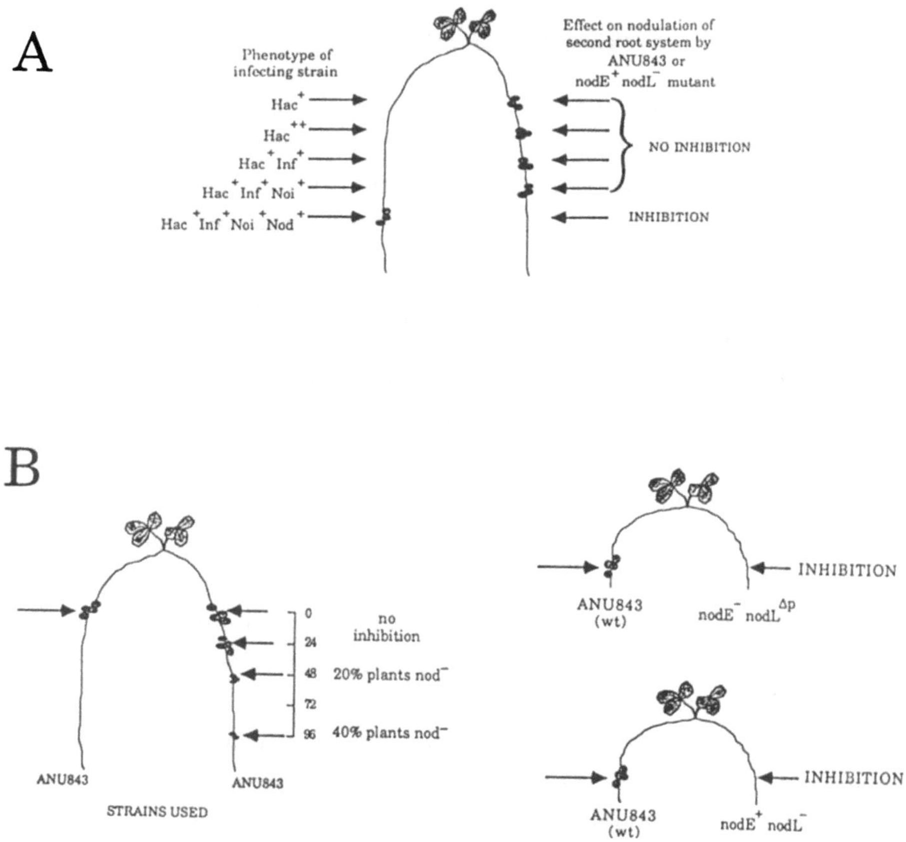

Figure 3. Diagrammatic representation of Induced Plant Systemic Inhibition of Nodulation (IPSIN). Subterranean clover seedlings were induced to develop two equally robust root systems following wounding of the tap root (Sargent *et al.*, 1987). Inoculation of a root system is indicated by an arrow. Delayed inoculation is indicated by additional arrows lower on the root system with the delay (in hours after the first inoculation) given on the scale to the side. Successful nodulation is shown by the presence of four nodules on the root system adjacent to the point of inoculation. Degrees of IPSIN are indicated by a reduction in the number of nodules adjacent to the point of inoculation. As appropriate, DHF (7,4'-dihydroxyflavone) was added to cultures of mutant strains to a 5.5 x 10^{-5} M concentration 4 h prior to inoculation. DHF, apiginin, and apiginin monoglucoside were added to rootlets as 20µl inoculations of a 5.5 x 10^{-5} M solution. Panel A illustrates the genetic requirement for a *Rhizobium* induced IPSIN; B, inoculation of strain ANU843 onto one root causes an IPSIN response on delayed ANU843, or concomitant *nodE*- or *nodL*- inoculation; C, IPSIN induced by *nodL* mutants; D, relief of IPSIN on a *nodL* mutant through preinduction of the mutant strain with DHF; E, IPSIN induced by flavonoid addition alone. Δp indicates a gene lacking its promoter.

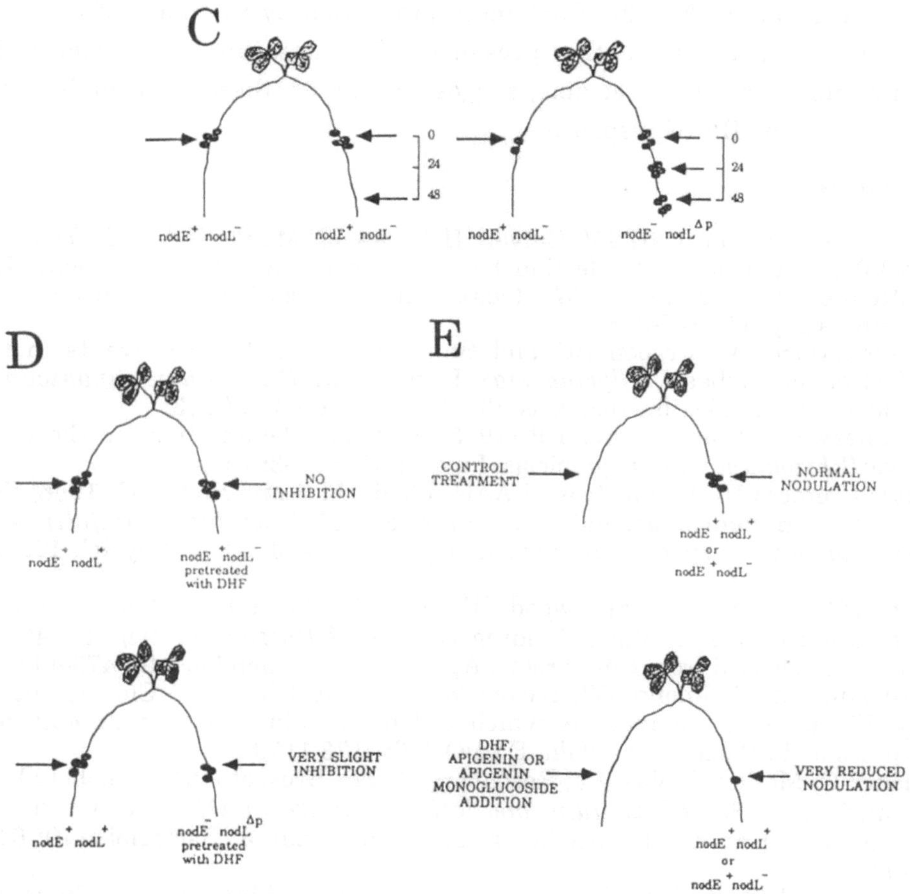

Flavonoid molecules are clearly established as plant signal molecules which induce the expression of nodulation genes. To test the hypothesis that competitive rhizobia are capable of more rapid responses to the presence of plant flavonoid molecules, we exposed *nodL* and *nodE* mutants to a single dose of the potent *nod* gene inducing flavone, DHF, prior to their inoculation onto split root clover plants. The opposing rootlet was simultaneously inoculated with strain ANU843 which normally inhibits nodule development by *nodL* or *nodE* mutants. The single exposure to DHF for 4h prior to inoculation enhanced the nodulation capacity of both these mutants to near normal levels (Fig. 3D) and abolished the ability of ANU843 to inhibit nodulation by these strains.

We have also demonstrated (Fig. 3E) that subterranean clover can mount an IPSIN response in the absence of an inoculating strain. The addition of DHF, apigenin, or apiginin monoglucoside will trigger an IPSIN response which can be observed on a separate rootlet. This result, together with the earlier observation (Fig. 2) of enhanced production by the plant of *nod* gene inducing compounds in the presence of compatible, and especially incompatible, *Rhizobium* strains, suggests a role for these compounds in the regulation of the IPSIN response.

References

Bauer WD, Bhuvaneswari TV, Calvert HE, Law IJ, Malik NSA and Vesper SJ (1985) Recognition and infection by slow-growing Rhizobia. In: Evans HJ, Bottomley PJ, and Newton WE (eds). Nitrogen Fixation Research Progress. Martinus Nijhoff : 247-253

Bhuvaneswari TV,Turgeon BG and Bauer WD (1980) Early events in the infection of soybean (*Glycine max* L. merr) by *Rhizobium japonicum*. I Localization of infectible root cells. Plant Physiol. **66:** 1027-1031

Bhuvaneswari TV and Solheim B (1985) Root hair deformations in the white clover/*Rhizobium trifolii* symbiosis. Physiol. Plant. **63:** 25-34

Canter-Cremers HCJ, van Brussel AAN, Plazinski J and Rolfe BG (1986) Sym plasmid and chromosomal gene products of *Rhizobium trifolii* elicit developmental responses on various legume roots. J. Plant Physiol. **122:** 25-40

Dazzo FB, Truchet GL, Sherwood JE, Hrabak EM and Gardiol AE. (1982) Alteration of the trifolin A-binding capsule of *Rhizobium trifolii* 0403 by enzymes released from clover roots. Appl. Environ. Microbiol. **44:** 478-490

Djordjevic MA, Redmond JW, Batley M and Rolfe BG (1987) Clovers secrete specific phenolic compounds which either stimulate or repress *nod* gene expression in *Rhizobium trifolii*. EMBO J. **6:** 1173-1179

Faucher C, Maillet F, Vasse J, Rosenberg C, van Brussel AAN, Truchet G and Denarie J (1988) *Rhizobium meliloti* host range *nodH* gene determines production of an alfalfa-specific extracellular signal. J. Bacteriol. **170:** 6489-5499

Firmin JL, Wilson KE, Rossen L and Johnston AWB (1986) Flavonoid activation of nodulation genes in *Rhizobium* reversed by other compounds present in plants. Nature **324**: 90-92

Kosslak RM and Bohlool BB (1984) Suppression of nodule development of one side of a split-root system of soybeans caused by prior inoculation of the other side. Plant Physiol. **75:** 125-130

Pierce M and Bauer WD (1983) A rapid regulatory response governing nodulation in soybean. Plant Physiol. **73:** 286-290.

Peters NK, Frost JW and Long SR (1986) A plant flavone, luteolin, induces expression of *Rhizobium meliloti* nodulation genes. Science **233:** 977-979

Redmond JW, Batley M, Djordjevic MA, Innes RW, Kuempel PL and Rolfe BG (1986). Flavones induce expression on nodulation genes in *Rhizobium*. Nature **323:** 632-635

Sargent CL, Huang SZ, Rolfe BG and Djordjevic MA (1987) Split-root assays using *Trifolium subterraneum* show that *Rhizobium* infection induces a systemic response that can inhibit nodulation of another invasive *Rhizobium* strain. Appl. Environ. Microbiol. **53:** 1611-1619

Weinman JJ, Djordjevic MA, Sargent CL, Dazzo FB and Rolfe BG (1988) A molecular analysis of the host range genes of *Rhizobium trifolii*. In: Palacios R and Verma DPS (eds) Molecular Genetics of Plant-Microbe Interactions. APS Press, St. Paul: 33-34

FIVE GENES ON THE SYM PLASMID OF RHIZOBIUM LEGUMINOSARUM WHOSE PRODUCTS ARE LOCATED IN THE BACTERIAL MEMBRANE OR PERIPLASM

A.W.B. Johnston, J.W. Latchford, E.O. Davis, A. Economou and J.A. Downie
A.F.R.C. Institute of Plant Science Research and
John Innes Institute,
Colney Lane, Norwich NR4 7UH, U.K.

INTRODUCTION

Various soil bacteria, collectively termed the rhizobia, can recognise specific leguminous plants and engage in a process in which the bacteria elicit a series of complex changes in the hosts and in which the rhizobia, during the process of nodule induction and occupation, alter in physiology, morphology and in their pattern of gene expression [see Long, 1989].

We are interested in genes in the species Rhizobium leguminosarum which are needed for nodulation and for determination of the host-range of a strain or biovar [bv.]. R.leguminosarum comprises three bv.'s [trifolii, phaseoli, and viciae], defined by their host ranges. These three bv.'s respectively nodulate clovers, Phaseolus beans and peas, lentils, vetches and field beans. nod and host range genes are on [Sym]biotic plasmids [Long, 1989] whose transfer between bv.'s of this species results in the acquisition of the ability to nodulate the hosts of the strains from which the Sym plasmid was derived. In several species of Rhizobium, including R.leguminosarum, genes required for nodulation and/or host range have been identified, mutated, sequenced and for some, their roles in the symbiosis have been elucidated or some clues on their function have been deduced from comparisons of their sequences with homologous genes of known function.

A priori, it seems likely that cell surfaces of the two symbionts would be important in the cell-to-cell signalling

NATO ASI Series, Vol. H 36
Signal Molecules in Plants and Plant-Microbe Interactions
Edited by B. J. J. Lugtenberg

events that presumably predicate the interaction between the two partners. Several lines of evidence, obtained from genetic studies on rhizobia, have supported this hypothesis. For example, mutants defective in the synthesis of various of the extracellular or cell-surface polysaccharides and lipopolysaccharides that are normally made by rhizobia, have been shown to be defective in the induction of normal nitrogen-fixing nodules [see Chapter by Walker et al this volume; see below; N.J. Brewin, pers. comm.]. Also, there is good evidence that the product of nodC [a gene that is essential for the early steps in the infection] is an unusual trans-membrane protein [John et al 1988]. Further, de Maagd et al [1989] showed that in R.l bv. viciae, there is a gene on the Sym plasmid of this strain which, when it is induced [see below] by exposure to flavonoids that activate nod genes, is required for the synthesis of an exported protein.

Here, we describe other genes that have been implicated in the early steps of nodulation and which encode proteins that are processed so that they are located either in the bacterial membrane or in the periplasm. These genes were identified by using the transposon Tn5phoA. This transposon contains a truncated version of the gene phoA of Escherichia coli which lacks the signal peptide sequence of alkaline phosphatase that is required for the enzyme to be transported into the periplasm where it is active. Thus, if Tn5phoA inserts into a gene, alkaline phosphatase activity is observed only if the insertion is correctly oriented and is in frame, thus giving a hybrid protein and if the gene product has sequences that target it to the periplasm or either of the bacterial membranes such that in the fusion product, the alkaline phosphatase is in the periplasmic space [Manoil and Beckwith, 1986]. Thus, several genes whose products are associated with the cell surfaces of Gram negative bacteria have been identified including some in R.meliloti which are needed for the synthesis of the acidic exopolysaccharide [Long et al 1988].

RESULTS

The product of the psi gene is associated with the membrane

A gene, psi, on the Sym plasmid of R.l bv. phaseoli, is close to other genes involved in nodulation or nitrogen fixation. psi mutants induce non-fixing nodules on Phaseolus beans and these are virtually devoid of bacteria [Borthakur et al 1985; unpublished]. Also, strains containing multiple copies of psi, cloned in plasmid vectors, are non-mucoid [ie. they fail to make acidic exopolysaccharide; also, they are severely reduced in their ability to form nodules on beans [Borthakur et al 1985]. psi encodes a small polypeptide of M_r 10kD which had an extremely hydrophobic N-terminus and a hydrophilic C-terminus suggesting that it might be inserted into one of the bacterial membranes [Borthakur et al 1987].

A recombinant plasmid, containing the cloned psi gene plus sequences upstream of psi was mutagenised with Tn5phoA, the mutant plasmids were transferred to R.l bv. phaseoli and the transconjugant colonies were screened on media containing X-P [a chromogenic indicator for alkaline phosphatase] for any that were blue. Three such colonies were identified; significantly, all were mucoid, indicating that these insertions abolished psi activity. Restriction mapping showed that two of the Tn5phoA insertions were in psi in the region that specified the hydrophobic part of its product. The third insertion was located in an ORF upstream of and in the same orientation as psi. Previous genetic evidence indicated that this ORF is in the same transcriptional unit as psi [Borthakur et al 1987]. The fact that the insertion in this upstream ORF abolishes the effects on EPS production of multiple copies of the psi gene is explained by a polar effect of the insertion on transcription of psi.

Thus Psi appears to be associated with the bacterial membrane and that psi is preceded by another gene whose product is also located in the membrane or periplasm. The precise mechanisms whereby mutations in this gene cause aberrant, non-fixing nodules are not clear nor is it known why over-expression of psi in strains containing multiple copies of this gene are defective both in nodulation and in EPS production.

nolE, which precedes nodD1 of bv. phaseoli specifies a periplasmic protein

In several rhizobia, it has been shown that most nod genes are not transcribed in cells grown in normal media. However, addition of particular flavonoid molecules, present in legume root exudates, causes active transcription of several nod operons. This induction requires the regulatory gene nodD whose product binds to conserved sequences [nod boxes] that precede such inducible nod operons. In some rhizobia there is a single copy of nodD but in others there are multiple copies of the gene [see Long, 1989 for review of nod gene regulation].

In R.l bv. phaseoli, there are three copies of nodD on the Sym plasmid pRP2JI. All three nodD genes are able to activate other nod operons in the presence of specific flavonoids [E.O.D., unpublished]. Two of these, nodD2 and nodD3, are linked and are transcribed towards each other [unpublished]. The third, nodD1, is approximately 25kb from these two other nodD genes and is located close to, but not adjacent to the nodA, B and C genes. nodD1 has some unusual features. Upstream of nodD1 is a regulatory nod box sequence [see Long,1989]. Transcription of nodD1 is activated by nodD1 itself, this induction being enhanced by addition of certain isoflavones such as genistein [E.O.D., unpublished]. Of relevance to the theme of this paper is the fact that, between nodD1 and the nod box upstream of this gene, is an ORF that defines a gene, nolE, with a high probability of encoding a protein. It was shown, using lac fusions in nolE and nodD1 that these two genes are in the same transcriptional unit, and are both under the control of the nod box preceding nolE.

The deduced polypeptide sequence of NolE has no similarity to that of any known nod genes. It was noted, though, that the N-terminus of NolE was similar to the concensus sequences of signal peptides that direct the transport of proteins through the cytoplasmic periplasm of Gram negative bacteria. To examine if NolE was indeed processed, a recombinant plasmid containing nolE was mutagenised with Tn5phoA. Two derivatives expressed alkaline phosphatase activity in Rhizobium and restriction mapping showed that these were located in nolE, with phoA in the same orientation as nolE. Since NolE has no obvious features that indicate that it might be associated with the bacterial membrane and because it is so small, we propose that it is a periplasmic protein and although its role in the nodulation process is not yet defined, it is interesting to speculate that it may serve to enhance the transport and/or modification of inducer flavonoids that are required for nodD1-mediated activation of nod gene transcription in R.l bv. phaseoli.

Further evidence that NodC specifies a protein associated with the bacterial surface

The nodulaton gene nodC is absolutely required for nodulation and, indeed, for the earliest steps of the infection process, namely root hair curling and cytokinesis in the host. John et al [1988] obtained antibodies against the NodC protein of R.meliloti and, using cell fractionation techniques, showed that the protein was located in the membrane fraction. These authors proposed that NodC was a transmembrane protein, held in place by a hydrophobic C-terminal region and that the N-terminus was exposed from the surface of the bacteria.

On the Sym plasmid pRL1JI of R.l bv. viciae, a cluster of nod genes has been identified, all of which are contained with a region that has been cloned in the recombinant plasmid pIJ1089 [Downie et al 1983]. pIJ1089 was mutagenised with Tn5phoA and the mutant derivatives of the plasmid were transferred to R.l bv. viciae. In two cases, the mutant plasmids specified alkaline phosphatase activity, but only when the colonies were plated on media containing flavonoid molecules that are nod gene inducers for this biovar. Restriction mapping of these two mutant derivatives of pIJ1089 showed that the Tn5phoA was in the 3' region of nodC, providing further evidence that NodC is associated with the bacterial membrane. Since insertion of the transposon disrupts the C-terminus of NodC, it is clear that this region of the protein is not required for the trafficking of the protein into a periplasmic/membrane location.

rhiC, a gene on the Sym plasmid of R.l. bv. viciae whose product is exported from the cytoplasm

During the course of Tn5phoA mutagenesis of pIJ1089 [see previous section] it was noted that when two of the mutant derivatives were transferred to Rhizobium and plated on media supplemented with X-P in the presence or absence of the nod inducer, hesperitin, the colonies were paler blue in the presence than in the absence of this flavanone indicating that transcription of the gene into which the transposon had inserted was reduced by exposure to nod gene inducers. This was reminiscent of the behaviour of the expression of a gene termed rhiA which is close to the nod genes on pRL1JI and which specifies the most abundant polypeptide made by strains of R.l bv. viciae [Dibb et al 1984; Economou et al 1989]. Indeed, in these two "down-regulated" pIJ1089::Tn5phoA derivatives, it was shown that the transposon was located downstream of rhiA and in the same orientation as that gene. It is possible that the gene, termed rhiC, into which these Tn5phoA insertions are located is in the same transcriptional unit as rhiA but this has not been formally proven. In this connection, it is relevant to note that Economou et al [1989] identified a lac fusion downstream of rhiA which had the same pattern of transcription as rhiA; its level of transcription was lower in cells exposed to nod gene inducers and its transcription depended on the presence of the regulatory gene rhiR. It is not yet clear if that lac fusion is in the same gene as that defined by the Tn5phoA insertion described here. Nevertheless, it is clear that close to the rhi gene cluster, there is a gene whose product is either exported into the periplasm or is associated with the one of the bacterial membranes. Mutations in this gene did not, however, lead to any clear inhibition of the nodulation of peas or Vicia.

CONCLUSIONS

These observations demonstrate the utility of the transposon Tn5phoA for the identification of gene products that are associated with the bacterial membrane or periplasm. By using this transposon, we have identified several genes, including some that are involved in EPS production, one that precedes the regulatory gene nodD, plus two that are in the nod gene region of R.l bv. viciae. Further work is required to define precisely the location of these gene products and, in some of the cases that we have described of the role of these genes in the nodulation of legumes.

ACKNOWLEDGMENTS

This work was supported by the Agricultural and Food Research Council. Further financial support was from the Agricultural and Genetics Company the EECand from Shell Research and we are grateful for their contributions. We are grateful to Graham Walker for supplying Tn5phoA.

REFERENCES

Borthakur, D., Downie, J.A., Johnston, A.W.B. and Lamb, J.W. [1985] psi, a plasmid-linked Rhizobium phaseoli gene that inhibits exopolysaccharide production and and which is required for symbiotic nitrogen fixation. Mol. Gen. Genet. 200: 278-282

Borthakur, D. and Johnston, A.W.B. [1987] Sequence of psi, a gene on the symbiotic plasmid of Rhizobium phaseoli which inhibits exoplysaccharide synthesis and nodulation and demonstration that its transcription is inhibited by psr, another gene on the Sym plasmid. Mol. Gen. Genet. 207:149-154

De Maagd, R.A., Spaink, H.P., Pees, E., Mulders, I.H.M., Wijfjes, A., Wijffelman, C.A., Okker, R.J.H. and Lugtenberg, B.J.J. [1989] Localization and symbiotic function of a region on the Rhizobium leguminosarum Sym plasmid pRL1JI responsible for a secreted, flavonoid-inducible 50-kilodalton protein. J. Bacteriol. 171:1151-1557

Dibb, N.J., Downie, J.A. and Brewin, N.J. 1984 Identification of a rhizosphere protein encoded by the symbiotic plasmid of Rhizobium leguminosarum. J. Bacteriol. 158:621-627

Downie, J.A., Hombrecher, G., Ma, Q.S., Knight, C.D., Wells, B. and Johnston, A.W.B. 1983 Cloned nodulation genes of Rhizobium leguminosarum determine host-range specicity. Mol. Gen. Genet. 190:359-365

Economou, A., Hawkins, F.K.L., Downie, J.A. and Johnston, A.W.B. 1989 Transcription of rhiA, a gene on a Rhizobium leguminosarum bv. viciae sym plasmid, requires rhiR and is repressed by flavanoids that induce nod genes. Mol. Microbiol. 3: 87-93

John, M., Schmidt, J., Wieneke, U., Krussman, H.-D. and Schell, J. 1988 Transmembrane orientation and receptor-like structure of the Rhizobium meliloti common nodulation gene nodC protein. EMBO J. 7:583-588

Long, S., McCune, S. and Walker, G.C. 1988 Symbiotic loci of Rhizobium meliloti identified by random TnphoA mutagenesis. J. Bacteriol. 170: 4257-4265

Long, S.R. 1989. Rhizobium-legume nodulation: life together in the underground. Cell 56:203-214

Manoil, C. and Beckwith, J. 1985 TnphoA transposon probe for protein export signals. Proc. Nat'l. Acad. Sci. USA 82:8129-8133

FACTORS CONTROLLING ROOT NODULE INDUCTION BY RHIZOBIUM MELILOTI

A. Kondorosi[1-3], E. Kondorosi[1,2], I. Dusha[1], Z. Banfalvi[1], J. Gyuris[1], M. John[2], J. Schmidt[2], A. Bakos[1], B. Hoffmann[2], E. Duda[1], F. de Bruijn[2] and J. Schell[2]
[1]Biological Research Center, Hungarian Academy of Sciences, H-6701 Szeged, Hungary, [2]Max-Planck Institut für Züchtungsforschung, D-5000 Köln 30, Germany, [3]Institut des Sciences Vegetales CNRS, 91198 Gif-sur-Yvette, France

INTRODUCTION

The establishment of nitrogen-fixing root nodules by Rhizobium requires multiple interactions between the bacterium and the plant. These interactions involve transmission of signal molecules, activating sets of genes determining symbiotic nodule development in the partner organism. This type of signaling between the host plant and its microsymbiont has been recently elucidated for the earliest interactions leading to the induction of root nodules (for reviews see Kondorosi and Kondorosi 1986; Rolfe and Gresshoff 1988). It was demonstrated that the host plants exude different flavonoids and isoflavones and some of these compounds activate (or inhibit) the expression of nodulation (nod) genes in Rhizobium. The activated nod genes are then involved in determining the production of diffusible signal molecules, evoking the earliest plant responses, root hair curling and division of cortical root cells. In this paper we discuss some aspects of the genetic control and specificity of nodule induction and signaling by R. meliloti.

POSITIVE AND NEGATIVE TRANSACTING FACTORS CONTROLLING NOD GENE EXPRESSION

We found that expression of common nodulation genes in R. meliloti is under positive as well as negative control. All the

NATO ASI Series, Vol. H 36
Signal Molecules in Plants and Plant-Microbe Interactions
Edited by B. J. J. Lugtenberg

three nodD gene copies (nodD1, D2 and D3) can activate the expression of the flavonoid-inducible nod genes (Göttfert et al 1986; Honma and Ausubel 1987). The products of the different nodD genes have distinct ability to interact with the various flavonoids exuded by the plant host, thus contributing to nod gene activation on a host dependent manner (Györgypal et al 1988).

Recently, an additional trans-acting factor, a repressor protein was found to be involved in the negative control of nod gene expression (Kondorosi et al 1989). This repressor protein allows only a lower level of nod gene induction. The repressor protein was detected in 18 out of 22 R. meliloti strains of different geographical origin, including AK631, a strain under detailed investigation in our laboratory. At the same time, in another widely studied strain (1021) this protein was not produced which correlated with the ability of this strain to express nod genes at significantly higher level.

The NodD protein was shown to bind to the conserved cis-regulatory element nod-box of the inducible nod gene promoters

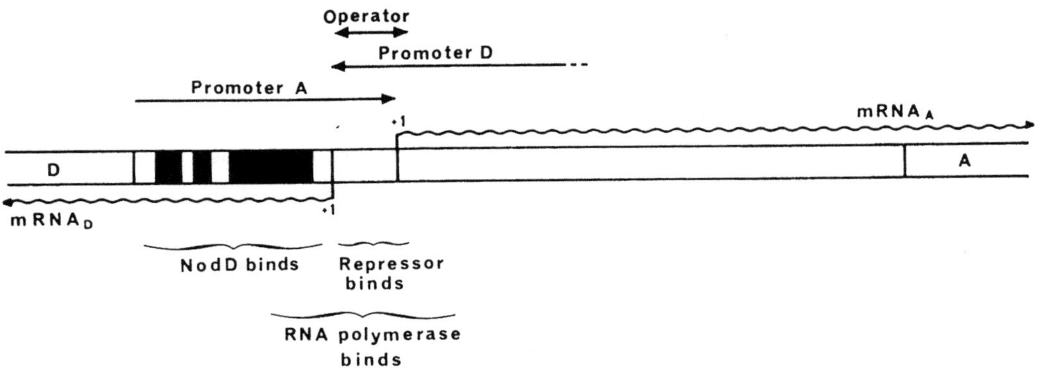

Fig. 1. Protein binding sites in the overlapping nodD1 and nodA promoter region.

(Fisher et al 1988; Kondorosi et al 1988), while the binding of the repressor to the overlapping *nodD1* and *nodA* promoter region at the RNA-polymerase binding site was demonstrated (Fig. 1; Kondorosi et al 1989). A model depicting the possible interaction of the flavonoid inducer, the NodD and the repressor was proposed (Kondorosi et al 1989). According to this model the repressor competes with RNA polymerase for the binding site. Therefore, binding of RNA polymerase in AK631 is occasional, resulting in the production of low amounts of the NodD protein. The low NodD concentration and the possible interfering effect of strong repressor binding cause weak binding of NodD to the *nod*-box. Upon addition of the flavonoid inducer the NodD protein interacts with the inducer, probably resulting in a conformational change of the NodD-DNA complex. The binding of RNA polymerase to this complex competes out the repressor leading to the transcription of the *nodABC* genes.

The DNA region encoding the repressor protein was cloned and by site-directed mutagenesis mutant derivatives of *R. meliloti* AK631 lacking the repressor were generated. These mutants exhibited delayed nodulation phenotype, indicating that fine tuning of *nod* gene expression, mediated by positive as well as negative transacting factors, is required for optimal nodulation of the plant host.

NITROGEN REGULATION OF EARLY NODULATION GENES

The presence of combined nitrogen in the soil generally suppresses the formation of root nodules by *Rhizobium*. We found that not only host plant genes involved in nodulation control are regulated by combined nitrogen, but the *nodABC* genes of *R. meliloti* are also under nitrogen (NH_4^+) control (Dusha et al 1989). Mutations in the nitrogen regulation (*ntr*) system of *R. meliloti* decreased the *ex planta* expression of *nodABC*-*lacZ* fusions (Table 1) as well as the nodulation ability of the host plant *Medicago sativa* . Using *nodD1*-*lacZ* and *nodD3*-*lacZ* fusions, it was found that while *nodD1* is not inducible by flavonoids of the plant host and its expression is independent

Table 1. Expression of nodD1-lacZ, nodC-lacZ and nodD3-lacZ fusions in wild type and mutant R. meliloti strains

STRAIN	GENE FUSION ON PLASMID	ß-GALACTOSIDASE ACTIVITY (Miller units)			
		nitrogen-limited medium		nitrogen-excess medium	
		-L	+L	-L	+L
R.m.1021	-	0.7	0.4	-	-
R.m.1021	nodD1-lacZ	714	577	550	388
R.m.5002(ntrC::Tn5)	nodD1-lacZ	786	591	552	344
R.m.1680(ntrA::Tn5)	nodD1-lacZ	758	675	687	550
R.m.399	nodD1-lacZ	740	575	708	391
R.m.1021	nodC-lacZ	13	354	9	64
R.m.5002(ntrC::Tn5)	nodC-lacZ	11	130	10	82
R.m.1680(ntrA::Tn5)	nodA-lacZ	8	114	9	72
R.m.399	nodA-lacZ	19	182	11	140
R.m.1021	nodD3-lacZ	36	130	15	16
R.m.5002(ntrC::Tn5)	nodD3-lacZ	40	69	22	51
R.m.1680(ntrA::Tn5)	nodD3-lacZ	43	84	23	80
R.m.399	nodD3-lacZ	42	98	19	44

Plasmids used: pRmM61(nodD1-lacZ)(Mulligan and Long 1985); pRmM57(nodC-lacZ)(Mulligan and Long 1985); pNID12(nodD3-lacZ)(constructed by N. Iyer). The ntr mutants were provided by F. Ausubel. R.m.399 exhibiting an elevated level of nodABC expression was isolated in this study. Nitrogen-limited medium contained 3mM $(NH_4)_2SO_4$, nitrogen-excess medium: 100 mM $(NH_4)_2SO_4$.
-L, +L: without or with luteolin induction.

of nitrogen control, the expression of nodD3 was induced by luteolin (the major inducer in M. sativa) and was repressed by NH_4^+ (Table 1). In addition, our experiments indicated that the control of nodD3 expression requires two genes located adjacent to nodD3. These genes were designated as syrM and syrA by Mulligan and Long (1989). It seems that the product of nodD3 may play a special role in the regulation of other nod genes, particularly in response to combined nitrogen. Possibly, under nitrogen starvation conditions, the NodD3 protein is the primary activating factor for nod gene expression. This suggestion is in line with the observation that under nitrogen limitation a nodD3::Tn5 mutant exhibited severely reduced ability to express the nodC-lacZ fusion.

R. meliloti mutants exhibiting an increased level of nodABC expression in the presence of ammonia were isolated (strain R.m.399 in Table 1). The altered sensitivity of mutants to ammonia regulation was accompanied by a more efficient nodulation ability of M. sativa in the presence of ammonia, compared to the parent wild type strain. The DNA region responsible for this altered sensitivity was cloned and it was found to be distinct from the ntr loci identified so far and from the nod repressor described above (Dusha et al 1989).

The demonstrated nitrogen control of nod gene expression and the involvement of the novel regulatory gene again contribute to the complex control mechanism for fine tuning of the expression of nodulation genes and, in this way, to a more sensitive modulation of nodule induction.

PRODUCTION OF ROOT HAIR DEFORMATION FACTORS BY R. MELILOTI NOD GENES IN E. COLI

As described above, fine tuning of nod gene expression in R. meliloti is required for optimal nodulation of the host plant suggesting that the products of these genes must be available at defined concentrations to have optimal activity in nodule induction. Studies from several laboratories have shown that the products of induced nod genes are involved in the synthesis of diffusible signal molecules evoking root hair curling and meristematic cell division and it is likely that these compounds must be present at optimal concentrations for efficient nodulation (Bhuvaneswari and Solheim 1985; van Brussel et al 1986; Zaat et al 1987; Schmidt et al 1988; Faucher et al 1988).

To study the involvement of particular _nod_ genes in these processes and to identify these signal molecules, we introduced and expressed _nod_ genes in different combinations in _E. coli_. The biological activity of diffusible compounds produced in these strains was tested on various leguminous plants by

Table 2. Root hair response of leguminous plants to _E.coli_ cells expressing the _nodABC_ and the _nodABC,hsnD_ genes of _R. meliloti_

GROUPS	PLANT SPECIES	_E. COLI_ CARRYING	
		nodABC	_nodABC,hsnD_
I.	_Trifolium repens_ (white clover)	+	+
	Trifolium pratense (red clover)	+	+
II.	_Macroptilium atropurpureum_ (siratro)	+/-	+
III.	_Melilotus albus_ (sweet clover)	*/-	+
	Medicago sativa (alfalfa)	-	+
IV.	_Vigna sinensis_ (cowpea)	-	-
	Desmodium intortum	-	-

E. coli was negative, while the appropriate microsymbionts of the various plant hosts were positive. The _nod_ genes were introduced into _E. coli_ on plasmids and the genes were expressed from plasmid promoters (Schmidt et al 1984; Horvath et al 1986; Banfalvi and Kondorosi 1989).

+: strong plant response, heavily deformed and curled root hairs

+/-: infrequent root hair deformation

*/-: very few deformed root hairs

-: no plant response

detecting root hair deformation (Had activity). In the first series of experiments E. coli strains carrying the nodABC or nodABC+hsnD(nodH) genes were used. Based on the response to these strains, four categories of plants were established (Table 2). Upon infection with E. coli carrying the nodABC construct, root hair deformation was detected on Trifolium repens or on T. pratense, while the hsnD gene was additionally needed for eliciting the same response on M. sativa and Melilotus albus. Weak root hair deformation was seen on Macroptilium atropurpureum with E. coli harboring the nodABC genes and was much more pronounced when the hsnD gene was also present. Vigna and Desmodium species did not respond to these E. coli strains. Exudates or cytosolic fractions of these strains elicited the same response as the intact bacteria (Banfalvi and Kondorosi 1989).

Subclones expressing only nodA, nodB or nodC genes or the same genes in pairs did not provide a compound comparable to the NodABC-factor, suggesting that all the three genes are required for the production of the Had-factor, which is active on T. repens. In the absence of any of the these common nod genes, even the cytosolic fraction of the E. coli carrying the other nod genes did not elicit significant Had response. This result suggests that all the three proteins (NodA, NodB and NodC) are enzymatically involved in the generation of the Had-factors.

It seems that a family of related compounds are produced by rhizobia induced for their nod genes. The basic Had-factor determined by the nodABC genes is modified by hsnD to enable its action on Medicago, Melilotus or Macroptilium. It is possible that hsn genes of other Rhizobium or Bradyrhizobium species which were shown to be involved in determining root hair curling ability, also modify the NodABC-factor, converting it into a signal molecule specific for the appropriate plant host.

REFERENCES

Banfalvi Z, Kondorosi A (1989) Production of root hair deformation factors by Rhizobium meliloti nodulation genes in Escherichia coli: HsnD (NodH) is involved in the plant host-specific modification of the NodABC-factor. Plant Mol Biol (in press)

Bhuvaneswari TV, Solheim B (1985) Root hair deformation in the white clover/Rhizobium trifolii symbiosis. Physiol Plant 63: 25-34

Dusha I, Bakos A, Kondorosi A, de Bruijn F, Schell J (1989) The Rhizobium meliloti early nodulation genes (nodABC) are nitrogen-regulated: Isolation of a mutant strain with efficient nodulation capacity on alfalfa in the presence of ammonium. (submitted)

Faucher C, Maillet F, Vasse J, Rosenberg C, van Brussel AAN, Truchet G, Denarie J (1988) Rhizobium meliloti host range nodH gene determines production of an alfalfa-specific extracellular signal. J Bacteriol 170: 5489-5499

Fisher RF, Egelhoff TT, Mulligan JT, Long SR. (1988) Specific binding of Rhizobium meliloti extracts containing NodD to DNA sequences upstream of inducible nodulation genes. Genes Dev 2:282-293

Göttfert M, Horvath B, Kondorosi E, Putnoky P,Rodriguez-Quinones F, Kondorosi A (1986) At least two nodD genes are necessary for efficient nodulation of alfalfa by Rhizobium meliloti. J Mol Biol 191: 411-420

Györgypal Z, Iyer N, Kondorosi A (1988) Three regulatory nodD alleles of diverged flavonoid-specificity are involved in host-dependent nodulation by Rhizobium meliloti. Mol Gen Genet 212: 85-92

Honma MA, Ausubel FM (1987) Rhizobium meliloti has three functional copies the nodD symbiotic regulatory gene. Proc Natl Acad Sci USA 84: 8558-8562

Horvath B, Kondorosi E, John M, Schmidt J, Török I,.Györgypal Z, Barabas I, Wieneke U, Schell J, Kondorosi A. (1986) Organization, structure and symbiotic function of Rhizobium meliloti nodulation genes determining host specificity for alfalfa. Cell 46: 335-343

Kondorosi E, Kondorosi A (1986) Nodule induction on plant roots by Rhizobium. Trends in Biochem Sci 11: 296-299

Kondorosi E, Gyuris J, Schmidt J, John M, Duda E, Schell J, Kondorosi A (1988) Positive and negative control of nodulation genes in Rhizobium meliloti strain 41. In: Palacios R and Verma DPS (eds). Molecular plant-microbe interactions. APS Press, St.Paul, Minn.: 73-78

Kondorosi E, Gyuris J, Schmidt J, John M, Duda E, Hoffmann B, Schell J, Kondorosi A (1989) Positive and negative control of nod gene expression in Rhizobium meliloti is required for optimal nodulation. EMBO J. (in press)

Mulligan JT, Long SR (1985) Induction of Rhizobium meliloti nodC expression by plant exudate requires nodD. Proc Natl Acad Sci USA 82: 6609-6613

Rolfe BG, Gresshoff PM (1988) Genetic analysis of legume nodule initiation. Ann Rev Plant Physiol Plant Mol Biol 39: 297-319

Schmidt J, John M, Kondorosi E, Kondorosi A, Wieneke U, Schröder G, Schröder J, Schell J (1984) Mapping of the protein coding regions of Rhizobium meliloti common nodulation genes. EMBO J 3: 1705-1711

Schmidt J, Wingender-Drissen R, John M, Wieneke U, Schell J (1988) Rhizobium meliloti nodA and nodB genes are involved in generating compounds that stimulate mitosis of plant cells. Proc Natl Acad Sci USA 85: 8578-8582

van Brussel AAN, Zaat SAJ, Canter-Cremers HCJ, Wijffelman CA, Pees E, Tak T, Lugtenberg BJJ (1986) Role of plant root exudate and Sym plasmid-localized nodulation genes in the synthesis by Rhizobium leguminosarum of Tsr factor, which causes thick and short roots on common vetch. J Bacteriol 165: 517-522

Zaat SAJ, van Brussel AAN, Tak T, Pees E, Lugtenberg BJJ (1987) Flavonoids induce Rhizobium leguminosarum to produce nodABC gene-related factors that cause thick, short roots and root hair responses on common vetch. J Bacteriol 169: 3388-3391

GENETIC ANALYSES SUGGESTING BACTERIAL-PLANT SIGNALLING DURING NODULATION

Jane Glazebrook, T. Lynne Reuber, and Graham C. Walker
Department of Biology, Massachusetts Institute of Technology, Cambridge, Massachusetts, 02139,
USA

INTRODUCTION

Bacteria of the genus *Rhizobium* fix nitrogen in symbiotic association with leguminous plants. The establishment of symbiosis probably requires many signalling interactions between the bacteria and the plant. The bacterial *nod* genes are induced by plant flavones (Peters et al 1986) and are involved in synthesis of a compound which has been reported to induce plant cell division (Schmidt et al 1988). In our laboratory, we have studied three classes of *Rhizobium meliloti* mutants which may be involved in signalling interactions. These are mutants affected in exopolysaccharide synthesis, *trpE(G)* mutants, and a group of Tn*phoA* insertion mutants.

EXOPOLYSACCHARIDES ARE REQUIRED FOR NODULE INVASION

Mutants of *R. meliloti* which fail to synthesize an acidic Calcofluor-binding exopolysaccharide (EPS I) form ineffective nodules on alfalfa (Finan et al 1985; Leigh et al 1985). These nodules have aborted infection threads, and do not contain any bacteroids (Leigh et al 1987). The symbiotic block of these *exo* mutants thus appears to be at the stage of nodule invasion and development. We have performed a fairly extensive genetic analysis of EPS I biosynthesis (Leigh et al 1985; Finan et al 1986; Long et al 1988a). There are two *exo* loci on the chromosome and a cluster of thirteen *exo* loci on the second symbiotic megaplasmid that affect the synthesis of EPS I. Mutations in the *exoH* and *exoG* loci are particularly interesting because they result in structural alterations of EPS I. *exoH* mutants fail to succinylate EPS I and have symbiotic phenotypes similar to those of mutants which make no exopolysaccharide at all (Leigh et al 1987). The simplest explanation for these results is that the succinyl group in EPS I is required for its function. Strains

NATO ASI Series, Vol. H 36
Signal Molecules in Plants and Plant-Microbe Interactions
Edited by B. J. J. Lugtenberg

with mutations in the *exoG* locus fluoresce slightly on Calcofluor plates, indicating production of a reduced amount of Calcofluor-binding material, but there is no detectable high molecular weight EPS I in their culture supernatants (Long et al 1988a). It appears that *exoG* mutants produce a low molecular weight form of EPS I. These strains are able to form nitrogen-fixing nodules on alfalfa, but do so less efficiently than the wild-type. These phenotypes suggest that, while high molecular weight EPS I has a symbiotic function, a low molecular weight form of EPS I may be sufficient for some nodule invasion to occur.

The symbiotic defects of *exo* mutants can be suppressed by the presence of a mutation, *expR101*, which causes overproduction of a second exopolysaccharide, EPS II (Glazebrook and Walker 1989). Genetic analyses have shown that the products of a cluster of at least six *exp* genes located on the second symbiotic megaplasmid, as well as the product of the *exoB* gene, are required for EPS II synthesis. The presence of the *expR101* mutation causes increased transcription of the *exp* genes, resulting in overproduction of EPS II. As a consequence of this genetic analysis, we were able to construct strains which produced EPS I or EPS II exclusively, or neither EPS. *Medicago sativa* plants inoculated with an EPS II producing strain grew about as well as plants inoculated with the EPS I producing strain. However, the EPS II producing strain formed empty, ineffective nodules on four other plant species which were effectively nodulated by the EPS I producing strain. Thus, it appears that exopolysaccharides are involved in determining host range at the level of nodule invasion, rather than at the level of primary recognition of the host.

The structure of EPS II has been determined (Glazebrook and Walker 1989; G.-R. Her, J. Glazebrook, G.C. Walker, and V.N. Reinhold, submitted), and is shown next to the structure of EPS I (Aman et al 1981) in the figure below. Note that EPS I contains approximately one succinyl and one acetyl modification per octasaccharide subunit at as yet undetermined positions. Both exopolysaccharides are acidic, contain glucose and galactose, and have acetyl and pyruvate (1-carboxyethylidene) modifications. However, the structures differ in many respects: i) EPS II does not contain any succinate groups, ii) EPS II contains more galactose than EPS I, iii) EPS II is unbranched, iv) EPS II has both α and β glycosidic linkages, while EPS I has only β glycosidic linkages, and v)

The pyruvate group in EPS I is linked to glucose, while in EPS II it is linked to galactose. The only structural motif that the two exopolysaccharides have in common is the dimer Glc-β(1,3)-Gal.

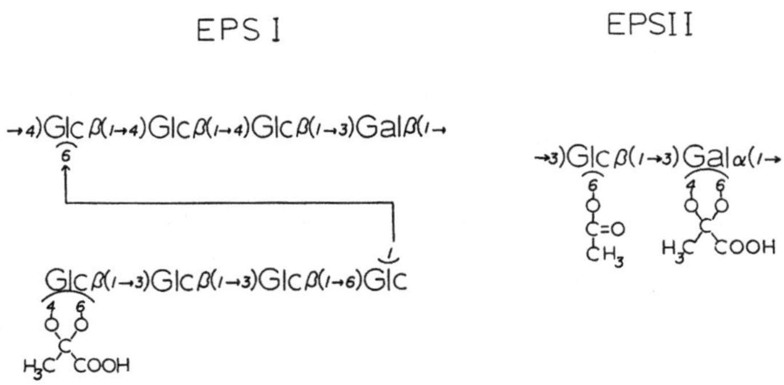

Exopolysaccharides may have more than one function in nodulation. Since *exoG* mutants, which make low molecular weight EPS I, but not high molecular weight EPS I, form effective nodules at reduced efficiency relative to wild-type, but much better than mutants which make no exopolysaccharide, it is possible that both the high and low molecular weight forms of EPS have symbiotic functions. One of these functions may be to enable the bacteria to avoid plant defense responses. This suggestion is based on a report that β(1,3)-glucanases are among the proteins induced when plants are exposed to pathogens (Kombrink et al 1988). Perhaps the modification around the Glc-β(1,3)-Gal linkage in EPS II prevents its cleavage by β(1,3) glucanases, thereby protecting the bacteria from plant defense responses (Glazebrook and Walker, 1989). This dimer could be similarly modified in EPS I, since the positions of the acetyl and succinyl groups in EPS I are not known. The observation that a crude enzyme preparation derived from lysates of phage ϕM12, which cleaves EPS II at the Glc-β(1,3)-Gal linkage, also cleaves EPS I at this position (G.R. Her, J. Glazebrook, V.N. Reinhold, and G.C. Walker, personal communication), is consistent with this idea. The host range effects of exopolysaccharides could be explained if some plants can degrade EPS II but not EPS I, while *Medicago sativa* cannot degrade either exopolysaccharide. Other possible roles for exopolysaccharides include serving as a carrier for extracellular enzymes or signal molecules, or forming part of the infection thread matrix. One of the more popular models is that exopolysaccharides are themselves signal molecules.

Carbohydrates have been shown to function as signal molecules in plants. Oligosaccharides derived from the breakdown of plant cell walls resulting from fungal attack induce synthesis of anti-fungal molecules (for review see McNeil et al 1984). Several lines of evidence suggest that oligosaccharide fragments of EPS may have a symbiotic function. First, *exoG* mutants, which produce mainly low molecular weight EPS I can invade nodules, albeit at reduced efficiency (Long et al 1988a). Second, Leigh and Lee (1988) have shown that *exoH* mutants produce less low molecular weight EPS I than the wild-type. Perhaps the symbiotic defects of these mutants result from a failure to process the non-succinylated EPS I into an oligosaccharide signal molecule. Third, *exo* mutants of *Rhizobium* sp. NGR234 and *Rhizobium trifolii* can form effective nodules if either high molecular weight EPS or oligosaccharide subunits of EPS are supplied exogenously (Djordjevic et al 1987). To date, we have been unable to suppress the symbiotic defects of *R. meliloti exo* mutants by addition of exogenous EPS.

If oligosaccharide fragments of EPS were to function as signal molecules, how could two exopolysaccharides with as many structural differences as EPS I and EPS II substitute for each other? One possibility is that *Medicago sativa* can recognize either of two signals, one derived from EPS I and one derived from EPS II. If this is the case, the plants which can only be nodulated by EPS I producing strains would be expected to recognize only the EPS I derived signal molecule. Alternatively, both exopolysaccharides could contain a common signal. This signal could consist of the Glc-β(1,3)-Gal dimer, and perhaps also of an acetyl group and an acidic group, depending upon the positions of the acetyl and succinyl groups in EPS I. This model predicts that the EPS I-requiring plants recognize a portion of the EPS I structure which is not contained in EPS II. In either of these models, the symbiotic defects resulting from non-succinylated EPS I could be due to either a defect in the structure of the signal molecule, or to the reduced amount of low molecular weight exopolysaccharide produced by the *exoH* mutant strains.

trpE(G) MUTANTS ARE BLOCKED AT A LATE STAGE IN NODULE DEVELOPMENT

The *R. meliloti trpE(G)* locus encodes anthranilate synthetase, which is the first enzyme in the tryptophan biosynthetic pathway. When alfalfa plants were inoculated with a *trpE(G)* deletion mutant, 20-50% of the plants grew no larger than uninoculated control plants (G. Barsomian, K. Lohmann, and G.C. Walker, in preparation). The remainder grew slightly larger than the controls, but not nearly as large as plants inoculated with wild-type bacteria. These plants were found to have low levels of nitrogenase activity. The nodules formed by the *trpE(G)* mutant were abnormally long and white. A small number of *trpE(G)* mutant nodules were sectioned longitudinally and examined by light microscopy. They seemed to consist of a nodule meristem, followed by a zone of cells resembling the early symbiotic zone of wild-type nodules, and a zone of senescent cells extending through the rest of the nodule to the plant. In some of the nodules, the cells closest to the root were packed with bacteroids. These nodules probably fix a small amount of nitrogen, and allow some of the plants inoculated with the *trpE(G)* mutant to grow slightly larger than the controls. This suggests that the symbiotic defects of *trpE(G)* mutants can be partially overcome by proximity to the plant root.

Strains with mutations in genes required for later steps in the tryptophan biosynthetic pathway (*trpA, B, C, D* and *F*) are symbiotically effective, so the reason for the symbiotic defects of *trpE(G)* mutants cannot be due to the failure to make tryptophan. This implies that a factor synthesized from anthranilate, or an alternate function of the anthranilate synthetase protein, is required for normal nodule development. The phenotypes of the nodules induced by the *trpE(G)* mutant suggest that the bacteria successfully invade nodules, but fail to become nitrogen-fixing bacteroids. It is possible that anthranilate synthetase is required for the synthesis of a signal molecule which acts at a late stage in nodule development.

RANDOM phoA FUSIONS ENRICH FOR NEW SYMBIOTIC LOCI

Many classes of gene products involved in signal reception and transduction in nodulation would be expected to be present on the cell surface or in the periplasmic space. Recognition of bacterial cell surface molecules by the plant may also be important in symbiosis. In the hope of enriching for mutants with symbiotic phenotypes, we undertook a mutagenesis which targeted exported proteins.

Genomic fusions were made by mutagenizing a Pho^- derivative of Rm1021 with Tn*phoA* (Long et al 1988b). Tn*phoA* is a derivative of Tn*5* which produces translational gene fusions to *E. coli* alkaline phosphatase (Manoil and Beckwith 1985). The alkaline phosphatase gene lacks a signal sequence, and the protein is inactive unless it is exported to the periplasmic space. Therefore, only in-frame fusions to exported proteins produce active alkaline phosphatase. These can be detected on plates containing the indicator XP (5-bromo-4-chloro-3-indolyl phosphate), which is cleaved by alkaline phosphatase to form an insoluble blue product. Approximately 1% of the Tn*phoA* insertion mutants were blue on XP plates. Two percent of these mutants were found to be Fix^-. By comparison, Meade et al (1982) reported that only 0.3% of random Tn*5* insertion mutants had symbiotic defects. Therefore, the use of Tn*phoA* to target exported proteins resulted in a six-fold enrichment for Fix^- mutants.

Of the twenty-five Fix^- fusions isolated, thirteen were found to be in genes necessary for production of the Calcofluor-binding exopolysaccharide (EPS I). Bacteria could be recovered from nodules induced by the other twelve mutants, indicating that the symbiotic defects of these mutants were at post-invasion stages. Two of the fusions were assigned to known genes. One was shown to be in a previously identified gene involved in lipopolysaccharide (LPS) biosynthesis (Kieber et al 1987), and the other was shown to be in *dctA*, a gene required for dicarboxylic acid transport. Of the remaining ten fusions, two pairs were found to be duplicates. Therefore, eight new symbiotic loci were identified. Five of the loci were chromosomal, one was on the first symbiotic megaplasmid, and two were on the second symbiotic megaplasmid. The gene products were localized by cell fractionation. Three of the fusions are periplasmic, and five are probably membrane proteins.

The common *nod* genes of *R. meliloti* are induced 2-3 fold in the presence of luteolin (Mulligan and Long 1985). Two of the eight *phoA* fusion proteins were modestly induced (1.4-fold) in the presence of luteolin. The LPS mutant was induced 2.4-fold. This suggests that, like the nod genes, some of the Tn*phoA* insertion mutants may be involved in responding to plant signals.

It is possible that the new loci we have identified may be involved in exchange of signals between the plant and the bacterium. Signals or specific recognition mechanisms may be involved at several post-invasion stages, including proliferation of the bacteria in the infection thread, release from the infection thread, bacteroid differentiation, initiation of nitrogen fixation, and maintainance of symbiosis. Our loci might be involved in reception or transmission of plant signals, or synthesis of bacterial signal molecules. They could also encode surface determinants which are recognized by the plant.

ACKNOWLEDGEMENTS

We wish to thank Guor-Rong Her, Vernon Reinhold, Gary Barsomian, and Karen Lohmann for communication of unpublished results. This work was supported by Public Health Service grant GM31030 to Ethan R. Signer and G.C.W.

REFERENCES

Aman P, McNeil M, Franzen L, Darvill AG, Albersheim P (1981) Structural elucidation, using HPLC-MS and GLC-MS, of the acidic polysaccharide secreted by *Rhizobium meliloti* strain Rm1021. Carbohydr Res 95:263-287

Djordjevic SP, Chen H, Batley M, Redmond JW, Rolfe BG (1987) Nitrogen fixation ability of exopolysaccharide synthesis mutants of *Rhizobium* sp. strain NGR234 and *Rhizobium trifolii* is restored by the addition of homologous exopolysaccharides. J Bacteriol 169:53-60

Finan TM, Hirsch AM, Leigh JA, Johansen E, Kuldau GA, Deegan S, Walker GC, Signer ER (1985) Symbiotic mutants of *Rhizobium meliloti* that uncouple plant from bacterial differentiation. Cell 40:869-877

Finan TM, Kunkel B, DeVos GF, Signer ER (1986) A second symbiotic megaplasmid in *Rhizobium meliloti* encodes exopolysaccharide and thiamine synthesis genes. J Bacteriol 167:66-72

Glazebrook J, Walker GC (1989) A novel exopolysaccharide can function in place of the Calcofluor-binding exopolysaccharide in nodulation of alfalfa by *Rhizobium meliloti*. Cell 56:661-672

Kieber J, Clover R, Finan TM, Signer ER (1987) Surface properties of *Rhizobium meliloti* associated with symbiosis. In: Verma DPS, Brisson LN (eds) Molecular genetics of plant-microbe interactions. Martinus Nijhoff Publishers Amsterdam, p182-184

Kombrink E, Schroeder M, Hahlbrock K (1988) Several "pathogenesis-related" proteins in potato are 1,3β-glucanases and chitinases. Proc Natl Acad Sci USA 85:782-786

Leigh JA, Lee CC (1988) Characterization of polysaccharides of *Rhizobium meliloti exo* mutants that form ineffective nodules. J Bacteriol 170:3327-3332

Leigh JA, Signer ER, Walker GC (1985) Exopolysaccharide-deficient mutants of *R. meliloti* that form ineffective nodules. Proc Natl Acad Sci USA 82:6231-6235

Leigh JA, Reed JW, Hanks JF, Hirsch AM, Walker GC (1987) *Rhizobium meliloti* mutants that fail to succinylate their Calcofluor-binding exopolysaccharide are defective in nodule invasion. Cell 51:579-587

Long S, Reed JW, Himawan JW, Walker GC (1988a) Genetic analysis of a cluster of genes required for synthesis of the Calcofluor-binding exopolysaccharide of *Rhizobium meliloti*. J Bacteriol 170:4239-4248

Long S, McCune S, Walker GC (1988b) Symbiotic loci of *Rhizobium meliloti* identified by random Tn*phoA* mutagenesis. J Bacteriol 170:4257-4265

Manoil C, Beckwith J (1985) Tn*phoA* transposon probe for protein export signals. Proc Natl Acad Sci USA 82:8129-8133

McNeil M, Darvill AG, Fry SC, Albersheim P (1984) Structure and function of the primary cell walls of plants. Ann Rev Biochem 53:625-663

Meade HM, Long SR, Ruvkun GB, Brown SE, Ausubel FM (1982) Physical and genetic characterization of symbiotic and auxotrophic mutants of *Rhizobium meliloti* induced by transposon Tn5 mutagenesis. J Bacteriol 149:114-122

Mulligan JT, Long SR (1985) Induction of *Rhizobium meliloti nodC* expression by plant exudate requires *nodD*. Proc Natl Acad Sci USA 82:6609-6613

Peters NK, Frost JW, Long SR (1986) A plant flavone, luteolin, induces expression of *Rhizobium meliloti* nodulation genes. Science 233:977-980

Schmidt J, Wingender R, John M, Wieneke U, Schell J (1988) *Rhizobium meliloti nodA* and *nodB* genes are involved in generating compounds that stimulate mitosis of plant cells. Proc Natl Acad Sci USA 85:8578-8582

LIPOPOLYSACCHARIDE: A SIGNAL IN THE ESTABLISHMENT OF THE RHIZOBIUM/LEGUME SYMBIOSIS?

Ruud A. de Maagd and Ben J.J. Lugtenberg
Department of Plant Molecular Biology, Leiden University,
Nonnensteeg 3, 2311 VJ Leiden, The Netherlands

INTRODUCTION

An essential role of lipopolysaccharides (LPS) in the nodulation of legumes by Rhizobia has long been suspected, and the recent isolation of mutants with altered LPSs has given research on this subject new momentum. In this paper we will attempt to give an overview of the data on this subject as well as speculate on the role of this surface molecule in symbiosis.

Lipopolysaccharide is an amphipathic molecule found exclusively in the outer membrane (OM) of Gram-negative bacteria. Its structure and functions have been best characterized for Enterobacteriaceae (Lugtenberg and van Alphen 1983). Basically it can consist of three regions: (i). Lipid A, which is a glycolipid consisting of a glucosamine disaccharide unit and approximately six acylresidues, which are embedded in the outer leaflet of the outer membrane bilayer. (ii). Linked to it is the core oligosaccharide, which contains a number of sugars, among which 2-keto-3-deoxyoctonate (KDO) and L-glycero-D-mannoheptose are almost LPS-specific. (iii). The outer part of the molecule, the O-antigen, consists of repeating units of 3 to 6 sugar residues. The number of repeating units may vary, even within one culture. This leads to a "ladder-like" appearance of the profile of LPSs separated according to their molecular weights by SDS-PAGE. The structure of the O-antigen is extremely variable, even within a single species. This property, together with the antigenicity of LPS, makes the O-antigen very useful for serotyping of strains. A large diversity of sugars, including neutral sugars, amino sugars and uronic acids is found in O-antigens. Moreover, sugars can be substituted with O-acetyl

NATO ASI Series, Vol. H 36
Signal Molecules in Plants and Plant-Microbe Interactions
Edited by B. J. J. Lugtenberg

groups, phosphate, amino acids or ethanolamine triphosphate.

In Enterobacteriaceae, the presence of LPS is thought to be an important determinant for the permeability properties of the outer membrane. Presence of LPS exclusively in the outer leaflet of the OM, is probably responsible for the high resistance against detergents and antibiotics. Mutants with altered LPS structure, in particular when they are affected in their inner core structure (deep rough mutants), have become much more sensitive to these agents. LPS/LPS-interactions, stabilized by divalent cations, are an important factor in the stabilization of the outer membrane and its permeability barrier. Moreover, LPS may be required for the activity of outer membrane proteins and influence outer membrane biogenesis. Lipopolysaccharides and their functions in other genera are much less documented. In the field of plant-microbe interactions it has been reported that mutants of certain plant-pathogenic bacteria with altered LPSs have lost their virulence (Hendrick and Sequiria 1984; Schoonejans *et al* 1987). Similarly, in plant-growth promoting *Pseudomonas* strains an intact LPS is required for efficient root colonization (de Weger *et al*, this volume).

The polysaccharide moieties of LPSs of various strains of *Rhizobium* have been partially characterized, because of their suspected role in symbiosis and, in particular host plant recognition (for review: Carlson 1982). Sugar analysis as well as SDS-polyacrylamide gel electrophoresis and immunological analysis have shown that the LPSs of fast-growing rhizobia vary as much among strains of a single species or biovar as they differ between species and biovars. SDS-PAGE and gel filtration studies have shown that the *R. leguminosarum* LPS can be divided into two species (Carlson 1984), which may be tentatively described as follows. The slower migrating, high molecular weight LPS I is often heterogeneous in size. Its differs from strain to strain and therefore can be used for strain identification (de Maagd *et al* 1988a). As it is also an immunodominant, surface-exposed antigen, this LPS species is considered to contain an O-antigen-like structure. The faster migrating, low molecular weight LPS II consists of one molecular species, which probably represents a "core" LPS, possibly an intermediate in synthesis of LPS I. Although heterogeneity in LPS chain length exists, it is often much less extensive than generally found in *Enterobacteriaceae*, and in several *Rhizobium* strains the LPS I seems to consist of only one molecular species (Carlson 1984; de Maagd *et al* 1988a). LPS of *R. meliloti* seems to be quite different, both in SDS-PAGE (de Maagd *et al* 1988a) as well as in that it has a very high KDO content and that it is highly cross-reactive with antisera against LPS from different strains, suggesting that it lacks strain-specific O-antigens.

Bacteroid cell envelopes seem to have a decreased polysaccharide content, suggesting either a decrease of total lipopolysaccharide or a specific conversion of "smooth" type LPS to "rough" type LPS. Using monoclonal antibodies to bacteroid LPS, Brewin *et al* (1986) showed that bacteroids of *R. leguminosarum* bv. *viciae* contain relatively less LPS with low electrophoretic mobility ("smooth" LPS), compared to free-living bacteria of the same strain. Comparing isolated cell envelopes, we also found that bacteroids contain relatively less LPS (de Maagd *et al* 1989a). The changes in LPS composition that occur during differentiation may differ for a given Rhizobium strain in different host plants (Sindhu *et al* 1988). Using the same monoclonal antibodies, Bradley *et al.* (1986) demonstrated that loss of LPS or outer membrane material in general from bacteroids may occur *in vivo*, as they observed close associations of the peribacteroid membrane with bacteroid LPS, often at a considerable distance from the bacteroid surface. LPS could even be detected in other plant cell membranes in the nodule.

RHIZOBIUM MUTANTS WITH ALTERED LIPOPOLYSACCHARIDES

In the last few years, a number of groups have isolated mutants of fast-growing Rhizobia with altered LPSs. At this moment, one or more LPS mutants are available for the different *R. leguminosarum* biovars as well as -although not yet thoroughly characterized- for *R. meliloti*. Sofar it appears that for all biovars of *R. leguminosarum* LPS mutants which almost or completely lack the LPS I species, show symbiotic defects. The exact stage of nodule development in which these mutants are blocked, is different for each biovar. All mutants are able to perform the first stages of nodule development, including attachment, root hair curling, meristem induction and infection thread initiation. Biovar *phaseoli* mutants lacking LPS I form abortive infection threads and are not released into bean plant cells (Noel *et al* 1986). Biovar *viciae* mutants lacking LPS I are released from the infection thread (possibly less frequently than the wild type) and form (some) bacteroid-filled *V. hirsuta* plant cells (de Maagd *et al* 1988b, 1989b; Pühler *et al* 1988). In all these cases the formed nodules are small, white and devoid of nitrogenase activity. LPS mutants of *R. leguminosarum* bv. *trifolii* lacking LPS I and forming non-nitrogen fixing nodules have been isolated (Brink *et al* 1988). Mutants of *R. meliloti* affected in LPS synthesis have been isolated (Signer *et al* 1988). Some, but not all of these mutants are symbiotically defective, non-nitrogen fixing, and appear to be blocked in infection thread development. Thus, it appears that LPS I is not essential for the early steps of symbiosis, but that, depending on the bacterium/host combination, some property of LPS I is needed for normal infection thread development, bacterial release and bacteroid differentiation.

POSSIBLE (SIGNAL) FUNCTIONS OF LIPOPOLYSACCHARIDES IN SYMBIOSIS

Although the evidence for requirement for intact LPS seems compelling, the exact role of LPS and the extent of this requirement remains unclear. The picture is complicated by the fact that LPS seems to play a role in more than one stage of the symbiosis. When speculating about the role of LPS in symbiosis, one may globally distinguish between three possibilities, which will be looked at in more detail below. 1. LPS may contain, either constitutively or induced during nodule development, a structure that acts as a signal to the plant or as a receptor for a plant signal. 2. The presence on the cell surface of a minimal density of LPS polysaccharide moieties may be essential, while the exact composition is of minor importance. 3. Either the mutations leading to defective LPS, or the defective LPS itself may have pleiotrophic effects causing symbiotic defects, whereas the LPS molecule itself plays no direct role. This distinction is somewhat arbitrary and does not exclude that more than one property of LPS is involved in symbiosis.

Ad 1. Lipopolysaccharide, or part(s) of it which might be released by plant enzymes, could form a signal towards the plant. This signal might be responsible for inducing the root cells to form an infection thread, or in later stages, induce the endocytosis of bacteria (bacterial release). Alternatively, the signal might inhibit plant defense reactions that would otherwise cause the death of invading bacteria. In the case that the intact LPS would form the signal, this would imply a close interaction between the bacterial surface and the walls or cytoplasmic membranes of the plant cells containing the developing infection thread or, in a later stage, the peribacteroid membrane. However, this interaction may also be mediated by outer membrane "blebs" released from the cells. LPS is readily released from free-living bacteria, for example as a result of calcium limitation (de Maagd *et al* 1989a).

Thinking along the same lines, LPS might be a bacterial receptor for a plant signal, that would trigger a next step in symbiosis in the bacterium. A possible candidate for a plant cell wall receptor for a LPS derived signal is the root lectin. A role for host root lectin/LPS interaction in symbiosis has been proposed (Wolpert and Albersheim 1976), but in spite of a large amount of data gathered since, conclusive evidence for the essential involvement of lectin in any step of nodule development has been lacking up to very recently. Diaz *et al* (1989) (see also J.W. Kijne *et al* this volume) have shown that pea lectin is a factor determining the host specificity of *R.leguminosarum* bv. *viciae*/legume interaction. Lectin binding to LPS might also act as a signal towards the bacterium, as it was shown that lectin binding to *B. japonicum* cells induces *de novo* protein synthesis

(Halverson and Stacey 1986).

Ad 2. The possible signal function of LPS suggests a high degree of specificity, which is contradicted by the observed heterogeneity in structure. Lipopolysaccharide may also have functions that are much less dependent on the exact molecular structure, especially where the structure of the O-antigen is concerned. (A). The presence of (O-antigen)polysaccharide on the bacterial surface may either act to protect the bacterial surface against bactericidal components of the host plant defense (Jones *et al* 1987), or to shield bacterial surface components that otherwise might induce such a defense reaction. A similar explanation has also been used for the avirulence of LPS mutants of plant-pathogenic bacteria (see above), although an intact O-antigen was not required in these cases. (B). The O-antigenic part of the LPS appears to be an important determinant of cell surface hydrophobicity and surface charge. We have shown that mutants of *R. leguminosarum* bv. *viciae* lacking the O-antigenic part of the wild-type LPS have increased surface hydrophobicity and electrophoretic mobility (de Maagd *et al* 1989b). Apparently the sugars of the O-antigen chain make the cell surface more hydrophylic and at the same time shield negatively charged groups near the outer membrane bilayer. Both properties may be important in determining the outcome of interactions between the rhizobial cell surface and the host cell surface, for example during bacterial release or bacteroid development. (C). Lipopolysaccharide may also act as a substrate for "decoration" with *nod* gene-dependent or otherwise (e.g. low oxygen pressure, succinate) induced specific signal structures. There is some evidence that *nod* gene activity has effects on LPS structure (Carlson *et al* 1987), and homology studies of *nod* gene nucleotide sequences also suggest that some of the *nod* genes may be involved in synthesis or modification of polysaccharides.

Ad 3. (A). Besides the possibilities given above, there remains the possibility that the LPS molecule itself has no direct function in symbiosis. The genes involved in LPS synthesis which were mutated in the examples mentioned above, may have a second function in the synthesis of another essential component, just as for example some *R. meliloti exo* genes also have a function in LPS synthesis (Leigh and Lee 1988). In that case, similar LPS mutants with no symbiotic defects could occur. Such mutants have not been found. (B). An intact LPS, or at least some features of it, may be of importance for outer membrane integrity and function, which in their turn should be important for normal symbiosis. We found no clear evidence for changes in membrane stability in LPS mutants. However, the possibility of some effect of LPS mutations on membrane stability or function, particularly in the host plant, cannot be ruled out yet. It is for example not yet clear why many LPS mutants are not motile.

CONCLUSIONS

The summary of possible functions of LPS in symbiosis given above, although probably not complete, gives an impression of the complexity of this problem, and the amount of work that faces us. An important question is whether LPS mutants are killed, either by constitutively present plant factors or even by elicited defense responses. Another question is whether LPS contains specific (signal-)structures, possibly even determining host-specificty, or that the requirement for intact LPS is only a requirement for a certain length and/or surface density of any O-antigen. Illustrative in this respect are the results of Brink *et al* (1988), who show that a *Rhizobium leguminosarum* containing a "hybrid" LPS is capable of normal symbiosis. Construction of a large group of these hybrids and the isolation of more mutants with LPSs that more gradually differ from their parent LPS (instead of mutants that lack the complete O-antigen) may lead to an answer to this question.

REFERENCES

Bradley DJ, Butcher GW, Galfre G, Wood EA, Brewin NJ (1986) Physical association between the peribacteroid membrane and lipopolysaccharide from the bacteroid outer membrane in *Rhizobium* infected pea root nodule cells. J Cell Sci *85*:47-61

Brewin NJ, Wood EA, Larkins AP, Galfre G, Butcher GW (1986) Analysis of lipopolysaccharide from root nodule bacteroids of *Rhizobium leguminosarum*

using monoclonal antibodies. J Gen Microbiol *132*:1959-1968
Brink BA, Cava JR, Noel KD (1988) Rhizobium *leguminosarum* lipopolysaccharide genetics and role in clover and bean nodulation. In: Palacios R and Verma DPS (eds). Proceedings of the 4th International Symposium on Molecular Genetics of Plant-Microbe Interactions. APS Press, St.Paul, p 55-56
Carlson RW (1982) Surface Chemistry. In: Broughton WJ (ed). Nitrogen Fixation: *Rhizobium*. Clarendon Press, Oxford, p 199-234
Carlson RW (1984) Heterogeneity of *Rhizobium* lipopolysaccharides. J Bacteriol *158*:1012-1017
Carlson RW, Shatters R, Duh JL, Turnbull E, Hanley B, Rolfe BG, Djordjevic MA (1987) The isolation and partial characterization of the lipopolysaccharides from several *Rhizobium trifolii* mutants affected in root hair infection. Plant Physiol *84*:421-427
De Maagd RA, de Rijk R, Mulders IHM, Lugtenberg BJJ (1989a) Immunological characterization of *Rhizobium leguminosarum* outer membrane antigens using polyclonal and monoclonal antibodies. J Bacteriol *171*:1136-1142
De Maagd RA, Rao AS, Mulders IHM, Goosen-de Roo L, van Loosdrecht M, Wijffelman CA, Lugtenberg BJJ (1989b) Isolation and characterization of mutants of *Rhizobium leguminosarum* biovar *viciae* strain 248 with altered lipopolysaccharides:possible role of surface charge or hydrophobicity in bacterial release from the infection thread. J Bacteriol *171*:1143-1150
De Maagd RA, Van Rossum C, Lugtenberg BJJ (1988a) Recognition of individual strains of fast-growing Rhizobia by using profiles of membrane proteins and lipopolysaccharides. J Bacteriol *170*:3782-3785
De Maagd, RA, CA Wijffelman, E Pees, BJJ Lugtenberg (1988b) Isolation and characterization of three classes of mutants of *Rhizobium leguminosarum* with altered lipopolysaccharides. In: Bothe H, De Bruijn FJ, Newton WE (eds.). Nitrogen Fixation: Hundred Years After. Gustav Fisher, Stuttgart, p. 473.
Diaz, CL, Melchers LS, Hooykaas PJJ, Lugtenberg BJJ, Kijne JW (1989) Root lectin as a determinant of host-plant specificity in the *Rhizobium*-legume symbiosis. Nature:in press
Halverson LJ, Stacey G (1986) Signal exchange in plant-microbe interactions. Microbiol Rev *50*:193-225
Hendrick CA, Sequeira L (1984) Lipopolysaccharide-defective mutants of the wilt pathogen *Pseudomonas solanacearum*. Appl Environ Microbiol *48*:94-101
Jones WT, Macdonald PE, Jones SD, Pankhurst CE (1987) Peptidoglycan-bound polysaccharide associated with resistance of *Rhizobium loti* strain NZP 2037 to *Lotus pedunculatus* root flavolan. J Gen Microbiol *133*:2617-2619
Leigh JA, Lee CC (1988) Characterization of polysaccharides of *Rhizobium meliloti exo* mutants that form ineffective nodules. J Bacteriol *170*:3327--3332
Lugtenberg B, Van Alphen L (1983) Molecular architecture and functioning of the outer membrane of *Escherichia coli* and other Gram-negative bacteria. Biochim Biophys Acta *737*:51-115
Noel KD, Van den Bosch KA, Kulpaca B (1986) Mutations in *Rhizobium phaseoli* that lead to arrested development of infection threads. J Bacteriol 168: 1392-1401
Pühler A, Enenkel B, Hilleman A, Kapp D, Keller M, Muller P, Niehaus K, Priefer UB, Quandt J, Schmidt C (1988) *Rhizobium meliloti* and *Rhizobium leguminosarum* mutants defective in surface polysaccharide synthesis and root nodule development. In: Bothe H, De Bruijn FJ (eds). Nitrogen Fixation: Hundred Years After. Gustav Fisher, Stuttgart, p 423-430
Schoonejans E, Expert D, Toussaint A (1987) Characterization and virulence properties of *Erwinia chrysanthemi* lipopolysaccharide-defective, 0EC2-resistant mutants. J Bacteriol *169*:4011-4017
Signer ER, Sharma SB, Williams MNV, Bent A, Clover R, Wacks DB, Kieber J, Klein S (1988) Search for new symbiotic genes in *Rhizobium meliloti*. In: Palacios R, Verma DPS (eds). Proceedings of the 4th International Symposium on Molecular Genetics of Plant-Microbe Interactions. APS Press, St.Paul, p 20-25
Sindhu SS, Kannenberg EL, Brewin NJ (1988) Lipopolysaccharide maturation in pea and bean bacteroids. In: Bothe H, De Bruijn FJ, Newton WE (eds). Nitrogen Fixation: Hundred Years After. Gustav Fisher, Stuttgart, p 480
Wolpert J, Albersheim P (1976) Host-symbiont interactions. I. The lectins of legumes interact with the O-antigen-containing lipopolysaccharides of their symbiont rhizobia. Biochem Biophys Res Commun. *70*:729-737

IDENTIFICATION AND ANALYSIS OF GENES INVOLVED IN LIPOPOLY-SACCHARIDE PRODUCTION AND SYMBIOTIC EFFICIENCY IN RHIZOBIUM LEGUMINOSARUM BIOVAR VICIAE VF39

U. B. Priefer, D. Kapp, S. Preisler, J. de Wall and A. Pühler
Department of Genetics
University of Bielefeld
4800 Bielefeld 1
Federal Republic of Germany

INTRODUCTION

Bacterial cell surface carbohydrates undoubtedly play an important role in signal exchange between *Rhizobium* and its host plant during the process of symbiotic nitrogen fixation. As a component of the outer membrane, the rhizobial lipopoly-saccharide (LPS) has been hypothesized to be involved in host-specific recognition by interacting with the lectins of legumes (Wolpert and Albersheim 1976). Recent reports suggest that the function of LPS is, at least in *Rhizobium leguminosarum*, not restricted to the early events of attachment and recognition (Noel et al 1986; Cava et al 1989; DeMaagd et al 1989).

In this report we describe the characteristics of Tn*5*-induced mutants of *R. leguminosarum* bv. *viciae* VF39, which fail to produce complete LPS. Genetic analyses revealed that the mutant phenotypes were caused by single Tn*5* insertions, clustered in a defined region of the VF39 genome.

ISOLATION AND PHENOTYPE OF LIPOPOLYSACCHARIDE MUTANTS

The strain *R. leguminosarum* bv. *viciae* VF39, used in this study, was isolated in Bielefeld from field-grown *Vicia faba* nodules; details have been published previously (Hynes et al

NATO ASI Series, Vol. H 36
Signal Molecules in Plants and Plant-Microbe Interactions
Edited by B. J. J. Lugtenberg

1988). Mutants were generated by random *Tn5* mutagenesis using the transposon delivery system described earlier (Simon et al 1983). Mutants defective in normal lipopolysaccharide production were selected by their rough colony surface on tryptone-yeast medium (TY). The four mutants described in more detail are listed in Table 1. All of them are additionally characterized by auto-agglutination in TY liquid medium and loss of motility on TY soft agar.

When total SDS extracts treated with proteinase K or LPS extracted by the hot phenol-water procedure were subjected to SDS-polyacrylamide gel electrophoresis, the mutants lacked a slow-migrating lipopolysaccharide portion (LPS I) present in the wildtype. For *R. leguminosarum* bv. *phaseoli* and bv. *trifolii*, it has been shown that this high molecular weight fraction corresponds to complete LPS containing the O-antigen side chain (Carlson et al 1987a, 1987b). Therefore, it is most likely, that all four mutants described in this study also lack or produce very reduced quantities of the O-antigen polysaccharide.

Also, the mutants behaved very similarly with respect to their symbiotic phenotype on *Vicia hirsuta*, where they were all able to induce numerous nodules. Light microscopic analysis of longitudinal semithin sections revealed the typical morphology of indeterminate nodules with infection threads and bacteria released into plant cells. However, the nodules remained very small and white, were completely devoid of nitrogenase activity and senesced prematurely. Thus, the defect occurs relatively late in symbiotic development, similar to the *R. leguminosarum* bv. *viciae* mutants described by DeMaagd et al (1989). In contrast, the LPS mutants isolated from *R. leguminosarum* bv. *phaseoli* show no bacterial release at all (Noel et al 1986; Cava et al 1989).

COMPLEMENTATION OF THE LIPOPOLYSACCHARIDE MUTANTS

One of the mutants was complemented by introducing a cosmid library of the VF39 wildtype genome. This cosmid gene bank was constructed in vector pSUP205 (Simon et al 1983) and maintained as a lambda phage lysate.

For "in trans" complementation, the cosmid gene bank was introduced via lambda phage infection into Escherichia coli containing an RP4 derivative. Due to DNA homology between the cosmid vector pSUP205 and the RP4 plasmid, single recombination events between the two molecules can occur leading to cointegrates. The principle of this method is described in more detail by Simon et al (1986). The RP4-cosmid cointegrates were transferred "en masse" into one of the R. leguminosarum mutants and complementation was indicated by smooth colony surface, lack of agglutination and restoration of motility. Complementing cointegrates were conjugally re-introduced into E.coli (λ cIts), where spontaneous resolution of the cointegrates can occur. The complementing cosmid was separated from the RP4 plasmid by "in vivo" packaging upon heat induction of the prophage and introduction (via phage infection) into another E.coli strain.

Accordingly, cosmid pCos4 was selected and found to restore wildtype characteristics in all the four mutants described here. The mutants carrying pCos4, either as RP4 cointegrate or integrated into their genomes, showed normal growth phenotypes, wild type LPS profiles in SDS-PAGE experiments and nitrogen fixing nodules on Vicia hirsuta plants.

Table 1. Characteristics of R. leguminosarum VF39 LPS mutants

Strain	Phenotype					compl.	Tn5 in
	R/S	Agg	Mot	LPS	Fix	cosmid	EcoRI fragment
VF39 (WT)	S	-	+	+	+		
VF39-23	R	+	-	-	-	pCos4	E1, 0.6 kb
VF39-32	R	+	-	-	-	pCos4	E6., 6.3 kb
VF39-C48	R	+	-	-	-	pCos4	E6, 6.3 kb
VF39-51	R	+	-	-	-	pCos4	E8, 9.5 kb

R/S=rough/smooth colony surface, Agg=agglutination in TY liquid medium, Mot=motility, LPS=presence of LPS I, Fix=fixation actity

MAPPING AND CLONING OF THE lps MUTATIONS

Cosmid pCos4 was used to localize the four lps mutations in the genome of VF39. Total DNA of wildtype and mutant strains was isolated, digested with EcoRI and hybridized against pCos4 DNA labelled with biotin-ATP (Bethesda Research Laboratory, Blue Gene System). Since Tn5 does not contain an EcoRI restriction site, the EcoRI fragments carrying the lps mutations should be enlarged by the size of Tn5. The hybridization patterns, shown in Figure 1, revealed that the Tn5 insertions occurred in three different EcoRI fragments. These results were confirmed by "in vivo" cloning of the Tn5 mutations from the mutant genomes. Details will be described elsewhere.

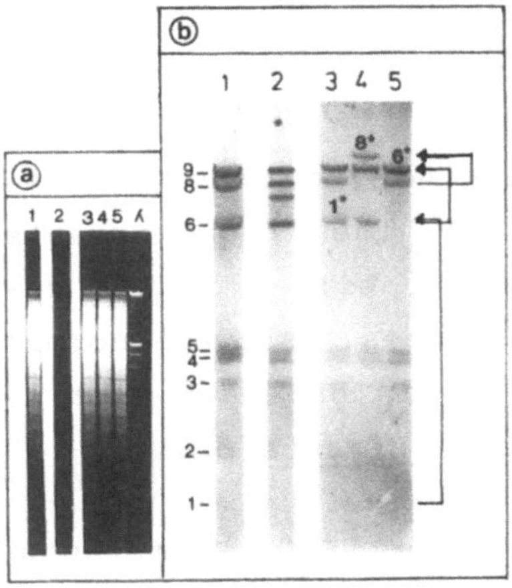

Figure 1. Localization of the Tn5 insertions in VF39 Lps mutants. Total DNA of wildtype (lane 1) and mutants VF39-23 (lane 3), VF39-51 (lane 4), and VF39-32 (lane 5) was digested with EcoRI (a) and hybridized to biotin-ATP labelled pCos4 DNA (b). EcoRI digested pCos4 DNA was loaded in lane 2. Numbers on the left refer to the VF39 EcoRI fragments (E1-E9) cloned in pCos4; E7 corresponds to the pSUP205 vector. Hybridizing bands marked with * represent the Tn5-carrying EcoRI fragments present in the mutant strains, i.e. E1 in VF39-23, E8 in VF39-51, and E6 in VF39-32. The pattern obtained for VF39-C48 is identical to that of VF39-32.

Fragments E6 and E8 were mutagenized with Tn5 by fragment-specific mutagenesis. The Tn5 insertions were introduced into the genome of VF39 by reverse mutagenesis (homogenotization) and their effect on growth characteristics and LPS synthesis was tested. In addition, cross-complementation studies were carried out. For this purpose, a Tn5-carrying fragment was introduced into the genome of a Tn5-induced mutant and the double mutant was tested for its phenotype. The principle of this method is outlined in Figure 2.

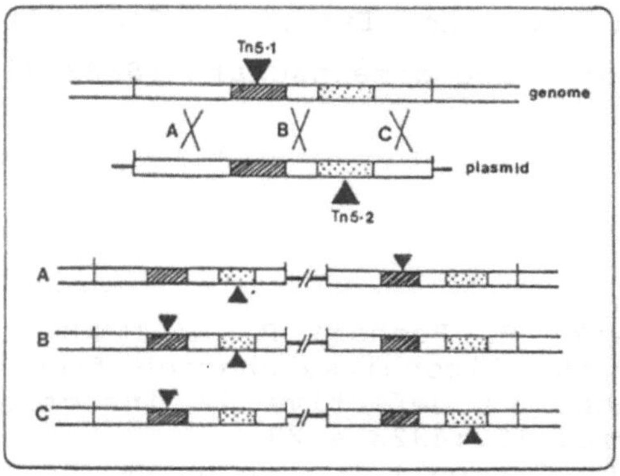

Figure 2. Scheme of a cross-complementation experiment to distinguish between independent transcriptional units. A cloned Tn5 mutation (Tn5-2) is introduced into the genome of a Tn5 mutant (Tn5-1). Single cross over events can occur at different positions relative to the Tn5 insertions sites (A, B,C), leading to partial diploids. If all transconjugants are complemented to wildtype phenotype, the two Tn5 insertions are located within independent transcriptional units.

These experiments indicated at least three transcriptional units on fragment E6 and four transcriptional units on fragment E8 (data not shown). Thus, at least eight transcriptional units involved in normal LPS synthesis are located within the VF39 region covered by pCos4.

CONCLUSION

The genetic data presented in this study clearly demonstrated that the pleiotropic phenotypes of LPS mutants are caused by defined single Tn5 insertions and therefore strongly correlate changes in the LPS structure with alterations in symbiotic ability. However, it remains unclear whether the symbiotic defect is a direct consequence of the altered LPS or whether secondary effects, such as general changes in the outer membrane architecture cause the symbiotic phenotype.

Interestingly, all genetically defined LPS mutants of *R. leguminosarum* published until now, including those described in this study, are able to induce nodules on their normal host plants. This would imply that the LPS of *R. leguminosarum* is not primarily required for host-specific recognition but for later steps of nodule development, possibly as a signal molecule or as a protection of the bacterium from plant defenses.

This work was supported by the European Community (BAP-0081-D) and the Deutsche Forschungsgemeinschaft (SFB223).

REFERENCES

Carlson RW, Kalembasa S, Turowski D, Pachori P, Noel KD (1987a) Characterization of the lipopolysaccharide from a Rhizobium phaseoli mutant that is defective in infection thread development. J Bacteriol 169:4923-4928

Carlson RW, Shatters R, Duh J-L, Turnbull E, Hanley B, Rolfe BG, Djordjevic MA (1987b) The isolation and partial characterization of the lipopolysaccharides from several Rhizobium trifolii mutants affected in root hair infection. Plant Physiol 84:421-427

Cava JR, Elias PM, Turowski DA, Noel DK (1989) Rhizobium leguminosarum CFN42 genetic regions encoding lipopolysaccharide structures essential for complete nodule development on bean plants. J Bacteriol 171:8-15

DeMaagd RA, Rao AS, Mulders IHM, Goosen-de Roo L, VanLoosdrecht MCM, Wijffelman CA, Lugtenberg BJJ (1989) Isolation and characterization of mutants of Rhizobium leguminosarum bv. viciae 248 with altered lipopolysaccharides: possible role of surface charge or hydrophobicity in bacterial release from the infection thread. J Bacteriol 171:1143-1150

Hynes MF, Brucksch K, Priefer U (1988) Melanin production encoded by a cryptic plasmid in a Rhizobium leguminosarum strain. Arch Microbiol 150:326-332

Noel KD, VandenBosch KA, Kulpaca B (1986) Mutations in Rhizobium phaseoli that lead to arrested development of infection threads. J Bacteriol 168:1392-1401

Simon R, Priefer U, Pühler A (1983) A broad host range mobilization system for in vivo genetic engineering: transposon mutagenesis in Gram-negative bacteria. Bio/Technology 1: 784-791

Simon R, O'Connell M, Labes M, Pühler A (1986) Plasmid vectors for the genetic analysis and manipulation of Rhizobia and other Gram-negative bacteria. Methods in Enzymology 118: 640-659

Wolpert JS, Albersheim P (1976) Host-symbionts interactions. I. The lectins of legumes interact with the O-antigen containing lipopolysaccharides of their symbiont rhizobia. Biochem Biophys Res Commun 70:729-737

ROLE OF LECTIN IN THE PEA-*RHIZOBIUM* SYMBIOSIS

Jan W. Kijne, Clara L. Díaz and Ben J.J. Lugtenberg
Department of Plant Molecular Biology
Botanical Laboratory, Nonnensteeg 3, 2311 VJ Leiden
The Netherlands

PEA LECTIN

Pea lectin (*Pisum sativum* lectin, Psl) is a 49 kDa protein with two high affinity binding sites for oligosaccharides containing a trimannoside core with a neighbouring fucosyl-α-1,6-N-acetylglucosamine group (Kornfeld et al 1981, Debray et al 1981). Psl is a tetramer and can dissociate into two α subunits and two β subunits with molecular weights of 5800 Da and 17000 Da, respectively (Trowbridge 1974). Amino acid residues in one αβ-subunit pair contribute to the structure of one sugar-binding site (Van Driessche 1988). Psl is encoded by a family of four genes of which only one gene appears to be functional (Kaminski et al 1987). The sequence of the functional *psl* gene (LecA, Gatehouse et al 1987; Psl1, Kaminski et al 1987) corresponds with the nucleotide sequence of cloned DNA complementary to Psl mRNA (Higgins et al 1983a). Furthermore, the deduced amino acid sequences correspond with previously reported amino acid sequences of parts of Psl (Van Driessche et al 1976, Foriers et al 1977, Richardson et al 1978). Psl is synthesized as a single translation product and post-translational cleavage of the Psl precursor yields one α and one β subunit (Higgins et al 1983a, 1983b, Lauwereys et al 1983, Van Driessche et al 1988).

In pea seeds, Psl constitutes a mixture of two isolectins, Psl1 and Psl2, and a small amount of the unprocessed Psl precursor (Van der Schaal 1983). Psl1 and Psl2 have identical molecular weights and subunit structures and display the same specificity of sugar-binding. However, the isolectins differ in isoelectric point, pI 7.2 for Psl1 and pI 6.1 for Psl2

NATO ASI Series, Vol. H 36
Signal Molecules in Plants and Plant-Microbe Interactions
Edited by B. J. J. Lugtenberg

(Entlicher and Kocourek 1975). This charge difference corresponds with a difference in electrophoretic migration of the α subunits of the isolectins, and probably results from the additional removal of a small peptide at the COOH-terminus of the Psl1-α subunit, yielding the α subunit of Psl2 (Van Driessche et al 1988).

Interestingly, Psl1 is only found in pea seeds whereas a lectin similar to Psl2 is produced in pea roots and in other parts of the pea plant and is also secreted into the pea rhizosphere (Díaz 1989). Detection in pea root extracts of mRNA cross-hybridizing with a copy DNA probe derived from seed Psl mRNA (Buffard et al 1988) and of polypeptides similar to the seed Psl precursor (Díaz 1989) points at synthesis of Psl in the root of this plant. However, subtle differences were found between the sugar-binding specificities of pea seed and root lectin. These differences correspond with subtle structural differences possibly at or near to the sugar-binding site, as determined by circular dichroism measurements (Díaz 1989). These differences indicate that pea root lectin is not identical to Psl2, which necessitates a detailed study of genetics, processing and structure of Psl from roots.

LOCALIZATION OF PEA ROOT LECTIN

Using indirect immunofluorescence, Psl could be demonstrated on the surface of pea roots. Remarkably, the pattern of lectin location entirely corresponds with the pattern of susceptibility of pea root epidermal cells to infection by Rhizobium leguminosarum biovar viciae (Rl viciae) bacteria (Díaz 1989). Whereas the surface of the pea root tip appeared to be Psl-negative, lectin is present on the surface of epidermal cells at the time of their differentiation into trichoblasts. Trichoblasts with emerging root hair tips constitute the majority of Psl-positive root epidermal cells. At the tip of developing root hairs Psl appeared to be distributed in dense, small patches rather than uniformly (Díaz et al 1986). In contrast, the surface of the mature root hair zone and the zone between the root hairs and the cotyledons is characterized by absence of Psl and by an almost complete resistance to rhizobial infection. Yet, crude protein extracts of the various parts of pea roots all contain Psl (Díaz 1989). These observations point at developmental control of Psl secretion. The correspondence in patterns of susceptibility to rhizobial infection and of Psl location on the surface of pea roots suggests that both phenomena are regulated by the same developmental mechanism.

Recently it was shown that extracellular Psl on pea roots is present in a form allowing binding of specific saccharide receptor molecules (Díaz 1989). By using a novel binding assay with rat IgE antibodies as high-affinity receptors (see Kornfeld et al 1981), Psl could be localized on the pea root to the same extent and at sites similar to those observed after use of monospecific polyclonal antibodies, the indirect immunofluorescence method. These results indicate that most, if not all, Psl molecules at the outer surface of target cells for rhizobial infection are functional in their ability to bind specific saccharide sequences. Study of slightly plasmolyzed trichoblasts demonstrated that Psl is predominantly present at the external surface of the plasma membrane and that functional lectin molecules are absent from the root hair cell wall surface. This observation explains that direct binding of single Rhizobium cells to the pea root hair tip surface is not lectin-mediated, i.e. inhibited in the presence of the Psl-haptenic monosaccharide 3-O-methyl-D-glucose (Kijne et al 1988).

Psl secreted by pea roots enables enhanced accumulation of Rl viciae bacteria on pea root hair tips (Kijne et al 1988, Díaz 1989) after rhicadhesin-mediated attachment of single rhizobial cells to the root hair tip surface (Smit et al 1989 a,b, see Lugtenberg et al, this volume). Mutants of Rl viciae unable to produce or expose pea lectin receptor molecules are not yet available and should be isolated to test the relevance of lectin-enhanced accumulation for the nodulation process. Up till now, evidence is lacking that secreted Psl induces increased infectivity of Rl viciae cells like it appears to be the situation in Bradyrhizobium-soybean lectin interactions (Halverson and Stacey 1985, 1986).

PEA LECTIN IN TRANSGENIC CLOVER ROOTS

Recently, we tested the involvement of Psl in the establishment of the symbiosis between Rl viciae and its host plants by introducing a functional psl gene in the roots of a heterologous host plant, white clover, and by studying the infection of transgenic clover roots by Rl viciae cells (Díaz et al 1989). White clover was chosen as a test plant because the host-specificity barrier in the heterologous Rl viciae - white clover interaction is at the level of the functioning of rhizobial hsn (host specificity of nodulation) genes and infection thread formation rather than at the levels of specific induction of nod (nodulation) genes (Spaink et al 1987) and functioning of common nod gene related root hair deformation factors (Yao and Vincent 1969, Debellé et al 1986).

The complete genomic psl gene as well as a Psl cDNA construction were cloned into binary vectors (Hoekema et al 1983) and were introduced into the roots of white clover seedlings using Agrobacterium rhizogenes as a vehicle for transformation.

White clover hairy roots appeared to behave as normal roots in nodulation, and nodulation of these roots by the homologous R. leguminosarum biovar trifolii (Rl trifolii) symbiont was not in any way inhibited by the presence of the binary vector T-DNA with or without the psl gene. Moreover, nodulation appeared to be host-specific in that Rl viciae bacteria induced marked root hair curling but no infection threads, similar to the situation with normal clover roots.

Presence of the psl gene in white clover hairy roots appeared to enable nodulation by Rl viciae. Nodulation was delayed, infrequent and often abnormal with respect to nodule morphology. However, red functional nodules appeared on thirty percent of the nodulated transgenic clover plants, indicating that a significant number of infections by Rl viciae cells had been completely successful.

By using the rat IgE receptor binding assay, we were able to demonstrate indirectly the presence of Psl on white clover hairy roots transformed with the psl gene at sites similar to those on pea roots (Díaz 1989). Normal white clover roots and hairy roots without the psl gene were, as expected, Psl-negative. These results strongly suggest that root lectin somehow contributes to the legume-Rhizobium host plant specificity barrier at the level of induction of infection threads. It is up to future research to determine the molecular basis of Psl-Rl viciae interactions.

THE PEA LECTIN RECEPTOR

The molecular role of Psl will remain obscure until it is established which (rhizobial) product is its high affinity binding ligand. It is reasonable to expect that this ligand contains a specific saccharide sequence. Lectins from legumes representing members of different cross-inoculation groups differ in their sugar-binding specificity, and sequence homology between different lectin genes is low for those genomic regions considered to play a role in sugar-binding (Van Driessche 1988). In contrast, differences in the high affinity binding of hydrophobic molecules by different lectins have not been found yet and the hydrophobic pocket in lectin molecules is fairly well conserved.

If lectin molecules specifically induce increased infectivity of rhizobia (Halverson and Stacey 1986), one might expect the lectin-binding ligand to be located at the surface of the bacterial cell. Such a ligand could also contribute to lectin-enhanced accumulation of rhizobia to legume root hair tips. However, for Rl viciae any involvement of Sym plasmid-coded gene products in the process of root hair attachment could not be demonstrated (Smit et al 1987, Kijne et al 1988). Moreover, several Rhizobium species are able to bind Psl (Kamberger 1979). It may be envisaged that this "non-specific" binding of Psl masks the presence of a receptor in which production hsn genes are involved.

If, on the other hand, a specific rhizobial interaction with plasma membrane-located Psl molecules is essential for the induction of infection thread formation, the rhizobial ligand might well be an extracellular signal molecule. There is no evidence that physical contact of rhizobia with the plasma membrane of the host's cells is necessary for infection thread growth (Bakhuizen 1988). New experimental results indicate that host-range genes act by modifying factors encoded by the common nod genes (Faucher et al 1988). This modification might result in targeting of the rhizobial signal molecules to the perceptive site of the growing root hair, i.e. the tip growing area, at which Psl could act as a receptor molecule. Conclusive characterization of rhizobial signal molecules is necessary for testing this hypothesis. Whatever the character of the Psl ligand may be, it is likely that the rhizobial nodE gene is (in)directly involved in its production (see Spaink et al, this volume).

REFERENCES

Bakhuizen R (1988) The plant cytoskeleton in the Rhizobium-legume symbiosis. PhD Thesis, Leiden University, The Netherlands

Buffard D, Kaminski PA, Strosberg AD (1988) Lectin-gene expression in pea (Pisum sativum L.) roots. Planta 173: 367-372

Debellé F, Sharma SB, Rosenberg C, Vasse J, Maillet F, Truchet G, Dénarié J (1986) Respective roles of common and host-specific Rhizobium meliloti nod genes in the control of lucerne infection. In: Lugtenberg B (ed) Recognition in microbe-plant symbiotic and pathogenic interactions, pp. 17-28. Springer Verlag Berlin, Heidelberg

Debray H, Decout D, Strecker G, Spik G, Montreuil J (1981) Specificity of twelve lectins towards oligosaccharides and glycopeptides related to N-glycosylproteins. Eur J Biochem 117: 41-55

Díaz CL, Van Spronsen PC, Bakhuizen R, Logman GJJ, Lugtenberg EJJ, Kijne JW (1986) Correlation between infection by Rhizobium leguminosarum and lectin on the surface of Pisum sativum L. roots. Planta 168: 350-359

Díaz CL (1989) Root lectin as a determinant of host-plant specificity in the Rhizobium-legume symbiosis. PhD Thesis, Leiden University, The Netherlands

Díaz CL, Melchers LS, Hooykaas PJJ, Lugtenberg BJJ, Kijne JW (1989) Root lectin as a determinant of host-plant specificity in the Rhizobium-legume symbiosis. Nature (London) 338: 579-581

Entlicher G, Kocourek J (1975) Studies on phytohemagglutinins. XXIV. Isoelectric point and hybridization of the pea (Pisum sativum L.) isophytohemagglutinins. Biochim Biophys Acta 393:

165-169
Faucher C, Maillet F, Vasse J, Rosenberg C, Van Brussel AAN, Truchet G, Dénarié J (1988) Rhizobium meliloti host range nodH gene determines production of an alfalfa-specific extracellular signal. J Bacteriol 170: 5489-5499
Foriers A, Wuilmart C, Sharon N, Strosberg AD (1977) Extensive sequence homologies among lectins from leguminous plants. Biochem Biophys Res Comm 75: 980-986
Gatehouse JA, Bown D, Evans IM, Gatehouse LN, Jobes D, Preston P, Croy RRD (1987) Sequence of the seed lectin gene from pea (Pisum sativum L.). Nucl Acid Res 15: 7642
Halverson LJ, Stacey G (1985) Host recognition in the Rhizobium-soybean symbiosis. Evidence for the involvement of lectin in nodulation. Plant Physiol 77: 621-625
Halverson LJ, Stacey G (1986) Effect of lectin on nodulation by wild-type Bradyrhizobium japonicum and a nodulation-defective mutant. Appl Environ Microbiol 51: 753-760
Higgins TVJ, Chandler PM, Zurawski G, Button SC, Spencer D (1983a) The biosynthesis and primary structure of pea seed lectin. J Biol Chem 258: 9544-9549
Higgins TVJ, Chrispeels MJ, Chandler P, Spencer D (1983b) Intracellular site of synthesis and processing of lectin in developing pea cotyledons. J Biol Chem 258: 9550-9552
Hoekema A, Hirsch PR, Hooykaas PJJ, Schilperoort RA (1983) A binary plant vector strategy based on separation of vir and T-region of the Agrobacterium tumefaciens Ti-plasmid. Nature (London) 303: 179-180
Kamberger W (1979) Role of cell surface polysaccharides in the Rhizobium-pea symbiosis. FEMS Microbiol Lett 6: 361-365
Kaminski PA, Buffard D, Strosberg AD (1987) The pea lectin gene family contains only one functional lectin gene. Plant Mol Biol 9: 497-507
Kornfeld K, Reitman ML, Kornfeld R (1981) The carbohydrate-binding specificity of pea and lentil lectins. Fucose is an important determinant. J Biol Chem 256: 6633-6640
Kijne JW, Smit G, Díaz CL, Lugtenberg BJJ (1988) Lectin-enhanced accumulation of manganese-limited Rhizobium leguminosarum cells on pea root hair tips. J Bacteriol 170: 2994-3000
Lauwereys M, Van Driessche E, Strosberg AD, Dejaegere R, Kanarek L (1983) The α and β subunits of pea lectin are the result of a post-translational cleavage of a precursor chain. In: Lectins, biology, biochemistry, clinical biochemistry, Vol. 3, pp. 603-610, Bog-Hansen TC, Spengler GA (eds), W de Gruyter, Berlin New York
Richardson C, Behnke WD, Freisheim JH, Blumenthal KM (1978) The complete amino acid sequence of the α subunit of pea lectin, Pisum sativum. Biochim Biophys Acta 537: 310-319
Smit G, Kijne JW, Lugtenberg BJJ (1987) Both cellulose fibrils and a Ca^{2+}-dependent adhesin are involved in the attachment of Rhizobium leguminosarum to pea root hair tips. J Bacteriol 169: 4294-4301
Smit G, Kijne JW, Lugtenberg BJJ (1989a) Roles of flagella, lipopolysaccharide, and a Ca^{2+}-dependent cell surface protein in attachment of Rhizobium leguminosarum biovar viciae to pea root hair tips. J Bacteriol 171: 569-572
Smit G, Logman TJJ, Boerrigter METI, Kijne JW, Lugtenberg BJJ (1989b) Purification and partial characterization of the

Rhizobium leguminosarum biovar viciae Ca^{2+}-dependent adhesin, which mediates the first step in attachment of cells of the family Rhizobiaceae to plant root hair tips. J Bacteriol 171: 4054-4062

Spaink HP, Wijffelman CA, Pees E, Okker RJH, Lugtenberg BJJ (1987) Rhizobium nodulation gene nodD as a determinant of host specificity. Nature (London) 328: 337-340

Trowbridge IS (1974) Isolation and chemical characterization of a mitogeneic lectin from Pisum sativum. J Biol Chem 249: 6004-6012

Van der Schaal CAM (1983) Lectins and their possible involvement in the Rhizobium-leguminosae symbiosis. PhD Thesis, Leiden University, The Netherlands

Van Driessche E, Foriers A, Strosberg AD, Kanarek L (1976) N-terminal sequences of the α and β subunits of the lectin from the garden pea (Pisum sativum L.). FEBS Lett 71: 220-222

Van Driessche E (1988) Structure and function of leguminous lectins. In: Advances in lectin research, Vol. 1, pp. 73-134, Franz H. ed, Springer Verlag Berlin Heidelberg New York

Van Driessche E, Beeckmans S, Dejaegere R, Kanarek L (1988) Isolation of the pea lectin precursor and characterization of its processing products. In: Lectins, biology, biochemistry, clinical biochemistry, Vol. 6, pp. 355-362, Reed DLJ, Bøg-Hansen TC (eds), Sigma Chem Co St. Louis Missouri USA

Yao PY, Vincent JM (1969) Host specificity in the root hair curling factor of Rhizobium spp. Aust J Biol Sci 22: 413-423

THE *RHIZOBIUM* NODE PROTEIN AS A MAJOR DETERMINANT OF HOST SPECIFICITY

Herman P. Spaink, Henk Roest, Michel Vierboom, Rob J.H. Okker, Carel A. Wijffelman and Ben J.J. Lugtenberg
Leiden University, Department of Plant Molecular Biology, Nonnensteeg 3, 2311 VJ Leiden,
The Netherlands

INTRODUCTION

In the symbiosis between *Rhizobium* bacteria and their leguminous host plants each *Rhizobium* strain is only able to form nitrogen fixing root nodules on a limited number of host species. This has led to the definition of cross-inoculation groups, in which the bacterial species are classified according to their ability to establish an effective symbiosis with a particular group of host plants. The cross-inoculation groups of *R.leguminosarum*, with plants of the genera *Vicia*, *Lathyrus*, *Pisum* and *Lens* as hosts, and *R.trifolii*, with plants of the genus *Trifolium* as hosts, are the object of the study described in this paper.

In *R.leguminosarum* and *R.trifolii* a relatively small region of the large symbiosis (Sym) plasmid is crucial for the nodulation process, including its host-specific properties. These so-called nodulation (*nod*) regions of both *Rhizobium* species resemble each other in many respects. (i) The *nod* regions contain the same number of genes, eleven of which have homologous counterparts in both *R.leguminosarum* and *R.trifolii*. (ii) The genetic organization of the homologous *nod* genes is almost identical. (iii) The regulation of all identified genes shares a common principle in that they are transcribed as one regulon in which the constitutively transcribed *nodD* gene is the positive regulator. The *nod* regulon is only transcribed in the presence of an inducer which is produced by the host plant. The inducing compounds have been identified as flavones, flavanones or closely

NATO ASI Series, Vol. H 36
Signal Molecules in Plants and Plant-Microbe Interactions
Edited by B. J. J. Lugtenberg

related compounds. A strongly conserved regulatory domain, designated the nod box, is present upstream of all flavonoid-inducible operons and appears to be essential for the activity of the inducible promoters (for a review see Long, 1989).

The regulation of the genes in the nodulation region by the nodD gene has been shown to be a determinant of host specificity of R.leguminosarum and R.trifolii (see Wijffelman et al this volume). However, the specificity of R.leguminosarum and R.trifolii for the plants of their cross-inoculation group is also dependent on other nod genes than nodD. In this paper we will demonstrate that in isogenic R.trifolii strains the product of the nodE gene is the only factor which determines whether the host plants of R.trifolii or R.leguminosarum are efficiently nodulated. The other nod genes in this R.trifolii system can therefore be considered as functionally interchangeable between both cross-inoculation groups. We have studied the properties of a set of hybrid nodE genes derived of R.leguminosarum and R.trifolii. From this study we conclude that a central region, of 185 amino acids at the most, determines the difference in the host-specific characteristics of the two NodE proteins.

Using antibodies we have studied NodE protein in more detail. NodE protein appears to be localized in the cytoplasmic membrane and probably contains several transbilayer α-helices.

THE HOST-SPECIFIC PROPERTIES OF THE NODE PROTEIN

Djordjevic et al (1985) have shown that R.trifolii strain ANU843 acquires the ability to nodulate peas when it contains a Tn5 insertion in the nodE gene. However, as no other host plants of the R.leguminosarum cross-inoculation group are nodulated by such R.trifolii derivatives (Table 1), this is the only example of a host range-limiting influence of the nodE gene of R.trifolii. The results with the nodE mutant strains (Table 1) demonstrate that the nodE gene is crucial for the nodulation of several R.trifolii host plants.

To study the role of the nodE gene in host specificity in more detail we have constructed the plasmids pMP263 and pMP258

TABLE 1. Nodulation characteristics of R.trifolii nodE mutant strain K11 and derivatives containing a cloned nodE gene

Tested strains	Nodulation characteristics[1] with indicated test plants: T.pratense, T.arvense, T.campestre, T.lappaceum	T.repens	V.sativa, V.hirsuta, L.montanum, L.pratensis	P.sativum
ANU843 (wild type)	+	+	-	-
K11 (nodE::Tn5)	-	±	-	+
K11.pMP263 (nodE R.trifolii)	+	+	-	-
K11.pMP258 (nodE R.leguminosarum)	-	±	+	+
K11. nodE925, nodE912, nodE906, nodE910	-	n.t.	+	n.t.
K11. nodE921, nodE932, nodE919	+	n.t.	-	n.t.
K11. nodE911	±	n.t.	-	n.t.
K11. nodE913, nodE929	-	n.t.	±	n.t.

[1] -, 0 to 15 percent of plants nodulated; ±, 20 to 50 percent of plants nodulated; +, more than 80 percent of plants nodulated; n.t., not tested.

(Fig. 1) in which the nodE genes of R.trifolii and R.leguminosarum, respectively, are separated from the nodF genes, which are located upstream of nodE in the same operon. In pMP263 and pMP258 the NodE proteins are expressed under control of the flavonoid-inducible promoter of the nodABCIJ operon of R.leguminosarum. Since identical regulatory regions are present, these plasmids are well suited to compare the influence of the respective R.trifolii and R.leguminosarum nodE genes on the nodulation process. Moreover, the promoter of the nodABCIJ

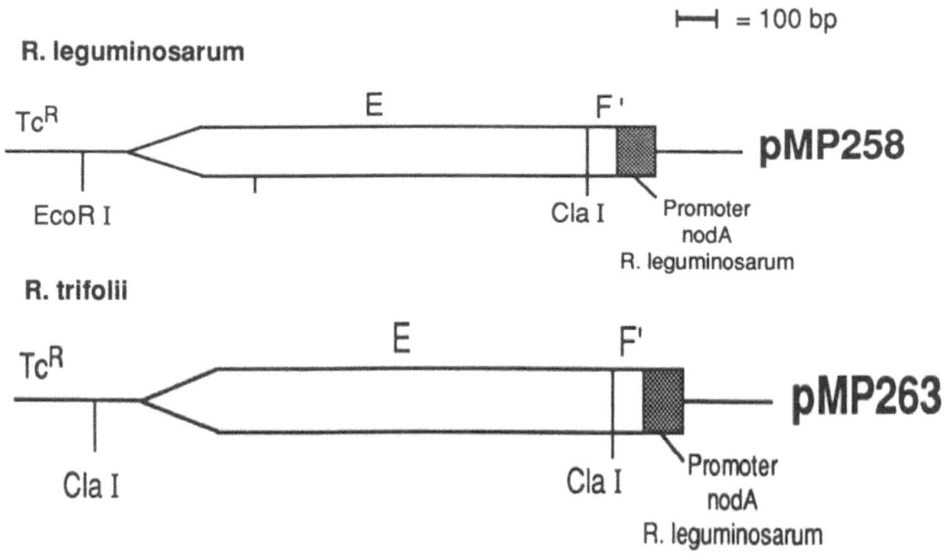

Fig. 1. IncP plasmid constructs containing the nodE genes of R.leguminosarum and R.trifolii. A 114 basepair BglII-BclI fragment was used as the source of the nodA promoter of R.leguminosarum (hatched box).

operon, which controls the expression of a common nodulation operon, was used to rule out a possible host-specific influence of the nodFE regulatory region. The plasmids pMP263 and pMP258 were introduced into the R.trifolii nodE mutant strain K11 and the resulting strains were tested for nodulation on several host plants of either R.trifolii or R.leguminosarum. As expected, plasmid pMP263 restores the mutant phenotype of strain K11 to that of the wild type strain ANU843. In strong contrast, the nodE gene of R.leguminosarum confers to strain K11 the ability to nodulate efficiently on the tested hosts of R.leguminosarum (Table 1). The nodulation phenotype of strain K11 on the R.trifolii host plants is not altered by the presence of the R.leguminosarum nodE gene. We therefore conclude that in isogenic R.trifolii strains the origin of the NodE product determines whether R.trifolii or R.leguminosarum host plants are efficiently nodulated.

LOCALIZATION OF THE HOST-SPECIFIC REGION OF NODE PROTEIN

Although the NodE proteins of R.trifolii and R.leguminosarum are clearly different in their host-specific properties the predicted sequence of these proteins is remarkably similar (Fig. 2). To localize one or more protein regions involved in the determination of the difference in host-specificity, we have studied the host-specific properties of hybrid nodE genes. These hybrid nodE genes were constructed by in vivo homologous recombination using the procedure described by Spaink et al (1989). In Fig. 2 the position of the fusion sites of several studied hybrid NodE proteins is indicated. All hybrids contain the 5′ part of the nodE gene of R.leguminosarum

```
       1
R.leg  MDRRVVITGLGGLCGLGTDASSIWTEMREGRSAIGPISNSEIHELKGMIG
R.tri  MDRRVVITGIGGLCGLGTNAASIWKEMREGPSAISPIITTDLYDLEGTVG
        925                            912

       51
R.leg  TEIKVLPQHDIDRKQLISMDRFSLLAVLAAKQAMLQAGLSCNEGNAHRFG
R.tri  LEIKAIPEHDIPRKQLVSMDRFSLLAVIAATEAMKQAGLSCDEQNAHRFG
        906                                          910

       101
R.leg  ATVGVGFGGWDATEKAYRTLLLGGATRTELFTGVKAMPSAAACQVSMNLG
R.tri  AAMGLGGPGWDTIEETYRSILLDGVTRARIFTAPKGMPSAAAGHVSIFLG
                                            911

       151
R.leg  LRGPVFGATSACASANHAIASAVDQIKLGRADVMLAGGSDAPLVWIVLKA
R.tri  LRGPVFGVTSACAAGNHAIASAVDQIRLGRADVMLAGGSDAPLTWGVLKS

       201
R.leg  WEAMRVLAPDTCRPFSADRKGLVLGEGAGMAVLESYEHAAARGATMLAEV
R.tri  WEALRVLAPDTCRPFSADRKGVVLGEGAGMAVLESYEHAAARGATMLAEV
                               913              929

       251
R.leg  AGIGLSADAYHIAAPAVHGPEAAMRACLVDASLNAEDVDYLNAHGTGTKA
R.tri  AGIGLSGDAYDIVMPSIEGPEAAMRSCLADAELNPDDVDYLNAHGTGTVA
                                                  921

       301
R.leg  NDQIETTAIKRVFGDHARSMSISSTKSTHAHCLGAASALEMIACVMAIQE
R.tri  NDEMETAAIKRVFGDHAFQMSVSSTKSMHAHCLGAASALEMIACVMAIQE
                                                     932

       351                                                 403
R.leg  GVVPPTANYREPDPDCDLDVTPNVPRERKVRVAMSNAFAMGGMNAVLAFKQVP
R.tri  GVIPPTANYREPDPQCDLDVTPNVPREQRC GSMSNAFAMGGTNAVLAFRQV
           919
```

Fig. 2. Comparison of the NodE proteins of R.leguminosarum and R.trifolii and localization of the fusion sites in NodE hybrids. All hybrid nodE genes contain the 5′ end of the R.leguminosarum nodE gene derived from pMP258 (Fig. 1). The non-conserved amino acids in both proteins are indicated by boxes. The underlined amino acids at position 27 to 31 are corrections on the data of Shearman et al (1986).

derived from pMP258 (Fig. 1).

The derivatives of R.trifolii nodE mutant strain K11, containing the representative nodE hybrid genes were tested for their nodulation ability on T.pratense, T.arvense, Vicia sativa and V.hirsuta. The results (Table 1) show that hybrids which contain a recombination site in the 5' prime part of the nodE gene, i.e. nodE925, nodE912, nodE906 and nodE910, cause a nodulation phenotype which is not significantly different from that caused by the R.trifolii parental nodE gene. Hybrids which contain a recombination site in the 3' part of the nodE gene, i.e. nodE921, nodE932, and nodE919, have a phenotype which is not significantly different from that of the R.leguminosarum parental nodE gene. The hybrids nodE911, nodE913 and nodE929, which resulted from a recombination in the central region of nodE, confer a phenotype which is different from that caused by both of the parental nodE genes. These results indicate that a central region of the NodE protein of 185 amino acids at the most, containing 44 non-conserved amino acids, is responsible for the difference in host specificity conferred by the studied NodE proteins.

IMMUNOLOGICAL STUDIES OF NODE PROTEIN

To obtain antibodies against NodE protein we constructed plasmid pMP1150 in which 90 percent of the 3' part of the R.trifolii nodE gene is fused in frame to the 5' terminal part of the E.coli lacZ gene. This plasmid expresses a fusion protein which was purified and subsequently used for immunization of a rabbit. The obtained serum reacts with flavonoid-inducible proteins produced by R.trifolii and R.leguminosarum whith apparent molecular weights of 51 kd and 45 kd, respectively, upon sodium dodecyl sulfate polyacrylamide gel electrophoresis (SDS-PAGE). Since these proteins are not produced by R.trifolii and R.leguminosarum strains in which the nodE gene is eliminated by transposon insertions it can be concluded that the detected proteins represent the respective NodE proteins. Using the NodE antiserum, flavonoid inducible protein could also be detected in

R.meliloti strains 102F28 and 102F51 as 44 kd and 43 kd proteins, respectively. No isoflavone-inducible proteins (Sanchez et al, 1988) could be detected in any of eight independent wild-type R.phaseoli strains tested.

INTRACELLULAR LOCALIZATION OF NODE PROTEIN

Cellular fractions of a wild type R.leguminosarum and a R.trifolii strain were tested for the presence of the NodE protein on Western blots. From the results we can conclude that NodE protein in R.leguminosarum and R.trifolii cells is present exclusively in the cytoplasmic membrane of the bacterial cell. Theoretical analysis of the predicted NodE proteins using the hydrophobicity scale of Engelman et al (1986) indicates that NodE protein has several potential transmembrane α-helical regions. Using the method described by Manoil and Beckwith (1985) we have constructed several fusions of the R.leguminosarum nodE gene with the alkaline phospatase (phoA) gene of E.coli. From the in vivo alkaline phosphatase activities of the resulting NodE-PhoA fusion proteins we conclude that the NodE protein is a membrane protein of complex topology with several regions present in the periplasmic space.

DISCUSSION

We have shown that NodE protein is a crucial factor in the determination of the host range of nodulation of R.trifolii and R.leguminosarum. The localization of NodE protein in the cytoplasmic membrane of the cell indicates that the host-specific interaction with the host plant is not determined directly by NodE protein. We therefore suggest that NodE protein is involved in the synthesis or modification of a signal towards the host plants. The nodF and nodE genes are translationally coupled (H. Spaink, unpublished results) and therefore NodF and NodE protein are probably involved in the same biochemical process. The predicted product of the nodF gene has been shown to be homologous with the E.coli acyl carrier protein (Shearman et al 1986). The E.coli acyl carrier protein has been shown also to function as a transglucosylase factor involved in the synthesis of β-1,2-glucans (Therisod et al 1986). These β-1,2-glucans constitute the backbone of the membrane-derived oligosaccharides of E.coli which very much resemble the cyclic β-1,2-glucans found in several rhizobia and Agrobacterium (e.g. Amemura et al 1983; Miller et al 1987). A role of the NodF and NodE proteins in the production of a specific cyclic β-1,2-glucan would be consistent with the presumed importance of such polysaccharides in the nodulation process (Abe et al 1982).

The bacterial signal which determines the host-specific characteristics of R.leguminosarum and R.trifolii, is likely to have some relation with the host plant lectin , another major determinant of host specificity (Diaz et al 1989; Kijne et al this volume). The function of the host plant lectins as sugar-binding proteins suggests that the bacterial signal towards the plant is of carbohydrate nature.

REFERENCES

Abe M, Amemura A and Higashi S (1982) Plant and Soil 64:315-324
Amemura A, Hisamatsu M and Mitani H (1983) Carbohydr Res 114: 277-285
Diaz CL, Melchers LS, Hooykaas PJJ, Lugtenberg BJJ and Kijne JW (1989) Nature 338:579-581
Engelman DM, Steitz TA and Goldman A (1986) Ann Rev Biophys Chem 15:145-168
Long SR (1989) Cell 56: 203-214
Manoil C and Beckwith J (1985) Proc Natl Acad Sci USA 82:8129-8133
Miller KJ, Reinhold VN, Weissborn AC and Kennedy EP (1987) Biochem Biophys Acta 901:112-118
Sanchez F, Quinto C, Vazquez M, Spaink HP, Wijffelman CA, Cevallos MA, De Las Peñas A, Campos F, Padilla J, Lara M (1988) In: Palacios R and Verma DPS (eds) Molecular genetics of plant-microbe interactions. APS Press, StPaul:370-375
Shearman CA, Rossen L, Johnston AWB and Downie JA (1986) EMBO J 5:647-652
Spaink HP, Wijffelman CA, Okker RJH and Lugtenberg BJJ (1989) Plant Mol Biol 12:59-73
Therisod H, Weissborn AC and Kennedy EP (1986) Proc Natl Acad Sci USA 83:7236-7240

IDENTIFICATION OF *RHIZOBIUM LEGUMINOSARUM* GENES AND SIGNAL COMPOUNDS INVOLVED IN THE INDUCTION OF EARLY NODULIN GENE EXPRESSION

Ben Scheres, Clemens van de Wiel, Andrei Zalensky+, Ann Hirsch*, Ab Van Kammen and Ton Bisseling

Dept. of Molecular Biology, Agricultural University, Dreijenlaan 3, 6703 HA Wageningen, The Netherlands
+All-Union Research Institute for Agricultural Microbiology, Leningrad, U.S.S.R.
*Biology Department, U.C.L.A., Los Angeles, Ca, U.S.A.

INTRODUCTION

The process of root nodule formation on legumes, induced by *Rhizobium*, can be looked upon as a sequence of several distinct steps. These steps have been defined by cytological studies on developing wild-type root nodules, and by analyses of nodules formed by either plant or bacterial mutants (Vincent 1980). Nowadays attachment of bacteria, root hair deformation and curling, induction of cortical cell division, infection thread formation, nodule development, bacterial release from infection threads, bacteroid development and effective nitrogen fixation are recognized as successive steps in root nodule formation (Vincent 1980). The multistep nature of root nodule formation has led to the hypothesis that at several stages in the *Rhizobium*-plant interaction signal molecules from either symbiontic partner are involved in inducing a process in the other partner Identification of the different bacterial and plant signals and analysis of the mode of action of each seperate compound would then significantly enlarge our knowledge about the establishment of symbiosis.

To date only two factors involved in the communication between the two symbionts have been identified. Using nod- lacZ reporter gene fusions, plant flavenoids have been identified as elicitors of the beginning of the communication. These flavenoids act in concert with the product of the constitutively expressed bacterial *nodD* gene to induce expression of the other nod genes (Mulligan and Long 1985). Upon induction of the *nod* genes the root hair deformation factor is formed by the bacteria. This factor has been identified using root hair deformation as qualitative bioassay (Bhuvaneswari & Solheim 1985). The relevant compound has however not been purified and the structure has not yet been determined due to the lacking of a quantitative assay. This demonstrates the importance of reporter genes for developing quantitative assays to be used for the purification of compounds, involved in different steps of root nodule formation.

We have set out to isolate nodule specific plant cDNA sequences representing genes of which the expression marks different stages of root nodule development. When we started, the already cloned plant nodulin sequences mainly represented genes expressed when nodule

NATO ASI Series, Vol. H 36
Signal Molecules in Plants and Plant-Microbe Interactions
Edited by B. J. J. Lugtenberg

development, according to cytological criteria, is completed (reviewed in Govers *et al* 1987a). We have now isolated cDNA clones representing genes expressed earlier in nodule development and examined whether the expression of these genes correlates to distinct steps in root nodule formation. The information that may be obtained with such marker genes is twofold. First, characterization of the proteins encoded by these genes will provide information about the type of plant proteins essential for different developmental steps. Second, if indeed the expression of a certain gene is characteristic for a step in root nodule development, the expression of such gene may be used to identify bacterial or plant signals necessary to invoke the developmental step in which the gene is involved. Here we shall briefly summarize the characteristics of a set of pea early nodulin cDNA clones. We will describe the correlation of the expression of different genes and different steps in root nodule formation. We further show how expression of these genes can be used to examine whether certain compounds can mimic the effect of signal molecules in the *Rhizobium*-plant interaction. We then focus on the question how one of the cloned cDNAs, pPsENOD12, may be used for developing an assay for purification of bacterial signal molecules.

EARLY NODULINS

Nodulin genes are differentially expressed during nodule development and therefore these genes have been divided into two groups, early and late nodulins (for review see Nap and Bisseling 1989). The late nodulin genes are first expressed shortly before or concomitantly with the onset of nitrogen fixation. The genes expressed at earlier stages of nodule development are named early nodulin genes.

pGmENOD2 was the first early nodulin for which a cDNA clone has been described (Franssen *et al* 1987). This cDNA clone was isolated from a soybean nodule cDNA library, and was found to encode a protein composed of 2 repeating penta- or hexapeptides, containing two or three prolines each. At the N-terminus a putative signal peptide is present. Later similar ENOD2 cDNA clones were isolated from alfalfa (Dickstein *et al* 1988), sesbania (F. de Bruijn, N-H.Chua, pers. comm.), and pea (our work). Besides pGmENOD2, two other soybean early nodulin clones have recently been described, pGmENOD13 and pGmENOD55 (Franssen *et al* 1988). ENOD13 has a structure similar to ENOD2. The 52 amino acid sequence of the polypeptide derived from the partial reading frame in the cloned sequence shows 70% homology to the C-terminal part of the ENOD2 protein. The structure of the ENOD55 protein differs from that of the ENOD2 protein, but ENOD55 is a proline rich protein as well. The proline rich region of the ENOD55 protein is confined to an internal domain of 32 amino acids and is composed of proline and serine residiues.

From *Pisum sativum*, the garden pea, we isolated five early nodulin clones pPsENOD2, pPsENOD12, pPsENOD5, pPsENOD3, and pPsENOD14, which could be involved in different developmental steps, judged by the different time courses of accumulation of the corresponding mRNAs during nodule development. pPsENOD12 encodes a protein with a structure similar to that of the pea ENOD2 protein , since both are built up of a series of two repeating pentapeptides each containing two prolines. The three other amino acids in the pentapeptide unit of pPsENOD12 differ from those in ENOD2. In conclusion, the pea ENOD2 and ENOD12 early nodulins are, like the soybean early nodulins ENOD2 and ENOD13, (hydroxy-)proline rich proteins. It is thought that (hydroxy-)proline rich proteins are often associated with cell walls. Therefore we assume that at least part of these proline rich early nodulins are components of cell walls of the different cell types that are formed during symbiosis. The homology of ENOD2, ENOD13 and ENOD12 to the soybean cell wall protein 1A10 (Averyhart- Fullard *et al* 1988) supports this hypothesis. pPsENOD5 encodes a protein with hydrophobic regions at both the C- and the N-terminus, the latter possibly forming a signal peptide. This early nodulin also has a proline rich domain, which moreover has a high content of glycine, alanine, and serine. The amino acid composition of this domain of ENOD5 is reminiscent of that of arabinogalactan proteins (Van Holst *et al* 1981). To date no amino acid sequences of arabinogalactan proteins have been published, which prevents a more detailed comparison of ENOD5 and arabinogalactan proteins. Upon searching different databases no significant homology between ENOD5 and other previously described proteins was found. pPsENOD3 and pPsENOD14 both encode small nodulins, with a molecular weight of about 6 kD. These two early nodulins are 60% homologuous and contain as most striking characteristic a cluster of four cysteins, arranged in such a way that a metal can be bound. This suggests that the ENOD3 and ENOD14 proteins contain a metal ion. Similar motifs of four cysteines are found in the late nodulins Ngm-20, Ngm-23, Ngm26b, Ngm-27, and Ngm-44, forming a small gene family in soybean (Jacobs *et al* 1987). However, there is no significant sequence homology between these or any other proteins from which the sequence is stored in the different databases, and the ENOD3 and ENOD14 proteins.

NODULIN GENE EXPRESSION DURING NODULE DEVELOPMENT

Pea nodules belong to the indeterminate nodule type. These nodules have a persistent meristem. Hence all nodules contain cells at different stages of development: the youngest cells adjacent to the apical meristem and the oldest cells at the basal root attachment point. This type of nodules have the advantage that the presence of transcripts related to particular stages of root nodule development is less dependent on the age of the nodule. A further advantage is that it is possible to compare the expresssion patterns of the nodulin genes in different cell types during nodule development by comparing the spatial distribution of different nodulin transcripts in serial sections from one nodule using *in situ* hybridization.

The different pea early nodulin clones were hybridized with sections of nodules from plants of various ages. The plants start to fix nitrogen from day 13 after inoculation and sowing, and at this day the first late nodulin transcripts are detectable by RNA blot analysis (Govers *et al* 1987b). On sections of these nodules the leghemoglobin (Lb) cDNA clone pPsLb was used as a marker for the expression of late nodulin genes, and the *R.leguminosarum nifH* gene as a marker for expression of bacterial genes involved in nitrogen fixation.

The transcripts corresponding to four of the five selected early nodulin clones are present in the central tissue of the nodule. For ENOD2, on the other hand, the transcript was specifically located in the nodule inner cortex. The ENOD12, ENOD5, ENOD3, and ENOD14 transcripts are present in successive, but partially overlapping zones of the central tissue. The ENOD12 mRNA is present in a small zone directly adjacent to the meristem, the invasion zone. The ENOD5 transcript starts to accumulate in the cells where ENOD12 mRNA is still present, but the concentration reaches a maximum in the zone of the central tissue that contains enlarging cells. This zone has been named the young symbiotic zone as in this zone the first plant cells that contain bacteria are present (Newcomb 1981). The ENOD5 transcript is present at strongly reduced levels in the cells of the central tissue that have reached their maximum size. This is the symbiotic zone. The ENOD3 and ENOD14 messengers start to accumulate in the cells where the ENOD5 transcript is present at a maximal level and the concentration of these transcripts reaches its maximum in the youngest 3-4 cell layers of the symbiontic zone. In the older cells of the symbiotic zone the ENOD3 and ENOD14 mRNA concentration decreases. In these older cells the level of the transcript of the late nodulin Lb is at its maximum. The ENOD5, ENOD3, and ENOD14 mRNAs are only present in the cells containing rhizobia, while ENOD12 mRNA is present in all the cells of the invasion zone. Therefore ENOD12 transcript is present in cells that will develop in either infected or uninfected cells. At the stage of development of the infected cells when the ENOD3 and 14 mRNA concentrations are decreasing, the nitrogenase mRNA can first be detected in the bacteroid . This proves that the infected cell becomes a functional nitrogen fixing cell at that stage. Therefore the decline of the ENOD3 and 14 mRNA concentrations marks the stage at which the infected cell is fully differentiated into a functional nodule cell.

In conclusion, the '*in situ*' hybridization studies show that the differentiation of meristematic cells into the infected cell type in the central zone of the nodule requires at least four successive steps of specific gene expression. These different steps are marked by the presence of ENOD12,ENOD5, ENOD3 and ENOD14, and Lb transcripts, respectively. Most likely several other late nodulin genes are expressed concomitantly with the Lb genes, but this has not yet been tested.

The availability of data on both the sequence and the localization of the early nodulin transcripts in some cases allows to speculate about functions of the encoded proteins. As for the site of expression of the ENOD2 gene, it has been shown by Witty *et al* (1986) that the nodule inner cortex is the major barrier for oxygen diffusion within a root nodule. It is conceivable that ENOD2, being a putative cell wall protein, contributes to the absence of intercellular spaces in the inner cortex which causes this tissue to be the major oxygen barrier.

In situ hybridization studies have now revealed that ENOD12 genes are expressed not only in the nodule invasion zone but also in root hairs and in the root cortex during the infection process. ENOD12 transcripts are found in the cells that become prepared for infection thread growth (Bakhuizen *et al* 1989), and in cells containing the infection thread tip. Therefore it was concluded that ENOD12 has a role in the "preparation" of cells for infection thread growth and maybe also in the formation of the infection thread. Sequence analyses suggest that ENOD12 is a cell wall component. In the "prepared" root cortex cells an additional cell wall is formed, an infection thread contains a wall. Whether these walls indeed contain ENOD12 protein remains to be established by immunocytological studies.

Combination of structural data of ENOD5, ENOD3, and ENOD14 early nodulins, and the location of the corresponding transcripts, does not yet allow predictions about the functions of these early nodulins during development. Immunocytological localization of the proteins might give more clues. More direct evidence for the function of these early nodulins would require the succesfull application of reverse genetics, e.g. antisense RNA inhibition.

In view of the possible function of nodulins we have wondered if the expression of nodulin genes is related to defense responses to plant pathogens. The study of signals and transduction mechansisms for activation of plant defense genes is a well developed area of plant molecular biology (Lamb *et al* 1989). Also the *Rhizobium*-legume symbiosis has been viewed as a modified defense response Djordjevic *et al* 1987). This hypothesis implies that during nodule development the expression of plant genes might be triggered as a result of bacterial signals and signal transduction pathways normally used as part of a response to pathogens. We tested whether pea early nodulin genes were expressed when pea roots were inoculated with the pathogen *Fusarum oxysporum* bv *pisi*. While we found accumulation of defense related hydroxyproline rich glycoprotein mRNAs in these roots, no accumulation of early nodulin transcripts could be detected. We conclude that early nodulin gene expression is not related to a general plant defense respons and that the signals and transduction mechanisms that trigger early nodulin gene expression therefore differ from those occuring during a pathogenic interaction.

EARLY NODULIN GENES AS REPORTERS FOR BACTERIAL OR PLANT SIGNALS

The differential accumulation of nodulin transcripts during the differentiation of the infected cell type points to the occurence of successive steps in nodulin gene expression. Successive induction of gene expression can be caused by different bacterial signals, by different kinetics of gene expression in response to the same bacterial signal, or by different second messengers that are formed in the plant as a result of a process induced by one bacterial signal. The mechanisms used by *Rhizobium* to establish the differential gene expression during nodule development can now be studied with the set of nodulin cDNA clones described here. We do not pretend to have cloned all early nodulin transcripts, and hence we cannot expect that the six genes studied here are markers for all steps of plant-bacterium communication and nodule development. Still the answers obtained on the expression of the available nodulin genes can substantially increase the insight into the way plant and bacterium communicate. Here we will demonstrate the use of the pea ENOD12 and ENOD2 genes as marker genes.

A. Nodulin genes as tools to study compounds that mimic *Rhizobium* signals.

More than forty years ago Allen and Allen (1940) showed that compounds that block the polar transport of auxin induce nodule-like outgrowths on the roots of several plants. Later it was shown that auxins and cytokinins induce cell division in the inner cortex of root explants, possibly in cooperation with a factor from the xylem (Libbenga *et al* 1973). In these studies processes were induced that resemble steps occurring in root nodule formation. Unfortunately these processes could only be studied on a cytological level at that time, and it remained undecided whether these compounds really mimiced part of the nodule formation process. Now probes for the expression of several early and late nodulin genes are available and the expression of specific marker genes can be used to mark different steps in the *Rhizobium*-legume interaction. It is now possible to reexamine the effect of such compounds and re-evaluate data that were already buried in the archaeology of science. The nodule structures formed by anti-auxins were recently studied at the molecular level by Hirsch *et al* (1989). In such nodules formed on alfalfa roots upon treatment with the anti-auxins 2,3,5-triiodobenzoic acid (TIBA) or N-(1-naphtyl)-phthalamic acid (NPA) the alfalfa early nodulin genes ENOD2 and Nms-30 appeared to be expressed. Moreover we demonstrated that the ENOD2 transcripts are located in a tissue at the periphery of these nodule structures. By position and gene expression this tissue is therefore very similar to the nodule inner cortex, which points to the existence of both cytological and molecular similarities between anti-auxin formed nodules and nodules formed by *Rhizobium*.

If anti-auxins can induce cortical cell division and development of a nodule structure similar to that induced by rhizobia, an obvious question is whether *Rhizobium* produces analoguous compounds. The occurence of a *Rhizobium* compound specifically involved in induction of cell division has been demonstrated, but the compound has not yet been characterized (Schmidt *et al* 1988). Two lines of evidence demonstrate a crucial role of the *nod* genes for the production of this factor. First, mutations in the common *nod* genes abolisch the ability of *Rhizobium* to induce cortical cell division (Dudley *et al* 1987). Second, 12 kb of the *R.leguminosarum* Sym plasmid, containing the *nod* genes, can confer to *Agrobacterium tumefaciens* the ability to form root nodules, expressing the ENOD2 gene (Nap *et al* 1987).

Based on the similarities in nodule development induced by anti-auxins and *nod* gene dependent bacterial compounds it is plausible that *Rhizobium nod* gene products interfere with the phytohormone distribution in the legume root. The subsequent change in hormone balance could then induce centres of mitotic activity at certain sites in the root cortex. The observation that *Rhizobium nod* gene mutants can be complemented with the *A.tumefaciens* zeatin gene, which is involved in cytokinin synthesis, is in accordance with this hypothesis (J.Cooper, pers.comm.).

Preliminary data suggest that, upon treatment of pea roots with anti-auxins, also the expression of ENOD12 is induced in root hairs already after 48 hrs. During the pea-*Rhizobium* symbiosis expression of this gene marks root hair-, root cortex-, and nodule cells involved in, or preparing for, infection thread growth. Specific microtubule rearrangement and nuclear migration in these cells also occurs in cells which, upon *Rhizobium* infection, become part of the centre of mitotic activity, the nodule primordium (Bakhuizen *et al* 1989). We demonstrated that in these cells ENOD12 gene expression is also induced. This would imply that cells preparing for infection thread passage and nodule primordium cells are similar as far as ENOD12 expression is concerned. Apparently anti-auxins can at least induce expression of the ENOD12 gene as a marker gene for these cell types. Since anti-auxins appear to be able to elicit both ENOD12 and ENOD2 gene expression, they appear to mimic two different processes. First, ENOD12 nodulin gene expression, related to cells involved in both infection thread growth and the establishment of a centre of meristematic activity, is induced. Second, meristematic activity indeed leads to differentiation into at least one nodule tissue where the proper nodulin gene, ENOD2, is expressed. Conclusively, anti-auxins induce a cascade of events, mimicing parts of both nodule morphogenesis and the *Rhizobium* infection process. Therefore signal molecules produced by *Rhizobium*, under the direction of a small set of *nod* genes, may similarly establish the morphogenesis of certain cell types and parts of the infection process by interfering with the plant hormone balance.

B. ENOD12 gene expression in root hairs as an assay for *Rhizobium* signal compounds.

Genes which are expressed in root hairs upon inoculation with *Rhizobium* seem most suitable to study bacterial signal compounds. First, these genes are expressed in a pre-existing cell type, and not in cells modified by *Rhizobium*. Hence the chain of events leading from a bacterial signal to expression of these plant genes might be less complicated. Second, root hairs directly can be treated with putative bacterial signal compounds, and then isolated and analyzed for gene expression.

By *in vitro*'translation of root hair RNA, two root hair specific pea transcripts, RH-42 and RH-44, have been shown to accumulate upon inoculation with *R.leguminosarum* (Gloudemans *et al* 1989). The accumulation of both transcripts was shown to be dependent on the bacterial *nodC* gene. Furthermore, the accumulation of RH-44 mRNA could be induced with a cell-free preparation of deformation factor, obtained from *R.leguminosarum* cultured in the presence of the *nod* gene inducer apigenin. Therefore, the appearance of RH-44 transcripts might be used as a molecular marker for the activity of the bacterial compound causing root hair deformation. Unfortunately the*in vitro* translation of root hair RNA followed by two dimensional gelelectrophoresis to detect this transcript, which has not yet been cloned, is too elaborate to use in a quantitative routine assay. Thus the use of RH-44 as marker has to wait untill more simple detection methods are available for this mRNA.

Another gene of which the expression could function as a possible marker for the action of *Rhizobium* signal compounds is the ENOD12 early nodulin gene. Also this gene is already expressed in root hairs 48 hrs after inoculation with *R.leguminosarum*, and we demonstrated that the expression of the bacterial common *nod* genes is essential for eliciting ENOD12 gene expression. The sequence data available from the pPsENOD12 were used in designing a specific assay for the presence of ENOD12 transcripts, using reverse transcription and polymerase chain reactions. This semi-quantitative detection method is now used to test the ability of different *R.leguminosarum* mutant strains and of soluble compounds excreted by the bacteria to induce ENOD12 gene expression. Using a variety of *R.leguminosarum* strains containing only small regions of the sym-plasmid or deletions spanning different *nod* genes (Spaink 1989) it was shown that both the common *nod* genes *nodABC* and the host specific *nodE* gene are essential for ENOD12 gene expression. This is consistent with the observation that both *R.leguminosarum nodABC* and *nodE* are necessary for infection thread formation, the process to which ENOD12 gene expression is correlated (Spaink 1989). Furthermore it was shown that ENOD12 gene expression is elicited by a soluble compound, excreted by *Rhizobium* upon induction of the *nod* genes with pea root exudate.

The common *nod* genes of *R.leguminosarum* are known to be sufficient for excretion of the soluble factor that establishes root hair deformation (Zaat *et al* 1987). We were able to partially purify this deformation factor using root hair deformation on *Vicia sativa* as a bio-assay. We are currently investigating whether the factor that elicits ENOD12 gene expression can be purified with a similar purification scheme, indicating that the compound has similar molecular properties. In this way we hope to establish whether the *nodE* gene product modifies a deformation factor made by a *nodABC* product in such a way that it is able to elicit ENOD12 gene expression. Alternatively, root hair deformation, established by the *nodABC* dependent deformation factor, might be a prerequisite for the abilitiy of a structurally unrelated *nodE* dependent factor to elicit ENOD12 gene expression. In the *R.meliloti*-alfalfa symbiosis not only the host specific nodulation genes *nod EF* and *nodG* (Horvath *et al* 1986) appear to be essential for infection thread formation, but also exopolysaccharide genes (Finan *et al* 1985). We also intend to investigate the role of exopolysaccharides in inducing expression of the infection related ENOD12 gene, and the possible link to *nod* gene products.

Summarizing the data on ENOD12 gene expression, soluble factors dependent on both *nod ABC* and *nodE* are required for induction of ENOD12 gene expression in root hairs. As stated in the previous section of this paragraph, preliminary data suggest that ENOD12 transcript also accumulates in root hairs upon treatment with anti-auxins. This leads us to the hypothesis that bacterial compounds, made either by the *nodE* product alone, or by the *nodABC* and *nodE* products together, alter the hormone balance in roots to allow bacterial infection and the development of a nodule structure.

CONCLUDING REMARKS

During the development of root nodules bacterial and plant signals play an important role in establishing an effective symbiosis. We have obtained a set of cDNA clones that can serve as probes for the expression of genes that mark different stages of root nodule development, demonstrating that progressive development is accompanied by differential plant gene expression. The induction of the expression of these specific genes might be due to the action of bacterial and/or plant signals effective during nodule development. We have demonstrated the use of ENOD2 and ENOD12 in showing that anti-auxins mimic several aspects of root nodule formation induced by *Rhizobium*. Using ENOD12, we have designed an assay for bacterial compounds produced by the bacterial *nodABC* and *nodE* genes, which are necessary to induce expression of the ENOD12 gene. The purification and characterization of these compounds are under way.

The use of ENOD5, ENOD3, and ENOD14 to identify signals involved in later steps of root nodule development will require a search for *R.leguminosarum* mutants unable to elicit expression of these genes, and characterization of the nature of the mutations. Possible candidate mutant strains for such a study are bacterial release (bar^-) and bacteroid development (bad^-) mutants.

REFERENCES

Allen, O.N., and Allen, E.K. (1940). Proc. Acad. Sci. Bernice P. Bishop Museum Spec. Publ. 35, 15-16

Averyhart-Fullard, V., Datta, K., and Marcus, A. (1988). A hydroxyproline-rich protein in the soybean cell wall. Proc. Natl. Acad. Sci. USA 85, 1082-1085

Bakhuizen, R., van Spronsen, P.C., Diaz, C.L., and Kijne, J.W. (1989). Rearrangement of cytoplasm and endoplasmic microtubules in cortical cells of *Pisum sativum* L. roots prior to infection by *Rhizobium leguminosarum* biovar viceae. Submitted

Bhuvaneswari, T.V., and Solheim, B. (1985). Root hari deformation in the white clover/*Rhizobium trifolii* symbiosis. Physiol. Plant. 63, 25-34

De Maagd, R.A., Rao, A.S., Mulders, L.H.M, Govers-de Roo, L., Van Loosdrecht, M.C.M., Wijffelman, C.A., Lugtenberg, B.J.J. (1989). Isolation and characterization of mutants of *R.leguminosarum* bv. Viciea strain 248 with altered lipopolysaccharides: possible role of surface charge or hydrophobicity in bacterial release from the infection thread. J. Bacteriol. 171, 1143-1150

Dickstein, R., Bisseling, T., Reinhold, V.N., and Ausubel, F.M. (1988). Expression of nodule-specific genes in alfalfa root nodules blocked at an early stage of development. Genes and Development 2, 677-687

Djordjevic, M.A., Gabriel, D.W., and Rolfe, B.G. (1987). *Rhizobium* - the refined parasite of legumes. Ann. Rev. Phytopathol. 25, 145-168

Dudley, M.E., Jacobs, T.W., and Long. S.R. (1987). Microscopic studies of cell divisions induced in alfalfa roots by *Rhizobium meliloti*. Planta 171, 289-301

Finan, T.M., Hirsch, A.M., Leigh, J.A., Hohansen, E., Kuldau, G.A., Deegan, S., Walker, G.C., and Singer, E.R. (1985). Symbiotic mutants of *Rhizobium meliloti* that uncouple plant from bacterial differentiation. Cell 40, 869-877

Franssen, H.J., Nap, J.P., Gloudemans, T., Stiekema, W., Van Dam, H., Govers, F., Louwerse, J., Van Kammen, A., and Bisseling, T. (1987). Characterization of cDNA for nodulin-75 of soybean: a gene product involved in early stages of root nodule development. Proc. Natl. Acad. Sci. USA 84, 4495-4499

Franssen, H.J., Scheres, B., Van De Wiel, C., and Bisseling, T. (1988). Characterization of soybean (hydroxy)proline-rich early nodulins. In: Molecular Genetics of Plant-Microbe interactions, R.Palacios and D.P.Verma, eds. (St.Paul: APS Press), 321-326

Gloudemans, T., Bhuvaneswari, T.V., Moerman, M., Van Brussel, T., Van Kammen, A., and Bisseling, T. (1989). Involvement of *Rhizobium leguminosarum* nodulation genes in gene expression in pea root hairs. Plant Mol. Biol. 12, 157-167

Govers, F., Nap, J.P., Van Kammen, A., and Bisseling, T. (1987a). Nodulins in the developing root nodule. Plant Physiol. Biochem. 25, 309-322

Govers, F., Nap, J.P., Moerman, M., Franssen, H.J., Van Kammen, A., and Bisseling, T. (1987b). cDNA cloning and developmental expression of pea nodulin genes. Plant Mol. Biol. 8, 425-435

Hirsch, A.M., Bhuvaneswari, T.V., Torrey, J.G., and Bisseling, T. (1989). Early nodulin genes are induced in alfalfa root outgrowths elicited by auxin transport inhibitors. Proc. Natl. Acad. Sci. USA 86, 1244-1248

Horvath, B., Kondorosi, E., John, M., Schmidt, J., Torok, J., Gyorgypal, Z., Barabas, I., Wieneke, U., Schell, J., and Kondorosi, A. (1986). Organization, structure and function of *Rhizobium meliloti* nodulation genes determining host specificity in alfalfa. Cell 46, 335-343

Jacobs, F.A., Zhang, M., Fortin, M.G., and Verma, D.P.S. (1987). Several nodulins of soybean share structural domains but differ in their subcellular locations. Nucl. Acids Res. 15, 1271-1280

Lamb, C.J., Lawton, M.A., Dron, M, and Dixon, R.A. (1989). Signals and transduction mechanisms for activation of plant defenses against microbial attack. Cell 56, 215-224

Libbenga, K.R., Van Iren, F., Bogers, R.J., Schraag-Lamers, M.F. (1973). The role of hormones and gradients in the initiation of cortex proliferation and nodule formation in *Pisum sativum* L. Planta 114, 29-39

Mulligan, J.T., and Long, S.R. (1985). Induction of *Rhizobium meliloti* nod C by plant exudate requires nod D. Proc. Natl. Acad. Sci. USA 82, 6609-6613

Nap, J.P., and Bisseling, T. Nodulin function and nodulin gene regulation in root nodule development. in: The Molecular Biology of Symbiotic Nitrogen Fixation, Gresshoff, P.M., Ed., CRC press, Florida, in press

Newcomb, W. (1981). Nodule morphogenesis and differentiation. in: Biology of of the Rhizobiaceae, Giles, K.L., and Atherly, A.G., Eds., Int. Rev. Cytol. Suppl 13, Academic Press, New York, 247-297

Schmidt, J., Wingender, R., John, M., Wieneke, U., and Schell, J. (1988). *Rhizobium meliloti* nodA and nodB genes are involved in generating compounds that stimulate mitosis of plant cells. Proc. Natl. Acad. Sci. USA 85, 8578-8582

Spaink, H. (1989). The *Rhizobium*-plant symbiosis: bacterial gene regulation and host specificity. Thesis, Leiden.

Van Holst, G-J., Klis, F.M., De Wildt, P.J.M., Hazenberg, C.A.M., Buijs, J., and Stegwee, D. (1981). Arabinogalactan protein from a crude cell organelle fraction of *Phaseolus vulgaris* L. Plant Physiol. 68, 910-913

Vincent, J.M. (1980). Factors controlling the legume-*Rhizobium* symbiosis. in: Nitrogen Fixation II, Newton, W.E., and Orme-Johnson, W.H.,Eds., University Park Press, Baltimore, 103-109

Witty, J.F., Minchin, F.R., Skot, L., and Sheehy, J.E. (1986). Nitrogen fixation and oxygen in legume root nodules. Oxford surveys of plant and cellular biology 3, 275-315

Zaat, S.A.J., Van Brussel, A.A.N., Tak, T., Pees, E., and Lugtenberg, B.J.J. (1987). Flavonoids induce *Rhizobium leguminosarum* to produce nodDABC gene-related factors that cause thick, short roots and root hair responses on common vetch. J. Bacteriol. 169, 3388-3391

THE COMMON *NODABC* GENES AND THE *NODH* AND *NODQ* HOST-RANGE GENES OF *RHIZOBIUM MELILOTI* DETERMINE THE PRODUCTION OF LOW MOLECULAR WEIGHT EXTRACELLULAR SIGNALS

Catherine Faucher[1*], Patrice Lerouge[2*], Philippe Roche[2*], Charles Rosenberg[1], Frédéric Debellé[1], Jacques Vasse[1], Emilio Cervantès[1], Shashi B. Sharma[1], Georges Truchet[1], Jean-Claude Promé[2] and Jean Dénarié[1]

1 Laboratoire de Biologie Moléculaire des Relations Plantes-Microorganismes, CNRS-INRA, BP27, 31326 Castanet-Tolosan Cedex,

2 Centre de Recherches de Biochimie et de Génétique Cellulaires, CNRS LP8201, 118 route de Narbonne, 31062 Toulouse Cedex, France

The interaction between fast-growing *Rhizobium* species and their leguminous hosts is highly specific. For example *R.meliloti* strains nodulate alfalfa while *R.leguminosarum* bv. *viceae* (= *R.leguminosarum*) form nodules on pea and vetch and *R.leguminosarum* bv. *trifolii* (= *R.trifolii*) on clover.

R.meliloti genes required for alfalfa infection and nodulation can be classified as follows: (i) the *nodABC* common genes which are structurally and functionally conserved in all *Rhizobium* species (see Long 1989), (ii) the regulatory *nodD* genes which modulate the expression of *nod* genes as a function of specific plant signals (Györgypal et al 1988) and (iii) the *nodFEG* and *nodH* host-range genes. In this paper we will describe recent results on the identification, structure and function of the *R.meliloti* host-range genes. In particular we will characterize a new host-range gene *nodQ* and will show that *nodH* and *nodQ* determine modifications of low molecular weight extracellular signals.

1. THE *R.MELILOTI NOD* GENES WHICH DETERMINE THE HOST-RANGE

Originally *R.meliloti nod* genes have been identified by isolation of mutants exhibiting an alteration of the nodulation pattern on the homologous host alfalfa. To distinguish among such mutants those which were altered in host-specific *nod* genes, Kondorosi et al (1984) introduced into them nodulation genes from another *Rhizobium* species and observed whether the normal nodulation of alfalfa could be restored. This approach allowed the common *nod* genes, which can be complemented, to be distinguished from the host-specific (or Hsn) *nod* genes, which cannot.

Molecular studies of the Hsn region revealed four putative host-range genes, *nodH* and *nodFEG*, which are divergently transcribed (Debellé et al 1986, Debellé and Sharma 1986, Horvath et al 1986). However, non-complementation is an indirect method of defining host-range genes. We thus looked for different approaches to identify and characterize host-range genes.

* The three first authors have made equal contributions to this work

NATO ASI Series, Vol. H 36
Signal Molecules in Plants and Plant-Microbe Interactions
Edited by B. J. J. Lugtenberg

1.1. Which genes are required to transfer the *R.meliloti* host-range?

We had already shown that the plasmid pGMI515, which carries *R.meliloti nod* region II, *nodFEG*, *nodH* and *nodD3* genes, enables *R.trifolii* to infect and nodulate alfalfa (Debellé et al 1986; see Fig.1). This raised the question of whether the *nodFEG* and *nodH* genes are sufficient to ensure the transfer of host-range. To answer this question we introduced into *R.leguminosarum* and *R.trifolii* (i) a series of

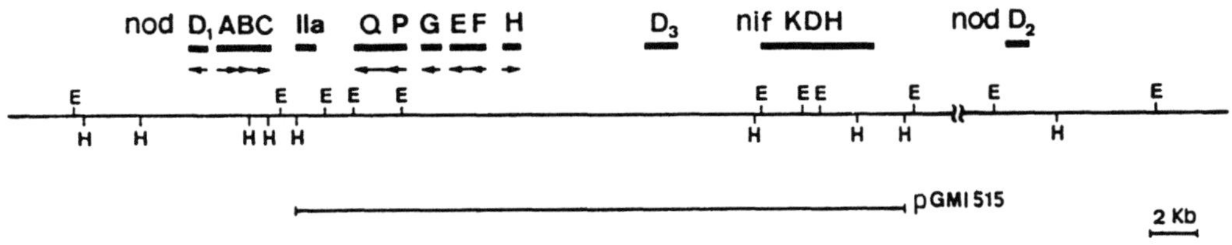

Fig.1 Genetic and physical map of the nodulation region of *R.meliloti* 2011. The horizontal line represents the restriction map (E, *Eco*RI; H, *Hin*dIII). The plasmid pGMI515 is shown below the physical map. The arrows indicate the direction of transcription of nod genes (data reviewed by Long (1989)).

pGMI515 derivatives carrying either deletions or Tn5 insertions in the *nod* region or (ii) subclones of the *nod* region which is present in pGMI515. In addition to *nodH* and *nodFEG*, the region IIb, a 3 kbp fragment located downstream of *nodFEG*, was found to be essential to transfer the ability to infect alfalfa (Cervantès et al 1989). Molecular studies of this region revealed the presence of two genes *nodP* and *nodQ* which are probably part of the same operon and are transcribed in the same direction as *nodFEG* (Cervantès et al 1989, Schwedock and Long 1989). Tn5 insertions in *nodQ* suppressed the ability of the *R.trifolii* (pGMI515) and *R.leguminosarum* (pGMI515) hybrid strains to infect alfalfa (Cervantès et al 1989, Faucher et al 1989). The regulatory *nodD3* gene was not required to transfer the host-range (Faucher et al 1989).

1.2. *R.meliloti* mutations resulting in host-range modifications

The wild-type strain *R.meliloti* 2011 is unable to infect (Hac^- Inf^-) and nodulate (Nod^-) non-host legumes such as common vetch and white clover. Deletions or Tn5-induced mutations in certain *nod* genes have been shown to result in host-range modifications. For example *nodH* mutants exhibit a shift of host-range: they are Hac^- Inf^- Nod^- on alfalfa, but Hac^+ on white clover (Debellé et al 1986) and Hac^+ Inf^+ (Faucher et al 1988) and Nod^+ on common vetch (Horvath et al 1986, Faucher et al 1988). Other types of mutants have an extended host-range, they can still infect and nodulate alfalfa (although with a reduced efficiency) but can also interact with non-homologous hosts. *nodF* and *nodE* mutants are Hac^+ on white clover (Debellé et al 1986) and *nodQ* mutants are Hac^+ Inf^+ on vetch (Cervantès et al 1989). It should be noted that (1) these changes are not due to point mutations which might result in slight modifications of gene products but are instead caused by deletions or Tn5 insertions (Debellé et al 1988, Faucher et al 1988), (2) all the mutations in a given gene result in the same host-range phenotype. These results show that the changes in host-range are due to the loss, and not a mere modification, of the gene products. Thus the loss of a host-range Nod protein results in the gain of the ability to elicit root hair reactions on new hosts or to infect them. The NodH, NodFE and NodQ proteins therefore restrict the ability of *R.meliloti* to interact with non-homologous hosts.

1.3. Symbiotic interference between *nod* genes of different *Rhizobium* species: the host-range genes of *R.meliloti* behave like avirulence genes

When the pGMI515 plasmid is introduced into *R.leguminosarum* and *R.trifolii*, the transconjugants not only gain the ability to nodulate alfalfa, but also exhibit a strong inhibition of their ability to infect and nodulate their normal host, vetch and white clover respectively (Debellé et al 1988, Faucher et al 1989). This *R.meliloti* "dominance" is essentially due to *nodH* and *nodFE* with *R.trifolii* and to *nodH* and *nodQ* with *R.leguminosarum*. The effect of *R.meliloti nodFE* genes in *R.trifolii* could be called dominance, since *nodFE* alleles do exist in both species. The effect of *nodH* and *nodQ* should rather be called epistatic, since no alleles of these genes have been detected among the *nod* genes of *R.leguminosarum* biovars. Thus the *R.meliloti* host-range genes *nodFE*, *nodH* and *nodQ*, which render *R.leguminosarum* and *R.trifolii* avirulent on their normal hosts by dominant or epistatic interactions, behave like avirulence (*avr*) genes of phytopathogenic bacteria (Keen and Staskawicz 1988).
The results of *nod* gene transfer experiments suggested that the *nodFE*, *nodH* and *nodQ* genes might determine host-range by directing the production of specific signals. The dominance effect could be interpreted by hypothesizing that these *R.meliloti* genes encode proteins which modify clover or vetch-specific signals into alfalfa-specific signals. These signals might be surface components or/and diffusible molecules.

1.4. Nucleotide sequence of host range genes

The sequence of *nod* region IIb revealed two ORFs, *nodP* and *nodQ*. The two genes are transcribed in the same direction and the stop codon of *nodP* overlaps with the start codon of *nodQ*; therefore it is likely that they belong to the same operon (Cervantès et al 1989). The *nodFE*, *nodG* and *nodPQ* genes are transcribed in the same direction and are all induced by luteolin (Schwedock and Long 1989; Fabienne Maillet, personal communication). The only detectable *nod* box in this region is located upstream of *nodF* which suggests that the five genes belong to the same operon. Nevertheless the intergenic distances between *nodE* and *nodG* and between *nodG* and *nodP* are long: 500 bp and 400 bp respectively.
Sequences of the *R.meliloti* host-range genes might provide some indications as to the possible role of their protein products. NodH protein is proline-rich and has a four contiguous proline domain which in eucaryotes has been observed in glycoproteins (Debellé and Sharma 1986). NodF shares homology

with acyl-carrier proteins (Debellé and Sharma 1986, Horvath et al 1986). NodG has strong homology along most of the chain to a bacterial ribitol dehydrogenase (Debellé and Sharma 1986). NodQ shares homology with translation initiation and elongation factors, possesses a GTP-binding domain and another nucleotide-binding domain (Cervantès et al 1989). If these analyses of the sequence data are compatible with the hypothesis that the host-range genes code for enzymes involved in the synthesis of specific signals, they do not allow a model to be proposed for the precise role of the gene products and the chemical nature of the signals.

2. EXTRACELLULAR SYMBIOTIC SIGNALS

2.1. Bioassays

A new approach to study the biochemical functions of *nod* genes was made possible by the discovery by Van Brussel et al (1986) and Zaat et al (1987) that common vetch (subsp. *nigra*) seedlings, treated with sterile supernatants of *R.leguminosarum* cultures, exhibit a "Thick and short root" (Tsr) reaction. The Tsr activity of the supernatants required the presence of *nodD* and *nodABC* genes in the *Rhizobium* strain, and an appropriate flavonoid inducer in the growth medium. The Tsr phenotype can be easily quantified by measuring root length. This assay is thus convenient for comparing the influence of various mutations on the production of extracellular symbiotic factors active on common vetch (Zaat et al 1987; Faucher et al 1988 and 1989).
However, a root hair deformation (Had) assay is of more general interest than the Tsr assay for the isolation of *nod* factors, since it is more sensitive and can be used with various legumes such as vetch (Zaat et al 1987), white clover (Bhuvaneswari and Solheim 1985 ; see the paper of Dazzo et al in this book) and alfalfa (Faucher et al 1988). We recently described a modified Had assay which is simpler to perform, requires smaller volumes of fluids and allows a quicker plant reaction (Faucher et al 1989). For both Tsr and Had assays low density *Rhizobium* cultures (around $5x10^5$ bacteria per ml) are grown with a flavonoid inducer (10 µM). After centrifugation the supernatants which had been sterilized by filtration were tested for activity on alfalfa or vetch. The activity was not altered by autoclaving, showing that the Tsr or Had factors are heat-stable (Zaat et al 1987; Faucher et al 1988).

2.2. *nodH* and *nodQ* genes determine modifications of symbiotic factors

The supernatant of a *R.meliloti* culture caused hair deformations on alfalfa ($HadA^+$) but not on vetch ($HadV^-$), whereas the supernatant of a *R.leguminosarum* culture was $HadA^-$ and $HadV^+$ (Faucher et al 1988). This suggested that host-range genes might be involved in the production of extracellular signals. We thus studied the activity of the filtrates from various *R.meliloti nod* mutants on alfalfa and vetch. *nodA* and *nodC* mutants are Had^- on both hosts. The *nodABC* operon is thus required for the production of active Had factors. A *R.meliloti* strain carrying only *nodD1* and the common *nodABC* genes is $HadV^+$ $HadA^-$, showing that these genes are sufficient to direct the production of a symbiotic

factor(s) which can be detected with vetch. This HadV factor(s) might be similar or identical to the compound(s) which are produced by *R.meliloti* and which stimulate mitosis of plant cells (Schmidt et al 1989). The filtrates of *R.meliloti nodH* mutants exhibit a shift of activity ($HadA^-$ $HadV^+$) and the filtrates of *nodQ* mutants an extended activity ($HadA^+$ $HadV^+$). Thus the *nodH* and *nodQ* host-range genes are involved in the modification of the specificity of extracellular heat-stable signals. Mutants in *nodFEG* are $HadA^+$ $HadV^-$ like the wild-type strain. It is worth noting that the filtrates of the various mutants which exhibit various spectra of activities were all produced in the presence of the same inducer, luteolin.
The supernatant of an induced *R.leguminosarum* wild-type culture is $HadA^-$ $TsrV^+$. The introduction of the pGMI515 plasmid carrying the *R.meliloti* host-range genes into *R.leguminosarum* results in a hybrid strain which produces a filtrate that is $HadA^+$ and has decreased TsrV activity. A Tn5 insertion in either *nodH* or *nodQ* restores the $HadA^-$ Tsr^+ phenotype (Faucher et al 1989). Thus the introduction of *nodH* and *nodQ* genes into *R.leguminosarum* results in the modification of the specificity of the extracellular Had factors (Faucher et al 1989).

2.3. HPLC identification of extracellular factors

The filtrate of the *nod*-induced culture of a *R.meliloti* strain carrying the wild-type nodulation region was extracted by ethyl acetate and butanol and the three fractions obtained were tested for Had activity on alfalfa seedlings. Most of the activity was detected in the butanol fraction.
The butanol fraction was fractionated by high performance liquid chromatography (HPLC) on a C18 reversed-phase column with a water/acetonitrile gradient. Two major compounds (peaks I and II) were detected, by UV-absorption, which could not be observed in the filtrate of a control strain deleted of the *nod* region (Fig. 2A). The introduction of a Tn5 insertion into either the *nodA* or the *nodC* gene suppressed simultaneously the two peaks, showing that the common *nod* genes are required for the production of compounds I and II (Fig.2A-B). Tn5 insertions in either *nodH* or *nodQ* resulted in a strong decrease of the area of both peaks I and II, and in the appearance of another set of more hydrophobic products (III; Fig.2C), providing a biochemical evidence that both of these host-range genes are involved in the modification of the extracellular factors. NMR and mass spectrometry data indicate that factors I and II are low molecular weight products. The complete structural analysis and the detailed biological activities of these factors are currently under study.

DISCUSSION

It had been previously shown that the nodulation genes of *R. meliloti* are induced when the bacterial cells are exposed to plant exudates (Mulligan and Long 1985). The inducing molecules have been identified as flavonoids (Peters and Long 1986). This activation requires the presence of the regulatory *nodD* genes and it has been shown that different *nodD* genes confer distinctive patterns of *nod* gene

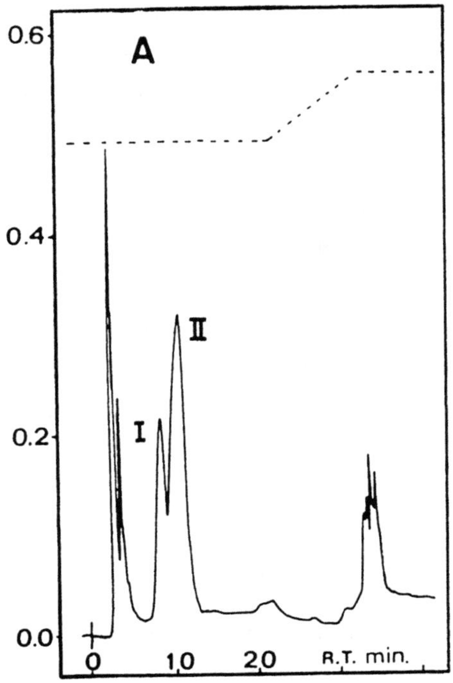

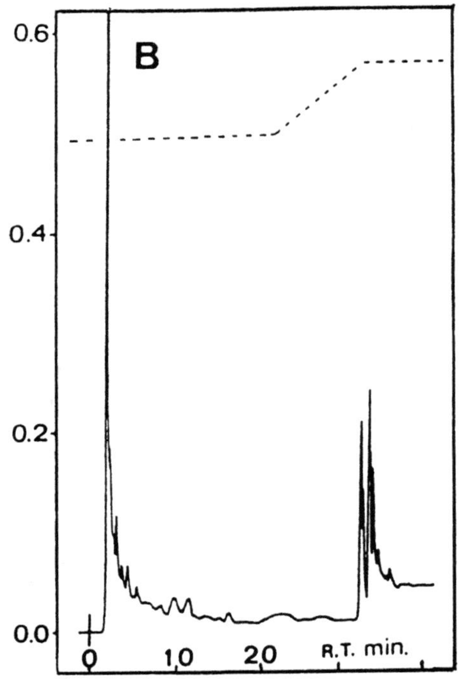

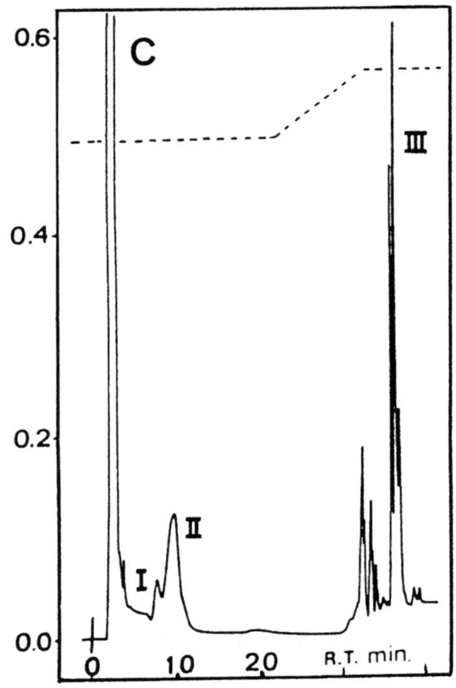

Fig.2 Chromatogram of butanolic extracts of *R.meliloti* derivatives with a reversed-phase C18 column. Solvent a: water-acetonitrile (80/20); solvent b: water-acetonitrile (50/50). Flow rate 1ml/min. (A) strain with the wild-type *nod* region, (B) isogenic strain with a *nodA*::Tn5 insertion, (C) strain with a *nodH*::Tn5 insertion.

response to a variety of flavonoids or plant exudates (Spaink et al 1987 ; Györgypal et al 1988). Thus plant signals activate in a somewhat specific manner the expression of *nod* genes.

In this paper we have shown that reciprocally the *nod* genes, once activated, determine the production of bacterial signals which elicit specific plant reactions, probably by regulating plant gene expression (see the paper of Bisseling et al in this book). Thus the early steps of the symbiosis seem to involve the exchange of low molecular weight signals between both partners.

To analyse the mechanisms by which the *R.meliloti nod* genes determine the specific infection of alfalfa various *R.meliloti* mutants, as well as *R.leguminosarum* derivatives carrying combinations of *R.meliloti* host-range genes, have been studied by three independent approaches: a cytological description of the symbiotic behaviour of bacteria, a study of the physiological activity of sterile filtrates in plant assays and a biochemical characterisation (HPLC) of extracellular products. The three approaches have revealed striking differences in the role of common and specific *nod* genes. Inactivation of the common *nodABC* genes results in a complete suppression of the ability of bacteria or their filtrates to elicit any plant reaction and in a disappearance of *nod*-related UV-absorbing products in the butanol extracts.

Inactivation of *nodH* and *nodQ* genes results in modifications of the specificity of bacterial infection and of the bacterial filtrate activity, and of the UV-absorbing compound pattern. *nodABC* genes are epistatic on *nodH* and *nodQ* for cytological, physiological and biochemical phenotypes.

In an attempt to interpret these data we propose the following model:

1. The common *nodABC* genes code for proteins which modify a precursor X, a cellular component synthesized independently from the *nod* genes, into a common Had factor(s) (detectable with the vetch Had assay);

2. Host specific genes, namely *nodH* and *nodQ* in *R.meliloti*, code for proteins which allow the modification of common Had factor(s) into specific Had signals; inactivation of individual host-range genes results in the accumulation of intermediate Had factors which have a modified or reduced host-specificity (resulting in a shift or extension of the host-range);

3. *R.meliloti nodABC*, *nodH* and *nodQ* genes determine alfalfa infection via these extracellular factors.

We hope that the determination of the structure of the Had factors produced by the wild-type strain and by the various *nod* mutants will allow deciphering the role of the various *nod* genes in the synthesis and export of these symbiotic signals.

ACKNOWLEDGEMENTS

We thank Fabienne Maillet and Sylvie Camut for their skillful technical assistance. We are grateful to Clare Gough and Les Barran for reviewing the manuscript.

REFERENCES

Bhuvaneswari TV and Solheim B (1985) Root hair deformation in the white clover /*Rhizobium trifolii* symbiosis. Physiol Plant 63:25-34

Cervantès E, Sharma S B, Maillet F, Vasse J, Truchet G and Rosenberg C (1989) The product of the host specific *nodQ* gene of *Rhizobium meliloti* shares homology with translation and initiation factors. Molec Microbiol (in press)

Debellé F, Rosenberg C, Vasse J, Maillet F, Martinez E, Dénarié J and Truchet G (1986) Assignment of symbiotic developmental phenotypes to common and specific nodulation (*nod*) genetic loci of *Rhizobium meliloti*. J Bacteriol 168:1075-1086

Debellé F and Sharma S B (1986) Nucleotide sequence of *Rhizobium meliloti* RCR2011 genes involved in host specificity of nodulation. Nucleic Acids Res 14:7453-7472

Debellé F, Maillet F, Vasse J, Rosenberg C, de Billy F, Truchet G, Dénarié J and Ausubel F M (1988) Interference beetween *Rhizobium meliloti* and *R.trifolii* nodulation genes: genetic basis of *R.meliloti* dominance. J Bacteriol 170:5717-5727

Faucher C, Maillet F, Vasse J, Rosenberg C, van Brussel A A N, Truchet G and Dénarié J (1988) *Rhizobium meliloti* host-range *nodH* gene determines production of an alfalfa-specific extracellular signal. J Bacteriol 170:5489-5499

Faucher C, Camut S, Dénarié J and Truchet G. (1989) The introduction of the *Rhizobium meliloti* host-range *nodH* and *nodQ* genes into *R.leguminosarum* results in changes in the production of plant-specific extracellular signals. Mol Plant-Microbe Interact, in press.

Györgypal Z, Iyer N and Kondorosi A (1988) Three regulatory *nodD* alleles of diverged flavonoid-specificity are involved in host-dependent nodulation of *Rhizobium meliloti.* Mol Gen Genet 212:85-92

Horvath B, Kondorosi E, John M, Schmidt J, Torok I, Györgypal Z, Barabas I, Wieneke U, Schell J and Kondorosi A (1986) Organisation, structure and symbiotic function of *Rhizobium meliloti* nodulation genes determining host specificity for alfalfa. Cell 46:335-343

Keen NT and Staskawicz BJ (1988) Host range determinants in plant pathogens and symbionts. Ann Rev Microbiol 4 2:421-440

Kondorosi E, Banfalvi Z and Kondorosi A (1984) Physical and genetic analysis of a symbiotic region of *Rhizobium meliloti*: identification of nodulation genes. Mol Gen Genet 193:445-452

Long SR (1989) Rhizobium-legume nodulation: life together in the underground. Cell 56:203-214

Mulligan JT and Long SR (1985) Induction of *Rhizobium meliloti nodC* expression by plant exudate requires *nodD*. Proc Natl Acad Sci USA 82:6609-6613

Peters NK and Long SR (1988) *Rhizobium meliloti* nodulation gene inducers and inhibitors. Plant Physiol 88:396-400

Schmidt J, Wingender R, John M, Wieneke U and Schell J (1988) *Rhizobium meliloti nodA* and *nodB* genes are involved in generating compounds that stimulate mitosis of plant cells. Proc Natl Acad Sci USA 85:8578-8582

Schwedock J and Long SR (1989) Nucleotide sequence and protein products of two new nodulation genes of *Rhizobium meliloti, nodP* and *nodQ*. Mol Plant Microbe Interact, in press

Spaink HP, Wijffelman CA, Pees E, Okker RJH and Lugtenberg BJJ (1987) *Rhizobium* nodulation gene *nodD* as a determinant of host specificity. Nature 328:337-340

van Brussel AAN, Zaat SAJ, Canter Cremers ACJ, Wijfellman CA, Pees E, Tak T and Lugtenberg BJJ (1986) Role of plant root exudate and Sym plasmid-localized nodulation genes in the synthesis by *Rhizobium leguminosarum* of Tsr factor, which causes thick and short roots on common vetch. J Bacteriol 165:517-522

Zaat SAJ, van Brussel AAN, Tak T, Pees EE and Lugtenberg BJJ (1987) Flavonoids induce *Rhizobium leguminosarum* to produce *nodDABC*-gene-related factors that cause thick, short roots and root hair responses on common vetch. J Bacteriol 169:3388-3391

ROOT HAIR DEFORMING AND NODULE INITIATING FACTORS FROM RHIZOBIUM TRIFOLII

R. Hollingsworth*, A. Squartini+, S. Philip-Hollingsworth+, and F. Dazzo+
Department of Biochemistry* and Microbiology+,
Michigan State University,
East Lansing, Michigan 48824
United States of America

Axenic legume root hairs typically remain straight. Early in the infection process, Rhizobium secretes compounds that induce various morphological changes in root hairs, including enhanced differentiation (leading to greater hair density per root surface area), elongation, various deformations including helices, branches, bulbs, and curlings including the 360° tight curl (so called "shepherd crooks") at hair tips (Yao & Vincent 1969). It has been proposed that these deformations may represent altered synthesis and/or stabilization of the root hair cell wall during growth (for reviews, see Bauer 1981; Dazzo & Hubbell 1982; Halverson & Stacey 1986; Rolfe & Gresshoff 1988).

Some types of root hair deformation can be induced by incubating homologous host roots with cell-free culture fluid or washed suspensions of rhizobia cultured without flavone inducers (Thornton & Nicol 1936; Yao & Vincent 1969; Hubbell 1970; Solheim & Raa 1973; Yao & Vincent 1976; Ervin & Hubbell 1985). Culture supernatants of rhizobia induced by appropriate flavone or root exudate to express pSym nod genes generally exhibit more exaggerated root hair deforming-ability. These supernatants also contain low molecular weight factors which can elicit other morphogenic responses in host plants, including a thickening and shortening of host roots with enhanced root hair differentiation (Canter-Cremers et al 1986; Van Brussel et al 1986), enhanced root hair branching (Bhuvaneswari & Solheim 1985), reduced root elongation (Faucher et al 1988), and mitosis in plant cells grown in vitro (Schmidt et al 1988).

Shepherd crooks on legume root hair tips are only induced by viable rhizobia specifically capable of infecting that legume host (Yao & Vincent 1969; Yao & Vincent 1976). Serial sectioning of shepherd crooks in white clover seedlings reveals that the rhizobia are enclosed in the pocket created by the overlapping wall of the root hair (Napoli & Hubbell 1975; Callaham & Torrey 1981). It has been proposed that the folds in curled root

NATO ASI Series, Vol. H 36
Signal Molecules in Plants and Plant-Microbe Interactions
Edited by B. J. J. Lugtenberg

hairs may, by entrapping the bacteria and their secretion products, facilitate wall erosion, bacterial penetration, and successful formation of infection threads (Napoli & Hubbell 1975; Callaham & Torrey 1981; Dazzo & Hubbell 1982). Computer simulation modeling of root hair growth (van Batenburg et al 1986) predicts that shepherd crooks form when hair growth is inhibited by attached Rhizobium on one side and is stimulated on the opposite side.

Several lines of evidence suggest that nodule initiation can result from diffusible substances excreted by Rhizobium during their interaction with host roots (Libbenga & Harkes 1973; Calvert et al 1984; Dudley et al 1987; Leigh et al 1987; Truchet et al 1985; Bauer et al 1986; Puhler et al 1988).

The identity of metabolites excreted by Rhizobium which are responsible for inducing the various morphological changes in legume roots during symbiotic development is unknown (for a review, see Halverson & Stacey 1986). Analyses of fractions active in root hair deformation have implicated bacterial oligosaccharides, polysaccharides, nucleic acids, and protein (Fahraeus & Ljunggren 1967; Hubbell 1970; Solheim & Raa 1973; Dazzo & Hubbell 1975; Yao & Vincent 1976; Bhuvaneswari 1984; Ervin & Hubbell 1985; Bhuvaneswari & Solheim 1985). In these studies, the biologically active compounds were not purified to a state sufficient for detailed structural analysis.

We have set out to isolate and identify the extracellular bacterial factors from R. trifolii which elicit morphological changes in root hairs and induce root nodules in white clover. Whenever possible, we compared these results to root responses on heterologous hosts. The seeding bioassays developed for this work require only small volumns of filter-sterilized test sample, are incubated vertically in a growth chamber (to allow geotropic development of roots under microbiologically controlled conditions), and are optically clear (thus permitting direct microscopic observations of the morphological changes without disturbing the root system or introducing microbial contamination) (Fig. 1). Root hair density and length are measured by video microscopy and image analysis. To measure the frequency of foci containing cortical cell divisions and cortical meristems resembling primordial nodules, roots are treated with test samples, incubated vertically, then cleared with dilute Chlorox under vacuum, stained with methylene blue, and examined by brightfield microscopy to distinguish the main histological features of root nodules from lateral roots (Truchet et al 1988). To assay for infection-related biological activity, roots of intact seedlings are preincubated vertically on plates with sterile test samples, then inoculated with R. trifolii. After 6 days incubation, the shepherd crooks and infected root hairs are counted. Sterile nitrogen-free Fahraeus medium was evaluated as a control in all bioassays.

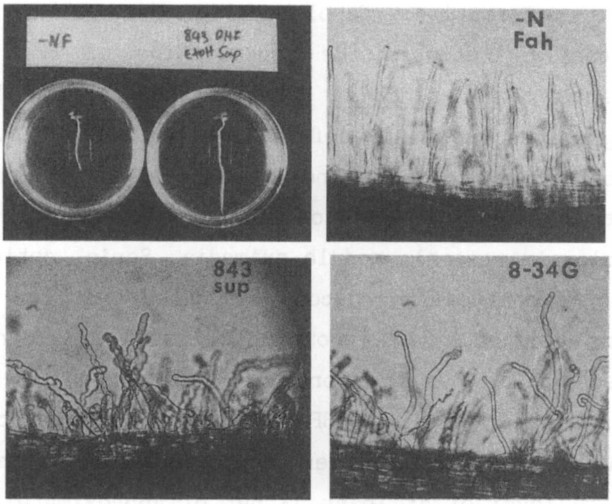

Fig. 1. Assay for root hair deformation using sterile metabolites from R. trifolii ANU843 applied to axenic seedlings of white clover. Samples used are indicated.

Using a variety of chromatographic procedures, we have fractionated the culture supernatant of wild-type R. trifolii ANU843 grown in defined BIII culture medium with and without 2 µM apigenin supplement (a flavone inducing pSym nod genes in R. trifolii, Redmond et al 1986), and have isolated several low molecular weight metabolites which elicit various morphological responses in white clover roots (Dazzo et al 1986; ibid 1988). Final purity was established by TLC and/or HPLC using different solvent systems. Based on results of our bioassays, we selected four metabolites covering a range of biological activities for further study. We will refer to these isolated metabolites as Bacterial Factors 1, 2, 3, and

4 (BF-1, BF-2, BF-3, and BF-4 for short). R. trifolii can produce BF-1 and BF-2 independent of added flavone, consistent with the pre-flavone era of Rhizobium research which documented the ability of R. trifolii culture supernatants to induce morphological changes in clover root hairs. BF-1 increases the length and density of root hairs in white clover seedlings, but not in birdsfoot trefoil (Dazzo et al 1986). Root hairs grown with BF-1 remain straight, but are more prone to develop shepherd crooks and infection threads when followed by inoculation with R. trifolii. BF-1 also elicits an altered cortical histology in white clover roots. BF-2 can induce root hair deformation. Supplementing the culture medium with apigenin significantly increases the accumulation of BF-3 and BF-4 in culture supernatants of R. trifolii ANU843. These metabolites induce a diversity of exaggerated root hair deformations in white clover, weak hair deformation in birdsfoot trefoil, and no hair deformations in alfalfa. White clover roots treated with BF-3 or BF-4 develop foci of cortical cell divisions which closely resemble nodule primodia. Serial optical sectioning using laser scanning confocal microscopy indicates that the foci of young cortical meristems induced by BF-3 consist of small, dividing cells which form a multilobed mound (resemble broccoli heads) and lack an internal vascular system. Neighter BF-3 nor BF-4 increase root branching in white clover. Incubation of white clover root meristems with BF-3 causes dramatic changes in level of at least 13 out of approximately 500 proteins examined by computer densitometry analysis of 2-dimensional O'Farrell gels. Of those 13 proteins, 6 new ones are detected and 7 others disappear upon exposure of meristems to BF-3.

Interestingly, BF-4 may enhance the ability of BF-3 to induce development of these foci of cortical meristems. When roots are treated with a mixture of both metabolites, the cortical meristems develop more fully, splitting the root epidermis and resembling very young nodules. The enhanced frequency of these structures resembling premature nodule meristems in white clover induced by BF-3 and BF-4 together was synergistic rather than additive.

The culture supernatant of apigenin-induced ANU843 also contains at least two other metabolites which can induce morphological responses in white clover roots. Although only partially purified, these have been separated from BF-1 through BF-4. One of these additional chromatographic peaks (peak B) can induce root hair branching. The other fraction (peak C) can cause root thickening and enhanced development of root hairs on white clover. The relative yield of peak B does not change with added apigenin, whereas peak C increases significantly with added apigenin.

In summary, our studies indicate that R. trifolii excretes several low molecular weight metabolites which can elicit some of the morphological responses in white clover roots that occur during infection and nodulation by the bacterial symbiont. These include various types of root hair deformation and induction of cortical meristems resembling nodule primordia. However, in no case yet have the 4 isolated metabolites tested been able to elicit true shepherd crooks or fully emerged and developed root nodules under the axenic conditions of our assay. R. trifolii also produces other metabolites capable of deforming white clover root hairs. An early indication of host selectivity is displayed by some of the isolated metabolites tested, but this needs more thorough evaluations. The active metabolites whose production is increased by culturing with inducer flavone are good candidates to examine whether certain pSym nod genes are required for their synthesis and/or export. Structural studies are in progress on the isolated bacterial metabolites. We are also interested in determining how these metabolites affect root physiology, and impact on host specificity, root hair infection, and development of root nodules in this nitrogen-fixing symbiosis.

ACKNOWLEDGMENTS

This work was supported by NIH Grant GM 34331-05 and the MSU Nitrogen Availability for Plants Center funded by the REED Biotechnology Program. We thank Georges Truchet and Barry Rolfe for their helpful comments.

REFERENCES

Bauer W, Bhuvaneswari T V, Calvert M E, Law I J, Malik N S A and Vesper S J (1985) Recognition and infection by slow-growing rhizobia. In: Evans H. Bottomley P, and Newton W (eds) Nitrogen fixation research progress. Martinus Nijhoff, Dordrecht, p 247

Bhuvaneswari T (1984) Preinvasion events in legume/Rhizobium symbiosis. In: Veegers C and Newton W (eds) Advances in nitrogen fixation research. Martinus Nijhoff, The Hague, p 389

Bhuvaneswari T and Solheim B (2985) Root hair deformation in the white clover/Rhizobium trifolii symbiosis. Physiol Plant **63**:25-34

Callaham D and Torrey J (1981) The structural basis for infection of root hairs of Trifolium repens by Rhizobium. Can J Bot **59**:1647-1664.

Calvert H, Pence M K, Pierce M, Malik N S A and Bauer W D (1984) Anatomical analysis of the development and distribution of Rhizobium infections in soybean roots. Can J Bot **62**:2375-2384

Canter-Cremers H van Brussel A A N, Plazinski J and Rolfe B G (1986) Sym plasmid and chromosomal gene products of Rhizobium trifolii elicit developmental responses on various legume roots. J Plant Physiol **122**:25-40

Dazzo F, Hollinsgworth R I, Philip S, Smith K B, Welsch M A, Salzwedel J, Morris P and McLaughlin L (1986) Involvement of pSym nodulation genes in production of surface and extracellular components of Rhizobium trifolii which interact with white clover roots. In Verma D and Brisson V (eds) Molecular genetics of plant-microbe interactions. Kluwer Academic Press, Dordrecht, p 171

Dazzo F, Hollingsworth R T, Salzwedel J, Philip-Hollingsworth S, Robeles M, Olen T A, Apenzeller L, Wang S, Toro I, Squartini A, Anderson S F, Chen J, Chapman K A, Maya-Flores J, Cargill L C, Baker D, Djordjevic M A and Rolfe B G (1988) Signal recognition responses in the Rhizobium trifolii-white clover symbiosis, In: Palacios R. Verma d (eds) Molecular genetics of plant-microbe interactions, American Phytopathological Society, St. Paul MN, p 35

Dazzo F and Hubbell D (1975) Cross-reactive antigens and lectin as determinants of symbiotic specificity in the Rhizobium trifolii-white clover assocation. Appl Microbiol **30**:1017-1033

Dazzo F and Hubbell D (1982) Control of root hair infection. In: Broughton W (ed) Nitrogen fixation, vol II Rhizobium Oxford University Press, Oxford, p 274

Dudley M, Jacobs W, and Long S R (1987) Microscopic studies of cell divisions induced in alfalfa roots by Rhizobium meliloti Planta **171**:289301

Ervin S and Hubbell D (1985) Root hair deformations associated with fractionated extracts from Rhizobium trifolii. Appl Environ Microbiol **49**:61-68

Fahraeus G and Ljunggren H (1967) Preinfection phases of the legume symbiosis. In: Grey T, Parkinson D (eds) Ecology of soil bacteria, University of Toronto Press, Toronto, p 396

Faucher C, Maillet F, Vasse J, Rosenberg C, Van Brussel A A N, Truchet J and Denarié J (1988) Rhizobium meliloti host range nodH gene determines production of an alfalfa-specific extracellular signal. J Bacteriol **170**:5489-5499

Halverson L and Stacey G (1986) Signal exchange in plant-microbe interactions. Microbiol Rev **50**:193-225

Hubbell D (1970) Studies on the root hair "curling factor" of Rhizobium. Bot Gaz **131**:137-342

Leigh J, Reed J W, Manks J F, Mirsch A M and Walker G C (1987) Rhizobium meliloti mutants that fail to succinylate their calcofluor-binding exopolysaccharide are defective in nodule invasion. Cell **51**:579-587

Libbenga K and Harkes P (1973) Initial proliferation of cortical cells in the formation of root nodules in Pisum sativum L. Planta **114**:17-28

Napoli C and Hubbell D (1975) Ultrastructure of Rhizobium-induced infection threads in clover root hairs. Appl Microbiol **30**:1003-1009

Napoli C, Dazzo F B and Hubbell D (1975) Production of cellulose microfibrils by Rhizobium. Appl Microbiol **30**:123-127

Puhler A, Enenkel B, Hilleman A, Kapp D, Keller M, Muller P, Niehaus K, Priefer U B, Quandt J and Schmidt C (1988) Rhizobium meliloti and Rhizobium leguminosarum mutants defective in surface polysaccharide synthesis and root nodule development. In: Bothe H, DeBruijn F, Newton W (eds) Nitrogen fixation: hundred years after, Gustav Fischer, Stuttgart, p 423

Redmond J, Batley M, Djordjevic M A, Innes R W, Kuempel P L and Rolfe B G (1986) Flavones induce expression of nodulation genes in Rhizobium. Nature 323:632-634

Rolfe B and Gresshoff P (1988) Genetic analysis of legume nodulation. Annu Rev Plant Physiol Plant Mol Biol **39**:297-319

Schmidt J, Wingender R, John M, Wieneke U and Schell J (1988) Rhizobium meliloti nodA and nodB genes are involved in generating compounds that stimulate mitosis of plant cells. Proc Natl Acad Sci (USA) **85**:8578-8582

Solheim B and Raa J (1973) Characterization of the substances causing deformation of root hairs of Trifolium repens when inoculated with Rhizobium trifolii. J Gen Microbiol **77**:241-247

Thornton M and Nicol H (1936) Stimulation of root hair growth in legumes by sterile secretions of nodule bacteria. Nature **137**:494-495

Truchet G, Debelle' F, Vasse J, Terzaghi B, Garnerone A M, Rosenberg, Batut J, Maillet K and Denarie J (1985) Identification of a Rhizobium meliloti pSym2011 region controlling host specificity of root hair curling and infection. J Bacteriol **164**:1200-1210

van Batenburg F, Jonker R, Kijne J W (1986) Rhizobium induces marked root hair curling by redirection of tip growth. Physiol Plant **66**:476-480

van Brussel A A N, Zaat S A J, Canter-Cremers H C J, Wijffelman C A, Pees E, Tak T and Lugtenberg B J J (1986) Role of plant root exudate and sym plasmid-localized nodulation genes in the synthesis by Rhizobium leguminosarum of Tsr factor, which causes thick and short roots on common vetch. J Bacteriol **165**:517-522

Yao P and Vincent J (1969) Host-specificity in the root hair 'curling-factor' of Rhizobium spp. Austr J Biol Sci **22**:413-423

Yao P and Vincent J (1976) Factors responsible for the curling and branching of clover root hairs by Rhizobium. Plant Soil **45**:1-16

DETERMINANTS OF HOST SPECIFICITY IN THE *BRADYRHIZOBIUM JAPONICUM*-SOYBEAN SYMBIOSIS

G. Stacey, M. G. Schell, and N. Deshmane
The University of Tennessee, Department of Microbiology, Knoxville, TN 37996-0845, U.S.A.

INTRODUCTION

One of the most fascinating, and still largely unexplained, aspects of *Rhizobium/Bradyrhizobium*-legume interaction is the host specificity exhibited. Investigations of *Rhizobium meliloti* and *R. leguminosarum* bv. *viciae* and *trifolii* have shown that host specificity is determined by the respective *nodD* gene as well as unique host specificity genes (reviewed in Long 1989). Against this background of work with various *Rhizobium* species, comparatively little is known about the determinants of host specificity in *Bradyrhizobium* species. Clearly, the specificity of the *nodD* gene product for specific isoflavone inducers from the plant plays an important role in legume infection by *Bradyrhizobium* (Kosslak *et al* 1988; Banfalvi *et al* 1988). In addition, a few loci have been identified by *mutagenesis* in *Bradyrhizobium* that appear to affect nodulation on only particular host species (Bender *et al* 1987; Nieuwkoop *et al* 1987; Hahn and Hennecke 1988).

In this report, two *hsn* regions identified in *B. japonicum* will be discussed. Figure 1 shows the location of these two regions in relation to the common *nod* genes.

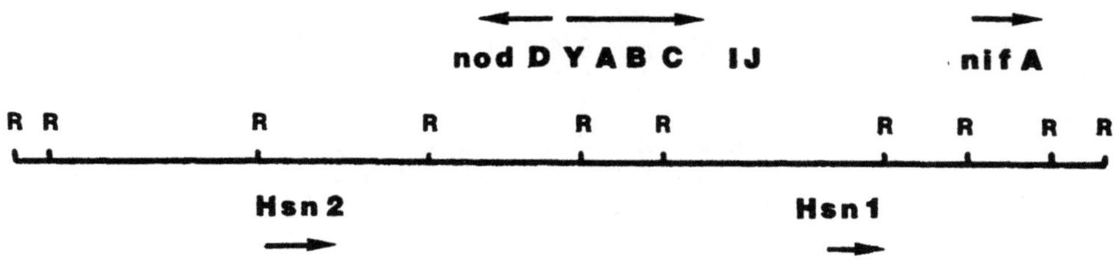

FIGURE 1. *Eco*RI Restriction map of the nodulation genes of *B. japonicum* showing the location of the Hsn-1 and 2 loci.

NATO ASI Series, Vol. H 36
Signal Molecules in Plants and Plant-Microbe Interactions
Edited by B. J. J. Lugtenberg

SIRATRO SPECIFIC NODULATION REGION (HSN-1)

This locus was previously defined in *B. japonicum* by its homology to a *hsn* locus isolated from *Rhizobium* sp strain MPIK3030 (Nieuwkoop *et al* 1988). Mutations in this region prevent *B. japonicum* from nodulating siratro, also a host of strain MPIK3030, but do not affect the ability to nodulate most soybean varieties. In order to further characterize this region, we determined the DNA sequence identifying an open-reading-frame encoding a 35 kilodalton protein. A protein of this size was later expressed from the cloned Hsn-1 locus in *E. coli*. Computer analysis of the ORF found did not reveal any significant homologies to known proteins. However, four possible metal coordination sites were found having the form of Cys-X_{2-4}-Cys or His-X_{2-4}-His (Harnisch *et al* 1985; Steppuhn *et al* 1987) suggesting the possibility that the protein may bind metals and participate in a redox reaction.

Since all known *nod* and *hsn* genes require *nodD* for expression and have the conserved 5′ *nod* box sequence, we also expected that the Hsn-1 locus would display these features. However, sequence analysis revealed the presence of a typical NtrA (RpoN)-dependent *nif/fix*-promoter (Beynon *et al* 1983) with the conserved sequence GG-N_9-TGC (Table 1).

TABLE 1. DNA sequence of the Hsn-1 promoter region compared to that of *B. japonicum nif/fix* gene promoters

*Bj*Hsn-1	G G G C A G G G T C G A T G C A
Bj nif DEN	T G G C A T G C C G G T T G C A
Bj nif H	T G G C A C G G C T G T T G C T
Bj nifBfrxA	C G G C A T G C A A G T T G C T
Bj glnA	T G G C A C G C T A A A T G C T
Bj fixA	T G G T A C A A G A C T T G C T
Bj fix BCX	T G G T A C G A C A C T T G C T
Bj nifS	T G G C A C G T G A C T T G C T
Bj fixRnifA	T G G C G C A A A T C C T G C T
Consensus	. G G . . . . T G C .

In addition, approximately 100 base-pairs 5′ of the transcriptional start is a *nif*-specific upstream activating sequence (TGT-N_{10}-ACA) (Buck *et al* 1986; Table 2). Northern hybridization to mRNA isolated from aerobic cells and

TABLE 2. The postulated upstream activating sequence of the Hsn-1 locus as compared to those of *B. japonicum nif/fix* genes.

Bj Hsn-1	A T A A T G T A C A A A G C G G A A C A T A G C
Bj nifH1	A G T T T G T T C G T T G T C T G A C A G C C G
Bj nifH2	A T C T T G T C A G A T C C A A A A G A G C C T
Bj nifD1	G C C T T G T T C C A T T T C G A A C A C C G C
Bj nifD2	C G A G T G T C G G A T T C G C A A C A A C A G
Bj nifB	C A T C T G T C G C T T T G C C A A C A A G C T
Bj nifS	G C C T T G T C A G G A A C C C G A C A T C C G
Bj fixB	C G G C T G T C G T A T T C C C G A C C C A T G
Consensus	. . . . T G T . . . . A C A . . . .

bacteroids of the wild type and a NifA-mutant showed that the Hsn-1 locus is only expressed in the bacteroid state and this expression is dependent on a functional *nifA* gene. Therefore, the Hsn-1 gene appears to be controlled similarly to *nif* or *fix* genes, but yet its observable phenotype is defective nodulation.

Most recently, we have found that mutants in Hsn-1 lack the ability to grow under microaerobic conditions in liquid sparged with 1% oxygen or in agar deeps. All of the above traits lead us to suggest that the Hsn-1 locus encodes a protein essential for the growth of the bacteria under the reduced oxygen conditions present within the plant host. Previously, NifA- mutants of *B. japonicum* were shown to produce nodules with abnormal ultrastructure suggesting that NifA plays a role in nodule development or maintenance (Fisher *et al* 1986). The discovery of the Hsn-1 locus suggests that NifA regulates genes which are essential for the successful growth and development of the bacteroid under the growth conditions present within the plant host.

SOYBEAN-SPECIFIC NODULATION LOCUS (HSN-2)

As shown in Figure 1, linked to the *nodD* gene we identified an additional locus (Hsn-2) of particular interest since it is required for nodulation of most soybean cultivars but not for other hosts of *B. japonicum* (Deshmane and Stacey 1989). The host range of a mutant in the Hsn-2 locus is similar to that of *Rhizobium fredii*; that is, *Glycine max* cv Peking, *Glycine soja*, and siratro are nodulated, but not other cultivars of soybean. No hybridization of a DNA fragment encompassing the Hsn-2 locus was found to chromosomal DNA from several *R. fredii* strains. Similar experiments detected hybridization to chromosomal DNA from all *B. japonicum* strains, cowpea rhizobia, and *B. parsponiae*, all of which can nodulate soybean. These results lead us to suggest that the nodulation mechanism of certain hosts of *B. japonicum* (e.g. siratro and *G. soja*) may be appreciably different from that used on most *G. max* cultivars and cowpea. This suggestion is further indicated by the identification of gene loci necessary for nodulation of siratro, but not soybean (e.g. Hsn-1).

Since mutations in the Hsn-2 region result in a complete inability to form nodules on most soybean cultivars, we sought to define microscopically at which point infection is blocked. Two of the earliest observable steps in nodulation are the induction by the symbiont of root hair curling and subepidermal cortical cell division. Microscopic observation of soybean roots infected with a mutant in the Hsn-2 locus revealed that root hair curling did occur but cortical cell division could not be observed.

Unlike the Hsn-1 locus, regulation of the Hsn-2 locus requires a functional NodD and the presence of specific isoflavone inducers (i.e., genistein or daidzein, Banfalvi *et al* 1988). This was shown by the construction of *LacZ* fusions in the Hsn-2 region using the Tn3HoHol transposon (Stachel *et al* 1985). The addition of soybean seed exudate, 2 μm genistein or daidzein resulted in a 3-, 2.5-, and 2- fold induction, respectively. The mode of regulation of the Hsn-2 locus and its apparent involvement in one of the earliest steps of nodule formation lead us to suggest that it encodes a protein comparable to those encoded by *Rhizobium hsn* loci.

CONCLUSION

Previous research on the genetics of nodulation in *Rhizobium* species indicated that two general classes of *hsn* genes could be identified; that is, those that recognize specific factors released by the plant (e.g., *nodD*) and those that produce and/or modify factors which affect the plant host (e.g. *nodH*). It is likely that the Hsn-2 locus described above falls into the latter of these two classes. However, we must now add a third class of *hsn* gene; those that are essential for the growth and/or maintenance of the symbiont in specific host species. The 35 kilodalton protein encoded in the Hsn-1 locus of *B. japonicum* appears to be required for microaerobic growth, a trait apparently more critical for growth in siratro than in the soybean cultivars tested. It is likely that these three general classes of gene function will also be found in other bacteria in close association with plants.

REFERENCES

Banfalvi Z, Nieuwkoop, A, Schell M., Besl L and Stacey G (1988) Regulation of *nod* gene expression in *Bradyrhizobium japonicum*. Mol Gen Genet 214: 420-424

Bender GL, Goydych W, Rolfe BG and Nayudu M (1987) The role of *Rhizobium* conserved and host-specific nodulation genes in the infection of the non-legume *Parasponia andersonii*. Mol Gen Genet 210: 299-306

Beynon JL, Cannon MC, Buchanan-Wollaston V and Cannon FC (1983) The *nif* promoters of *Klebsiella pneumoniae* have a characteristic primary structure. Cell 34: 665-671

Buck M, Miller S, Drummond M, and Dixon R (1986) Upstream activator sequences are present in the promoters of nitrogen fixation genes. Nature 320: 374-378

Deshmane N and Stacey G (1989) Identification of *Bradyrhizobium nod* genes involved in host specific nodulation. J Bacteriol 171: 3324-3330

Fischer H-M, Alvarez-Morales A and Hennecke H (1986). The pleiotropic nature of symbiotic regulatory mutants: *Bradyrhizobium japonicum nifA* gene is involved in control of nif gene expression and formation of determinate symbiosis. EMBO J 5: 1165-1173

Hahn M and Hennecke H (1988) Cloning and mapping of a novel nodulation region from *Bradyrhizobium japonicum* by genetic complementation of a deletion mutant. Appl Environ Microbiol 54: 55-61

Harnisch U, Weiss H and Siebald W (1985) The primary structure of the iron-sulfur subunit of ubiquinolcytochrome C reductase from *Neurospora*, determined by cDNA and gene sequencing. Eur J Biochem 149: 95-99

Kosslak RM, Bookland R, Barkei J, Paaren HE and Appelbaum ER (1987) Induction of *Bradyrhizobium japonicum* common *nod* genes by isoflavones isolated from *Glycine max*. Proc Natl Acad Sci 84: 7428-7432

Long SR (1989) Rhizobium-legume nodulation: Life together in the underground. Cell 56: 203-214

Nieuwkoop AJ, Banfalvi Z, Deshmane N, Gerhold D, Schell MG, Sirotkin KM and Stacey G (1987) A locus encoding host range is linked to the common nodulation genes of *Bradyrhizobium japonicum*. J Bacteriol 169: 2631-2638

Stachel SE, An G, Flores C and Nester EW (1985) A Tn*3 LacZ* transposon for the random generation of β-galactosidase gene fusions: application to the analysis of gene expression in *Agrobacterium*. EMBO J 4: 891-898

Steppuhn J, Rother C, Hermans J. Jansen T, Salnikow, J, Hauska G and Herrmann RG (1987) The complete amino acid sequence of the Rieske FeS-precursor protein from spinach chloroplasts deduced by cDNA analysis. Mol Gen Genet 210: 171-177

pica, C. Rosenberg, J. Dénarié,

ologie Moléculaire des Relations Plantes Microorganismes, Centre National
Scientifique-Institut National de la Recherche Agronomique, BP27, 31326
Cedex, France

Fleury, M. Maille, B. Message, D. Tepfer

iologie de la Rhizosphère, Institut National de la Recherche Agronomique,
rsailles, France

N

associate with plants are generally saprophitic. They compete with other ganisms for the products of plant photosynthesis. In certain cases their h plants are commensalistic or symbiotic. Transfer of energy from plants to ace primarily in the soil through the release of exudates by roots and the l and subterranean plant parts: leaves and roots (e.g. root cap). The ability to nces produced by the plant is surely crucial to the survival of soil bacteria, ces are likely objects of intense competition (Nutman 1965). Plants produce ndary metabolites that are potential carbon and/or nitrogen sources for soil l seem logical that bacteria would evolve catabolic functions to degrade and tabolites, and that exclusive nutritional relationships could co-evolve, i.e. duce exotic metabolites, not generally catabolized by microorganisms, and would benefit from these substances, which we have called nutritional er et al 1988), by evolving the corresponding catabolic functions. These ators can be thought of as signals that at the same time trigger and fuel ng to growth. It is also logical that selective microbial growth would be specificity of plant-microorganism interactions.

the catabolism of plant secondary metabolites by bacteria has mainly focused ilized by *Agrobacterium* (Tempé and Goldmann 1982) and *Pseudomonas* Dion 1985; Bouzar and Moore 1987), and on several unrelated substances

NATO ASI Series, Vol. H 36
Signal Molecules in Plants and Plant-Microbe Interactions
Edited by B. J. J. Lugtenberg

degraded by *Rhizobium*, including rhizolotine (Scott et al 1987), catechin (Gajendiran et al 1988), homoserine (Economou and Johnston 1988), rhizopine (Murphy et al 1987) and a family of novel compounds called calystegins (Tepfer et al 1988). The latter could be responsible for the nutritional selection of rhizosphere bacteria. We localized the bacterial genes responsible for calystegin catabolism (termed *cac*) to a plasmid from *Rhizobium meliloti* strain 41 (pRme41a), and we suggested that other couples of synthetic and catabolic functions might exist for other plant/microorganism relationships.

In the following paper we review and continue our study of genes in *R. meliloti* responsible for calystegin catabolism. We also continue the search for other nutritional mediators and present recent results showing that the Sym plasmid of *Rhizobium meliloti* encodes the catabolism of trigonelline, a betaine present in the seeds and seedlings of the legume hosts of *R. meliloti*. We speculate on the significance of the placement of these functions and on the role of trigonelline in nodule development.

CALYSTEGINS

The search for nutritional mediators began by screening a variety of plants for substances produced by subterranean organs in sufficient quantities and with sufficient specificity to make them probable candidates. Crude extracts of roots and leaves were analysed using high voltage paper electrophoresis and silver staining. The substances we call calystegins caught our attention because they appeared abundantly in the roots of only 3 species (out of the 105 species, representing 23 families, that we screened, see Tepfer et al 1988): these are two species of morning glory (*Calystegia sepium* and *Convolvulus arvensis*) and *Atropa belladonna*. We estimate their abundance to be on the order of several percent of the dry weight. Furthermore they are present in root exudates. In order to have a constant source of calystegins we cultured *C. sepium* roots transformed by *Agrobacterium rhizogenes*. These cultures produce kilogram quantities of roots in the space of a few weeks (Jung and Tepfer, 1987). Calystegins can be prepared from transformed roots or from rhizomes from pot-grown plants by methods previously used to prepare opines (Tepfer et al 1988).

We elected to first search for calystegin catabolic functions in genetically defined soil bacteria previously shown to associate with plants. We tested 44 bacterial species for the ability to degrade calystegins (Tepfer et al 1988). Only one of these, *Rhizobium meliloti* strain 41, was able to catabolize calystegins. Thus there was reason to suspect that calystegin catabolism would be limited to specific soil bacteria. We determined that *R. meliloti* 41 can use calystegins as a sole carbon and nitrogen source (Tepfer et al 1988). Calystegins are thus selective agents in the laboratory.

Calystegin catabolism in *R. meliloti* 41 is conferred (Tepfer et al 1988) by a large, previously cryptic plasmid, pRme41a. The organization of the genes encoding calystegin catabolism (*cac*) is under study. Using insertional and deletional mutagenesis we have determined that they are numerous and spread out over a 30 to 40 kb region of pRme41a (Fig. 1). Mutations affecting calystegin catabolism fall into three categories: Cac⁻ (no degradation of calystegins); Cac^{d} (delay in calystegin degradation; and Nca (calystegin degradation with the accumulation of neutral compound (s), likely resulting from a block in the catabolic pathway. Insertions conferring this third phenotype were mapped within a 10 kb region. No Tn5 insertion resulted in a Cac⁻ phenotype, suggesting a possible reiteration of some of the calystegin catabolism functions.

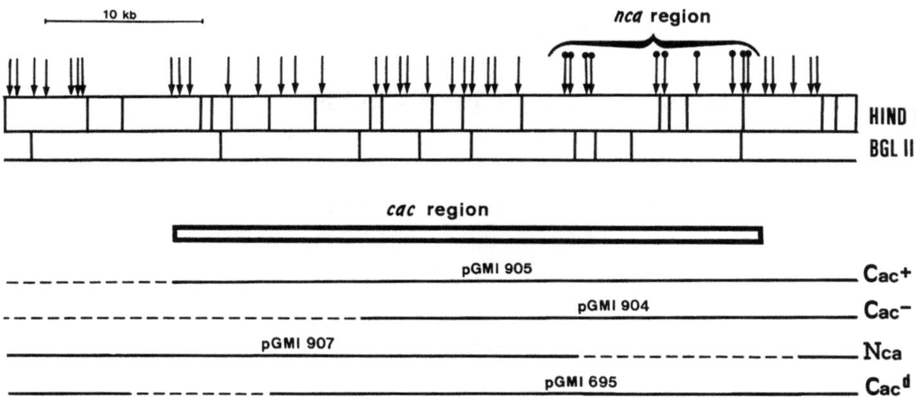

Fig. 1: Insertions and deletions into the *cac* region of pRme41a that alter calystegin catabolism. Deleted regions in plasmids pGMI905, pGMI904 and pGMI695 are represented by interrupted lines. The phenotypes of strains carrying the various plasmids are indicated on the right. Arrows represent Tn5 insertions: simple arrow = Cac^{+}; arrow beginning with a dot = Nca (Neutral compound accumulation).

If calystegins are nutritional mediators in the nature, one would expect to find Cac^{+} bacteria in the rhizospheres of plants that synthesize calystegins (Cas^{+}). We have searched for such bacteria both in the rhizospheres of plants that produce and do not produce them. Cac^{+} bacteria were found in the rhizospheres of plants that synthesize calystegins and not in the rhizospheres of other plants. The wild Cac^{+} bacteria are a mixture of Gram positives and negatives, and since their *cac* genes do not cross-hybridize with those of *R. meliloti*, we presume that they are only distantly or not at all related to those found in *R. meliloti* strain 41. Therefore, the Cac^{+} phenotype is not simply the product of recent sharing of *cac* genes among similar soil bacteria. *Cac* genes probably originated long ago and diverged in diverse bacteria after interspecific transfer or, alternatively, they may have arisen independently in widely different soil bacteria. (Specificity in plant-microbe interactions is likely to be the product of coevolution over long periods of interaction.) It would be of interest to determine whether *cac* genes are plasmid borne in these wild Cac^{+}, as they are in *R. meliloti* 41.

TRIGONELLINE

Trigonelline is a nicotinic acid derivative present in large quantities in the seeds and seedlings of legumes (Tramontano et al 1986) (Figure 2). We reasoned that it could be a substrate for bacterial growth, and thus a possible nutritional mediator in interactions between *Rhizobium* and legumes. We first tested the ability of a number of strains of soil bacteria to use trigonelline as a carbon and nitrogen source. The results (not shown) show that trigonelline catabolism is common in the *Rhizobiaceae*.

choline: $(CH_3)_3N^{+}CH_2CH_2OH$

glycine betaine : $(CH_3)_3N^{+}CH_2COO^{-}$

trigonelline :

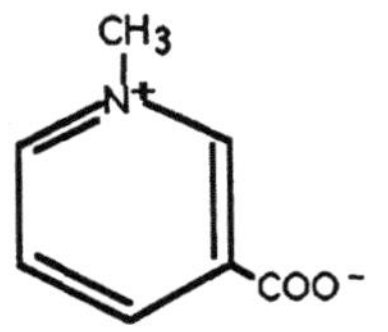

Fig. 2: Structures of the betaines: choline, glycine betaine and trigonelline. An important common feature among the three substances is the quarternized nitrogen atom, shown as N^{+}.

Efforts to localize the trigonelline catabolism functions in *R. meliloti* strain 2011 (not shown) showed that they are encoded on the Sym plasmid in a region linked to nodulation and nitrogen fixation functions. The evidence is as follows. Strain GMI766, a 2011 derivative carrying a large deletion that begins approximately 200 kb to the left of *nif HDK* and extends rightward for more than 300kb (Renalier et al 1987), is unable to grow on trigonelline either as the sole carbon and or nitrogen source. Introduction into this strain of a plasmid (not shown), carrying a 70 kb region to the right of nif HDK, restores the ability to grow on

In order to more precisely localize the *trc* (trigonelline catabolism) genes, we introduced into the Trc^{-}strain, GMI 766, R' plasmids containing inserts of variable length included in the 70 kb region. One of these, designated pGMI471, restored the ability to catabolize

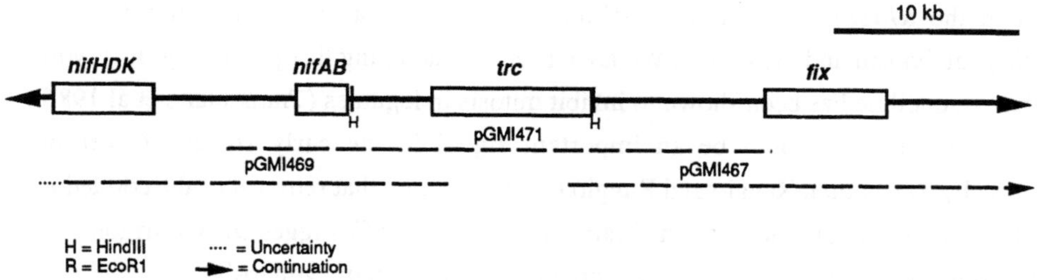

Fig. 3: Localization of the *trc* region on pSym. pSym is shown as a solid line, with the positions of *nif HDK nifAB, trc* and *fix.* given as boxes. R' plasmids (Renalier et al 1987) covering the region between *nifHDK* and *fix* are shown as dotted lines. The middle of these, pGMI471, complements a trigonelline⁻ (Trc ⁻), deletion in the symbiotic region of pSym (strain GMI766).

The genetic organization and regulation of the *trc* genes are under study. We are also analyzing the symbiotic properties of Trc⁻ mutants on various legume hosts and the ability of these mutants to colonize host plants.

Two other betaines, choline and glycine betaine (Fig. 2) are also catabolized by *R. meliloti* 2011. Preliminary results suggest that the corresponding catabolic genes could be encoded by the right most part of the deletion carried by strain GMI766. Plasmid control of betaine catabolism is not an obligatory feature of the *Rhizobiaceae*, since an *Agrobacterium tumefaciens* strain cured of detectable plasmids (Rosenberg and Huguet 1984) is still able to catabolize trigonelline.

DISCUSSION

We have examined the catabolism of calystegins and trigonelline among the *Rhizobiaceae*. The former were degraded by only one strain and the latter by the majority of the strains tested. On the other hand, a survey of soil bacteria associated with the roots of *C. sepium* showed that a large percentage of them were able to catabolize calystegins. Thus it would be premature to suggest that calystegins are better candidates to be nutritional mediators because of greater specificity of bacterial catabolism. However, the calystegins would seem more obvious candidates because they are produced in large quantities throughout the life of the plant and their synthesis is rare among plant species (Tepfer et al 1988). The possibility that calystegins and trigonelline serve as a nutritional mediators in nature remains open.

The function of the calystegins in the plant is not known. Trigonelline is a weak osmoticum (le Rudulier and Bernard 1986; Smith et al 1988) and is capable of inducing *nod* functions (Schmidt et al 1986). It's chemical structure is different from that of choline and glycine betaine, being derived from nicotinic acid rather than from glycine. It is perhaps significant that the region encoding trigonelline catabolism (*trc*) is surrounded by genes involved in the *Rhizobium*-legume symbiosis, suggesting a functional link between trigonelline catabolism and symbiosis. We are currently examining this possibility. It is worth noting that trigonelline has been shown to inhibit mitosis in legumes (Tramontano et al 1986). Trigonelline catabolism could be an important signal in the early stages of seedling colonization by *Rhizobium*. Raggio and Raggio (1956) reported that removal of the cotyledons from pea seedlings inhibits nodulation. Tramontano et al (1986) suggested that trigonelline accumulated in the cotyledons might be essential to the initiation of symbiosis.

Calystegin catabolism appears to not influence symbiosis because calystegins are not detected in legumes and *R. meliloti*, cured of pRme41a, the plasmid carrying the calystegin catabolism genes (*cac*) shows normal symbiotic properties, at least under laboratory conditions. A possible ecological role for these catabolic functions could be to favour *Rhizobium* in saprophytic associations with morning glory, in the absence of a legume host. Large cryptic plasmids are a general feature of the *Rhizobiaceae*. We suggest that some of these could be involved in the catabolism of nutritional mediators.

ACKNOWLEDGEMENTS

We wish to thank Marie-Helène Renalier and Pierre Boistard for permission to use plamids pGMI 467, pGMI469 and pGMI471.

REFERENCES

Bouzar H, Moore L (1987) Isolation of different *Agrobacterium* biovars from a natural oak savanna and tallgrass prairie. Appl. Environ. Microbiol 53:717-721

Economou A, Johnston A. W. B. (1988) Catabolism of L-homoserine by strains of *Rhizobium leguminosarum* biovar*viceae*, In: Annual Report of the AFRC Institute of Plant Science Research and the John Innes Institute. AFRC Institute of Plant Science Research and the John Innes Institute, Norwich, U.K., 116

Gajendiran N, Mahadevan A (1988) Utilization of catechin by *Rhizobium* sp. Plant and Soil 108:263-266

Jung G, Tepfer D (1987) Use of genetic transformation by the Ri JT-DNA of *Agrobacterium rhizogenes* to stimulate biomass and tropane alkaloid production in *Atropa belladonna* and *Calystegia sepium* roots grown *in vitro*. Plant Sci. 50: 145-151

le Rudulier D, Bernard T (1986) Salt tolerance in *Rhizobium*: a possible role for betaines. FEMS Microbiol. Rev. 39:67-72

Murphy PJ, Heycke N, Banvalvi Z, Tate ME, de Bruijn F, Kondorosi A, Tempé J, Schell J (1987) Genes for the catabolism and synthesis of an opine-like compound in *Rhizobium meliloti* are closely linked and on the pSym plasmid. Proc. Natl. Acad. Sci. 84:493-497

Nutman PS (1965) The relation between nodule bacteria and the legume host in the rhizosphere and in the process of infection, In: Baker KF, Synder WC (Eds) Ecology of soil-born plant pathogens. University of California Press, Berkeley, 231-247

Raggio M, Raggio N (1956) Relación entre cotiledones y nodulación y factores que la afectan. Phyton 7:103-119

Renalier M-H, Batut J, Ghai J, Terzaghi B, Gherardi M, David M, Garnerone A-M, Vasse J, Truchet G, Huguet T, Boistard P (1987) A new symbiotic cluster on the pSym megaplasmid of *Rhizobium melilot* 2011 carries a functional *fix* gene repeat and a *nod* locus. J. Bact. 169:2231-2238
Rosenberg C, Huguet T (1984) The pAtC58 plasmid of*Agrobacterium tumefaciens is not essential for tumour induction*. Mol. Gen. Genet. 196:533-536
Rossignol G, Dion P (1985) Octopine, nopaline and octopinic acid utilization in *Pseudomonas*. Can. J. Microbiol. 31:68-74
Schmidt J, John M, Wieneke U, Krussmann H, Schell J Expression of the nodulation gene *nodA* in *Rhizobium meliloti* and localization of the gene product in the cytosol. Proc. Natl. Acad. Sci. USA 83:9581-9585
Scott D, Wilson R, Show GJ, Petit A, Tempé J (1987) Biosynthesis and degradation of Nodule-Specific *Rhizobium loti* compounds in lotus nodules. J. Bact. 169:278-282
Smith LT, Pocard J-A, Bernard T, le Rudulier D (1988) Osmotic control of glycine betaine biosynthesis and degradation in *Rhizobium meliloti*. J. Bact. 170:3142-3149
Tempé J, Goldmann A (1982) Occurrence and biosynthesis of opines, In: Kahl G, Schell J (Eds) Molecular Biology of Plant Tumors. Academic Press, New York, 451-459
Tepfer D, Goldmann A, Pamboukdjian N, Maille M, Lepingle A, Chevalier D, Dénarié J, Rosenberg C (1988) A plasmid of *Rhizobium meliloti 41* encodes catabolism of two compounds from root exudate of *Calystegia sepium*. J. Bact. 170:1153-1161
Tramontano W, McGinley PA, Ciancaglini EF, Evans LS (1986) A survery of trigonelline concentrations in dry seeds of the *Dicotyledoneae*. Environmental and Experimental Botany 26:197-205

SIGNALS AND DEFENSE RESPONSES ASSOCIATED WITH THE RACE-CULTIVAR SPECIFIC INTERACTION BETWEEN TOBACCO AND PHYTOPHTHORA PARASITICA NICOTIANAE

M.T. Esquerré-Tugayé, J. Fournier, M. Rickauer and M.L. Pouénat
Centre de Physiologie Végétale
Université Paul Sabatier
118, route de Narbonne
31062 Toulouse cédex
France

Cell surface interactions between plants and pathogens induce defense responses in plants. In the past twenty years, it has been established that cell wall fragments of either pathogen or plant origin have the ability to induce the same defense reactions as pathogen attack. Such fragments, also called elicitors, are supposed to be among the earliest signals of the interaction (Darvill and Albersheim 1984). Distinct classes of elicitors (oligosaccharides, glycans, glycopeptides, peptides, lipids) have thus been identified. Similarly, a wide array of defense responses has been characterized, and several relevant genes and products isolated (Collinge and Slusarenko 1987). Despite such advances, our knowledge of transduction events leading from elicitor or pathogen recognition to defense gene expression remains scanty. Understanding signal transduction pathways requires the availability of model systems on which cellular and molecular approaches can be performed. The tobacco-*Phytophthora parasitica nicotianae* system is presented in this paper. A biochemical study of host responses, possibly related to signal processing in this system, is then reported.

THE TOBACCO-PHYTOPHTHORA PARASITICA NICOTIANAE SYSTEM

The fungal pathogen *Phytophthora parasitica nicotiane* interacts in a race-cultivar specific manner with its host plant *Nicotiana tabacum* (De Zoeten et al 1982). Interaction of cultivar Burley L8 with race 0 gives an incompatible response, while it is compatible between L8 and race 1, as well as between cultivar Burley KY 15 and the two races. Zoospore inoculation of the root system leads to the appearance of symptoms two to three days after inoculation. Browning of the tissues then rapidly extends in compatible interactions while it remains localized in the incompatible one.
Preliminary reports have shown that the interaction can be partly reproduced

NATO ASI Series, Vol. H 36
Signal Molecules in Plants and Plant-Microbe Interactions
Edited by B. J. J. Lugtenberg

by using tobacco cell culture and elicitors of Phytophthora (Pelissier et al 1986). The glycopeptide elicitors prepared from Phytophthora induce a slight browning of tobacco cells upon incubation.

As a prerequisite to studying signal transduction in this system, we have looked for host responses that would be equally expressed in whole plants and cell suspensions.

HOST RESPONSES

A biochemical study of proteinase inhibitors and of lipoxygenase was undertaken.

1. Proteinase inhibitors

Proteins with proteinase inhibitory activity are thought to participate in the defense of plants against pathogens and insects. Their level is low in the tobacco plants and cell suspension used in this study. A strong proteinase inhibitory activity is induced in plants upon infection, and in cell suspension upon elicitor treatment. This increase starts early and is transient during the incompatible interaction between cultivar L8 and race 0 of the fungus, with a maximum three days after inoculation. It is delayed and more prolonged in the three other compatible interactions described above. The increase of proteinase inhibitors in cell suspension occurs after 10 to 12 hours of incubation in the presence of elicitors of Phytophthora parasitica nicotianae. The activity was measured against trypsin but other inhibitors with different specificities are also present in tobacco. A trypsin inhibitor was purified to homogeneity by affinity chromatography, and characterized as a single peptide of MW 10.5 kD (Rickauer et al 1989). This and other criteria, suggest that it is a type II proteinase inhibitor. Types I and II refer to the inhibitors initially identified in wounded tomato plants by Plunkett et al (1982). Although wound-induced proteinase inhibitors are well characterized at the protein and gene level (Graham et al 1986), almost nothing is known on elicitation of these defense genes in plant-microorganism interactions. The present study extends previous findings obtained by Peng and Black (1976) on Phytophthora-infected tomato, and by Roby et al (1987) on melon infected by Colletotrichum lagenarium. In the latter case it was found that the melon extract inhibits the proteases of Colletotrichum, which strongly suggests a role in plant defense.

2. Lipoxygenases

Lipoxygenases are active on polyunsaturated fatty acids (PUFAs). A number of reports indicate that PUFAs and derived products might play a significant role in plant-pathogen interactions. Thus, arachidonic acid induces phytoalexin biosynthesis in potato (Bostock et al 1981), through the possible involvement of lipoxygenase-mediated degradation products. Lipoxygenase might also favour the last step of ethylene biosynthesis under certain conditions (Bousquet and Thimann 1984). This hormone, in its turn, induces several defense responses in plants (Roby et al 1986 ; Ecker and Davis 1987 ; Rumeau et al 1988).

The level of lipoxygenase is low in tobacco cell suspensions and is highly increased upon elicitor treatment (Fournier et al 1986). Elicitation of the enzyme starts 1.5 hour after addition of the elicitor preparation, and shows a maximum at about 24 hours. At the whole plant level, lipoxygenase is increased upon infection. As in the case of proteinase inhibitors, the data show that induction appears earlier in the incompatible interaction than in the compatible ones, and that it is transiently expressed. Lipoxygenase activity is maximal two days after inoculation in the resistant cultivar, while it starts to increase at the time of symptom appearance in the three other compatible situations. The effect of infection on lipoxygenase activity has also been reported in wheat infected by Puccinia graminis (Ocampo et al 1986), and in tobacco infected by Erisyphe cichoracearum (Lupu et al 1980).

Lipoxygenase was purified from elicitor treated cells and characterized. It has a MW of about 96 kD, an optimum pH of 6.8 and a pI comprised between 5.1. and 5.3. It is mainly a 9-lipoxygenase, according to the hydroperoxides that it forms from linoleic and linolenic acids, which are the two main PUFAs in tobacco.

The hydroperoxides are damaging compounds by themselves : they affect membrane integrity. They can also give rise to a variety of products (Figure 1), which exert deleterious effects on membranes, or which have regulatory activities (Vick and Zimmerman 1987).

For these reasons, PUFA-derived compounds might be involved in the hypersensitive response which typifies incompatibility. Some of them might also be part of a cascade of secondary signals.

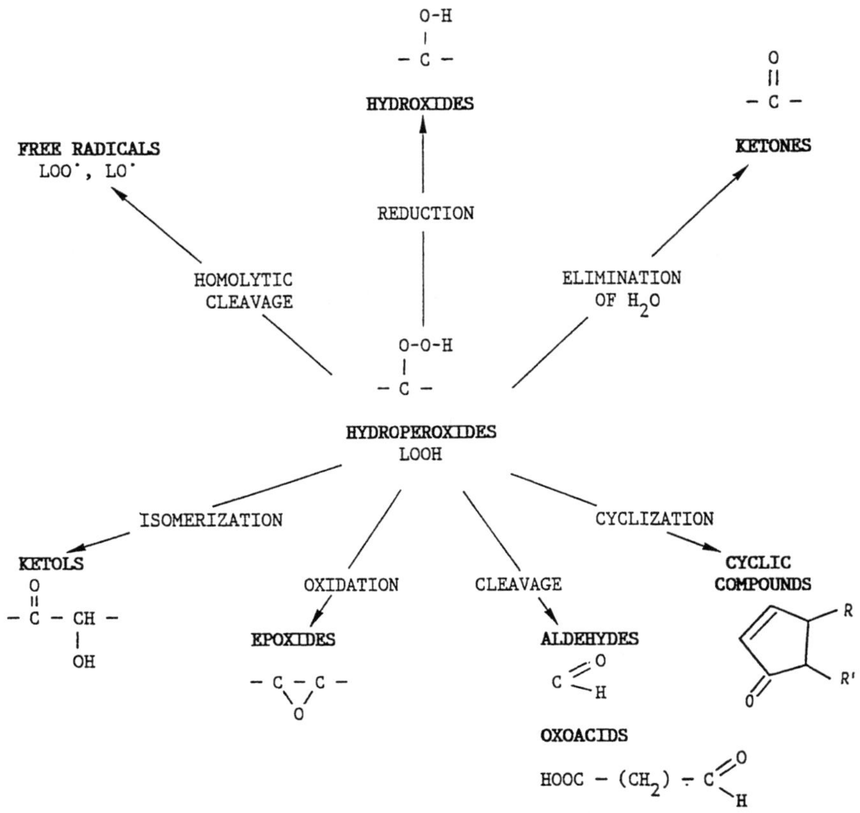

Figure 1. **SECONDARY CONVERSIONS OF LIPOXYGENASE-DERIVED PRODUCTS**

POSSIBLE SIGNAL PATHWAYS

In animals, PUFA-derived products are part of a phospholipid breakdown phenomenon which involves several enzymes, most notably phospholipases C and A_2, neutral lipases, lipoxygenases and other oxygenases. Whether or not all these membrane related effects are part of elicitor-induced signal cascade(s) is an open question. In the tobacco-Phytophthora system, we previously found that elicitors affect other membrane events, such as membrane potential and ethylene biosynthesis. The same elicitor also induces the methionine pathway of ethylene biosynthesis : after 15 minutes, there is an increase in ACC synthase activity, followed by ACC and ethylene production.

Considering the timing of elicitor-, or infection-induced events in tobacco, it appears that one can distinguish early membrane related steps and late defense responses.

As a working hypothesis (fig.2), we propose that elicitor recognition induces membrane depolarization, which in its turn affects several membrane related events, notably ethylene synthesis. Ethylene and membrane depolarization might play the role of rapidly exported, cell to cell, signals. Intracellular signals might involve phospholipid metabolites, such as compounds resulting from lipoxygenase action and/or inositol phosphates.

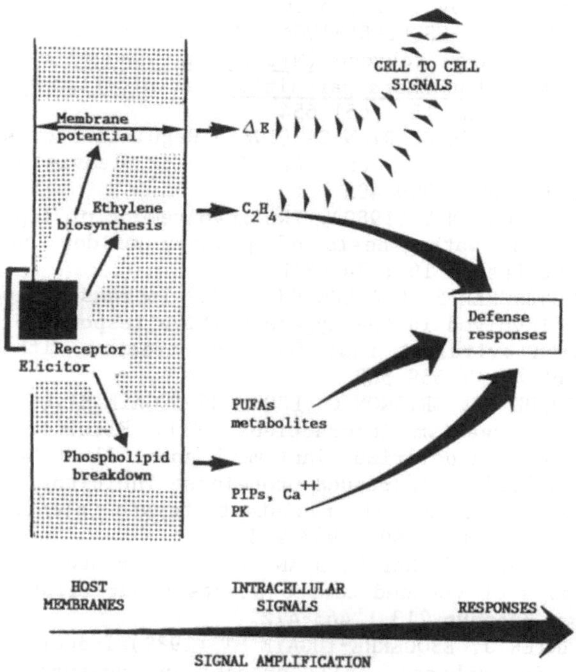

Figure 2. **SIGNAL TRANSDUCTION IN RESPONSE TO ELICITORS: A PROPOSED SCHEME**
(ΔE = variations of membrane potential ; PUFAs = polyunsaturated fatty acids ; PIPs = phosphoinositides ; PK = protein kinases)

Preliminary data suggest that several cascades are induced by elicitors, with complex cross regulatory mechanisms.

In conclusion, the interaction between tobacco and Phytophthora parasitica nicotianae appears as a model system for studying the recognition, transduction and expression phenomena which underly elicitor effects on plants.

REFERENCES

BOSTOK RM, KUC JA, LAINE RA (1981) Eicosapentaenoic and arachidonic acids from Phytophthora infestans elicit fungitoxic sesquiterpenes in the potato. Science 212 : 67-69

BOUQUET JF, THIMANN KV (1984) Lipid peroxidation forms ethylene from 1-aminocyclopropane-1-carboxylic acid and may operate in leaf senescence. Proc Natl Acad Sci USA 81 : 1724-1727

COLLINGE DB, SLUSARENKO AJ (1987) Plant gene expression in response to pathogens. Plant Mol Biol 9 : 389-410

DARVILL AG, ALBERSHEIM P (1984) Phytoalexins and their elicitors. A defense against microbial infection in plants. Ann Rev Plant Physiol 35 : 243-275

De ZOETEN GA, GAARD G, HABERLACH GT, HELGESON JP (1982) Infection of tobacco callus by Phytophthora parasitica nicotianae. Phytopathol 72 : 743-746.

ECKER JR, DAVIS RW (1987) Plant defense genes are regulated by ethylene. Proc Natl Acad Sci 84 : 5202-5206

FOURNIER J, PELISSIER B, ESQUERRE-TUGAYE MT (1986) Induction of lipoxygenase activity in cultured tobacco (Nicotiana tabacum) cells by elicitor of ethylene from Phytophthora parasitica var nicotianae. CR Acad Sc Paris t 303, serie III, n° 15 : 651-654

GRAHAM JS, HALL G, PEARCE G, RYAN CA (1986) Regulation of synthesis of proteinase inhibitors I and II mRNAs in leaves of wounded tomato plants. Planta 169 : 399-405

LUPU R, GROSSMAN S, COHEN Y (1980). The involvement of lipoxygenase and antioxydants in pathogenesis of powdery mildew on tobacco plants. Physiol Plant Pathol 16 : 241-248

OCAMPO CA, MOERSCHBACHER B, GRAMBOW HJ (1986) Increased lipoxygenase activity is involved in the hypersensitive response of wheat leaf cells infected with avirulent rust fungi or treated with fungal elicitor. Z Naturforsch 41c : 559-563

PELISSIER B, THIBAUD JB, GRIGNON C, ESQUERRE-TUGAYE MT (1986) Cell surfaces in plant-microorganism interactions. VII. Elicitor preparations from fungal pathogens depolarize plant membranes. Plant Sci 46 : 103-109

PENG JH, BLACK LL (1976). Increased proteinase inhibitor activity in response to infection of resistant tomato plants by Phytophthora infestans. Phytopathol 66 : 958-963

PLUNKETT G, SENEAR DF, ZUROSKE G, RYAN CA (1982) Proteinase inhibitors I and II from leaves of wounded tomato plants : purification and properties. Arch Biochem Biophys 213 : 463-472

RICKAUER M, FOURNIER J, ESQUERRE-TUGAYE MT (1989) Induction of proteinase inhibitors in tobacco cell suspension culture by elicitors of Phytophthora parasitica var. nicotianae. Plant Physiol (in press)

ROBY D, TOPPAN A, ESQUERRE-TUGAYE MT (1986) Cell surfaces in plant-microorganism interactions. VI. Elicitors of ethylene from Colletotrichum lagenarium trigger chitinase activity in melon plants. Plant Physiol 81 : 228-233

ROBY D, TOPPAN A, ESQUERRE-TUGAYE MT (1987) Increased proteinase inhibitor activity in melon plants in response to infection by Colletotrichum lagenarium or to treatment with an elicitor fraction from this fungus. Physiol Mol Plant Pathol 30 : 453-460

RUMEAU D, MAZAU D, PANABIERES F, DELSENY M, ESQUERRE-TUGAYE MT (1988) Accumulation of hydroxyproline-rich glycoprotein mRNAs in infected or ethylene treated melon plants. Physiol Mol Plant Pathol 33 : 419-428

VICK BA, ZIMMERMAN DC (1987) Oxidative systems for modification of fatty acids : the lipoxygenase pathway. In STUMPF PK (ed) The Biochemistry of Plants : a comprehensive treatise. Academic Press, New-York, pp 53-90

GENES FOR CULTIVAR SPECIFICITY IN THE RICE BLAST FUNGUS, *MAGNAPORTHE GRISEA*

Barbara Valent and Forrest Chumley
E. I. Du Pont de Nemours and Co., Inc.
Central Research and Development Department
Experimental Station, P. O. Box 80402
Wilmington, Delaware 19880-0402 U.S.A.

Introduction

Field isolates of the heterothallic Ascomycete, *Magnaporthe grisea* Barr (anamorph, *Pyricularia oryzae* Cav. or *P. grisea*), include pathogens of many grasses. Individual isolates, however, have a limited host range, parasitizing one or a few grass species (MACKILL and BONMAN 1986). Strains of the fungus that parasitize rice (*Oryza sativa*) are subdivided into races, depending on the rice cultivars they can successfully infect. The rice blast fungus shows a high degree of variability in the field; new races frequently appear with the ability to attack previously resistant rice cultivars (OU 1985).

Many genes in the pathogen must act together to execute the metabolic and developmental pathways that comprise the disease cycle, the ordered series of events leading from spore germination through infection, lesion development, and conidiation. Genes that control the development and function of infection structures are examples of general pathogenicity determinants, those required for the pathogen to infect any host. Genes that determine cultivar specificity, however, appear to provide important signals for a complex recognition system in the host that can trigger an effective disease resistance response.

Classic studies of race-cultivar specificity that defined the "gene-for-gene" relationship were conducted by Harold Flor, who analyzed the interactions between flax (*Linum usitatissimum*) and flax rust (*Melampsora lini*) (FLOR 1971). Flor's studies revealed that a flax cultivar is resistant to a particular race of the pathogen only if the cultivar carries a dominant resistance gene that corresponds to a dominant gene for aviru-

NATO ASI Series, Vol. H 36
Signal Molecules in Plants and Plant-Microbe Interactions
Edited by B.J.J. Lugtenberg

lence carried by the pathogen. Thus, there is a one-to-one (gene-for-gene) functional correlation between resistance genes in the host and avirulence genes in the pathogen. Flor's results showed the flax genome contains a number of genetic loci that can independently express rust resistance, and the genome of the pathogen contains a similar array of loci that can independently express avirulence. A particular race of *M. lini* is capable of causing disease on a particular flax cultivar only if the pathogen lacks all avirulence genes that correspond to the host's resistance genes. Although exceptions to this rule have been well documented (CRUTE 1986), it has been shown that the gene-for-gene model has great value in predicting the genetic basis of race-cultivar specificity in other systems (DAY 1974; CRUTE 1986).

Cultivars of rice that differ from one another by the presence or absence of dominant resistance genes have been developed (YAMADA et al 1976). The resistance of any one of these cultivars is effective only against certain races of the pathogen. The genetic basis for differences in cultivar specificity between races of the rice blast pathogen has been previosly unknown due to the infertility of field isolates of *M. grisea* that infect rice. Fertile *M. grisea* rice pathogens have now been developed. We have identified pathogen genes that control rice cultivar specificity, as well as genes that control pathogenicity toward a second host, weeping lovegrass (*Eragrostis curvula*). In this analysis, a gene was termed an "avirulence gene" if alleles determine virulence or avirulence in an all-or-nothing, cultivar specific fashion. Genes with no cultivar specific effects were termed "pathogenicity genes". Because obstacles to genetic analysis of rice pathogens have been overcome, the *M. grisea* system has properties that recommend it for undertaking a detailed analysis of the molecular basis for host specificity.

Avirulence Genes Identified in a Backcrossing Scheme

The interfertility of *M. grisea* strains with different host species specificities has permitted genetic analysis of those differences. A backcrossing regime was initiated by crossing the weeping lovegrass pathogen, 4091-5-8, a highly fertile hermaphrodite, and the rice pathogen, O-135, a female sterile field isolate that also infects weeping lovegrass (VALENT, FARRALL and CHUMLEY, submitted). All progeny from this cross infected weeping lovegrass, as expected, because both parents are weeping lovegrass pathogens. However, only 6 of 59 progeny were pathogens of rice, indicating that the two parental strains differ by several genes that control the ability to infect rice.

In the backcrossing regime, the rice pathogen was the recurrent parent. Female parents for the backcross generations were progeny strains chosen for having fertility comparable to that of 4091-5-8. The ratios of pathogenic to nonpathogenic (and virulent to avirulent) progeny through the backcross generations suggested that O-135 and 4091-5-8 differ in two types of genes that control the ability to infect rice. First, polygenically-inherited factors determine the extent of lesion development on rice. Second, single avirulence genes govern, in an all-or-nothing fashion, virulence toward specific cultivars of rice. Several crosses confirmed the segregation of three unlinked genes, *Avr1-CO39*, *Avr1-M201* and *Avr1-YM* that determine avirulence on rice cultivars CO39, M201, and Yashiro-mochi, respectively. *Avr1-M201* may be linked to the mating type locus, *Mat1*.

The avirulence alleles of the cultivar specificity genes identified in these crosses appear to have been inherited from the parent that is a nonpathogen of rice, because the rice pathogen parent is virulent on all three cultivars. We identified avirulence genes for each cultivar used in these experiments, suggesting a high potential for identification of other avirulence genes derived from the nonpathogen of rice. YAEGASHI and ASAGA (1981) reported similar results suggesting that a finger millet pathogen, WGG-FA40, carries an avirulence gene corresponding to the rice blast resistance gene, *Pi-a*.

Avirulence Genes Identified in Other Crosses

A separate series of crosses also identified single genes that control cultivar specificity (VALENT, FARRALL and CHUMLEY, in preparation). Both parents in Cross 4360 are virulent on rice cultivars CO39, M201 and Sariceltik, and all progeny are virulent on these three cultivars. The parents differ in specificity toward rice cultivars Yashiro-mochi and Maratelli. Data from five complete tetrads and from 30 random ascospore cultures suggest that two unlinked avirulence genes segregated among the progeny of this cross: *Avr2-YM* controls virulence or avirulence on cultivar Yashiro-mochi and *Avr-MARA* controls virulence or avirulence on cultivar Maratelli. The segregation of both genes has been confirmed in at least two subsequent crosses. These genes are not linked to the mating type locus. The parents of Cross 4360 differ in specificity toward five additional cultivars; further analysis of the progeny will probably reveal the segregation of additionial avirulence genes.

Genetic crosses were performed to determine if the two genes conferring specificity toward the rice cultivar Yashiro-mochi, *Avr1-YM* and *Avr2-YM*, are allelic. Data from three different crosses indicated that the two genes are not linked to one another.

Genes for Pathogenicity To Weeping Lovegrass

The parents of Cross 4360 described in the previous section also differ in ability to infect weeping lovegrass. Single gene segregation of pathogenicity toward weeping lovegrass was observed among the 30 random ascospore progeny and in the five complete tetrads. Subsequent crosses confirmed simple segregation of the gene now named *Pwl2*. Preliminary data suggest that *Pwl2* is not linked to the weeping lovegrass pathogenicity determinant, *Pwl1*, previously identified in a cross between a weeping lovegrass pathogen and a goosegrass pathogen (VALENT et al 1986; VALENT and CHUMLEY 1987).

The genes *Pwl1* and *Pwl2* must be considered "pathogenicity genes," with no cultivar specific effects, because cultivars of weeping lovegrass are not available. However, it remains an intriguing possiblity that these genes may function in a manner analagous to avirulence genes.

Unstable Genes in *M. grisea*

Two of the host specificity genes we have identified, *Pwl2* and *Avr2-YM*, appear to be unstable. That is, spontaneous pathogenic or virulent mutants derived from strains carrying the nonpathogenic or avirulent alleles of these genes appear frequently in standard assays. In a typical experiment, a pot of weeping lovegrass or rice seedlings is misted with a suspension containing 2 X 10^6 conidia. With *Pwl2* or *Avr2-YM*, such inoculations frequently yield one or more fully susceptible lesions. Fungus isolated from such a rare lesion devastates the host upon reinoculation. Markers such as mating type, fertility, other host specificities, and "DNA fingerprints" (HAMER et al SUBMITTED) have been used to confirm the origin of virulent mutants. Many virulent mutants of strains carrying *Pwl2* or *Avr2-YM* have been recovered, but no virulent mutants have been recovered from strains carrying *Avr1-CO39*, *Avr1-M201*, *Avr1-YM* and *Pwl1*.

Two other *M. grisea* genes appear to be unstable in several strains: *SMO*$^+$, a gene that controls spore morphology (HAMER, VALENT and CHUMLEY 1989), and *BUF*$^+$, a gene that encodes an enzyme involved in biosynthesis of the gray pigment, melanin (CHUMLEY, WEAVER and VALENT, IN PREPARATION). However, many other genes appear to mutate at normal frequencies, including *ALB*$^+$ and *RSY*$^+$, which encode melanin biosynthetic enzymes. Mutations that confer resistance to benomyl, sulfonylurea or 6-methyl purine also occur at normal low frequencies in strains that undergo frequent mutations at the named unstable loci. Cloning stable and unstable genes will yield clues to the molecular basis for genetic instability in *M. grisea*.

The mechanism by which new races of a pathogen arise is an intriguing problem. Although the extent of instability in race determinants has been debated (LATTERELL 1975; OU 1985), it is clear that new races rapidly appear in the field when new blast resistant rice cultivars are introduced.

Discussion

Seven host specificity genes of *M. grisea* have been confirmed by segregation in three or more crosses. Five of these genes determine virulence or avirulence on cultivars of rice, and two determine pathogenicity or nonpathogenicity on weeping lovegrass. LEUNG et. al. (1989) report evidence for additional genes. Our results suggest that nonpathogens of rice contain genes that function as cultivar specific avirulence genes when they are introduced into strains that infect rice. Similar observations have been published concerning bacterial plant pathogens (WHALEN, STALL and STASKAWICZ 1988; KOBAYASHI, TAMAKI and KEEN 1989). The ease with which single genes for cultivar specificity have been identified in the crosses reported here suggests that these genes are numerous.

Whether or not the cultivar specificity genes we have defined may be classical avirulence genes remains to be determined. We can provide no information on dominance relationships between alleles of these genes because *M. grisea* does not form stable vegetative diploids or heterokaryotic conidia (CRAWFORD et al 1986). Further genetic analysis of both the host and the pathogen will be required to define any functional correspondence that may exist between a particular pathogen gene and a particular resistance gene in rice. The rice cultivar Yashiro-mochi, one of the ten Japanese differential cultivars (YAMADA et al 1976), is reported to have the blast resistance gene, *Pi-ta*. Genetic crosses of Yashiro-mochi will determine if the pathogen genes *Avr1-YM* and *Avr2-YM* interact with the blast resistance gene, *Pi-ta*, with previously unidentified resistance genes, or with any identifiable gene for resistance. The cultivars CO39, M201 and Maratelli have not previously been reported to carry blast resistance genes. Genetic crosses involving these cultivars will determine if they contain unidentified blast resistance genes.

Our efforts now are focused on cloning avirulence genes as a first step toward understanding the molecular mechanisms that determine cultivar specificity. With cloned avirulence genes in hand, we plan to address the following questions. What gene products are encoded by avirulence genes? How might these gene products interact with the products of host resistance genes? How is the expression of avirulence genes controlled? What molecular genetic events accompany the appearance of new races in the field? What is the difference between stable and unstable avirulence genes? How do rice pathogen avirulence genes differ from those present in nonpathogens of rice? The goal of understanding the molecular basis for race-cultivar specificity remains a major challenge in biology.

Literature Cited

Chumley FG, Weaver CG, Valent B (In preparation) Genetic analysis of melanin deficient nonpathogenic mutants of *Magnaporthe grisea*.

Crawford MS, Chumley FG, Weaver CG, Valent B (1986) Characterization of the heterokaryotic and vegetative diploid phases of *Magnaporthe grisea*. Genetics 114:1111-1129

Crute IR (1986) The genetic basis of relationships between microbial parasites and their hosts. In: Fraser RSS (ed) *Mechanisms of Resistance to Plant Diseases*. Martinnus Nijhoff and W. Junk, Dordrecht pp 80-142

Day PR (1974) *Genetics of Host-Parasite Interaction*. WH Freeman and Co., San Fransisco

Flor HH (1971) Current status of the gene-for-gene concept. Annu Rev Phytopathol 9:275-296

Hamer JE, Farrall L, Orbach MJ, Chumley FG, Valent B, (Submitted) Host-species specific conservation of a family of repeated DNA sequences in the genome of a fungal plant pathogen

Hamer JE, Valent B, Chumley FG (1989) Mutations at the *SMO* genetic locus affect the shape of diverse cell types in the rice blast fungus. Genetics, in press

Kobayashi DY, Tamaki SJ, Keen NT (1989) Cloned avirulence genes from the tomato pathogen *Pseudomonas syringae* pv. *tomato* confer cultivar specificity on soybean. Proc Natl Acad Sci USA 86:157-161

Latterell FM (1975) Phenotypic stability of pathogenic races of Pyricularia oryzae, and its implication for breeding blast resistant rice varieties. In *Proceedings of the Seminar on Horizontal Resistance to Blast Disease of Rice*. Colombia Series CE-No. 9 Centro Internacional de Agricultura Tropical, Cali pp 199-234

Leung H, Borromeo ES, Bernardo MA, Notteghem JL (1988) Genetic analysis of virulence in the rice blast fungus *Magnaporthe grisea*. Phytopathology 78:1227-1233

Mackill AO, Bonman JM (1986) New hosts of *Pyricularia oryzae*. Plant Disease 70:125-127

Ou SH (1985) In *Rice Diseases*, Commonwealth Agricultural Bureaux, Slough, UK.pp. 109-201

Valent B, Chumley FG (1987) Genetic Analysis of Host Species Specificity in Magnaporthe grisea. UCLA Symp Mol Cell Biol (New Series) 48:83-93

Valent B, Crawford MS, Weaver CG, Chumley FG (1986) Genetic studies of pathogenicity and fertility of *Magnaporthe grisea*. Iowa State J Res 60:569-594

Valent B, Farrall L, Chumley FG (Submitted) *Magnaporthe grisea* genes for pathogenicity and virulence identified through a series of backcrosses

Valent B, Farrall L, Chumley FG (In preparation) Genes for host species and host cultivar specificity in *Magnaporthe grisea*

Whalen MC, Stall RE, Staskawicz B (1988) Characterization of a gene from a tomato pathogen determining hypersensitive resistance in non-host species and genetic analysis of this resistance in bean. Proc Nat Acad Sci USA 85:6743-6747

Yaegashi H, Asaga K (1981) Further studies on the inheritance of pathogenicity in crosses of *Pyricularia oryzae* with *Pyricularia* sp. from finger millet. Ann Phytopathol Soc Jpn 47:677-679

Yamada M, Kiyosawa S, Yamaguchi T, Hirano T, Kobayashi T, Kushibuchi K, Watanabe S (1976) Proposal of a new method for differentiating races of *Pyricularia oryzae* Cavara in Japan. Ann Phytopathol Soc Jpn 42:216-219

SUBJECT INDEX

(A page number marked* refers to the whole paper)

NATO ASI Series H

Vol. 1: **Biology and Molecular Biology of Plant-Pathogen Interactions.** Edited by J. A. Bailey. 415 pages. 1986.

Vol. 2: **Glial-Neuronal Communication in Development and Regeneration.** Edited by H. H. Althaus and W. Seifert. 865 pages. 1987.

Vol. 3: **Nicotinic Acetylcholine Receptor: Structure and Function.** Edited by A. Maelicke. 489 pages. 1986.

Vol. 4: **Recognition in Microbe-Plant Symbiotic and Pathogenic Interactions.** Edited by B. Lugtenberg. 449 pages. 1986.

Vol. 5: **Mesenchymal-Epithelial Interactions in Neural Development.** Edited by J. R. Wolff, J. Sievers, and M. Berry. 428 pages. 1987.

Vol. 6: **Molecular Mechanisms of Desensitization to Signal Molecules.** Edited by T. M. Konijn, P. J. M. Van Haastert, H. Van der Starre, H. Van der Wel, and M. D. Houslay. 336 pages. 1987.

Vol. 7: **Gangliosides and Modulation of Neuronal Functions.** Edited by H. Rahmann. 647 pages. 1987.

Vol. 8: **Molecular and Cellular Aspects of Erythropoietin and Erythropoiesis.** Edited by I. N. Rich. 460 pages. 1987.

Vol. 9: **Modification of Cell to Cell Signals During Normal and Pathological Aging.** Edited by S. Govoni and F. Battaini. 297 pages. 1987.

Vol. 10: **Plant Hormone Receptors.** Edited by D. Klämbt. 319 pages. 1987.

Vol. 11: **Host-Parasite Cellular and Molecular Interactions in Protozoal Infections.** Edited by K.-P. Chang and D. Snary. 425 pages. 1987.

Vol. 12: **The Cell Surface in Signal Transduction.** Edited by E. Wagner, H. Greppin, and B. Millet. 243 pages. 1987.

Vol. 13: **Toxicology of Pesticides: Experimental, Clinical and Regulatory Perspectives.** Edited by L. G. Costa, C. L. Galli, and S. D. Murphy. 320 pages. 1987.

Vol. 14: **Genetics of Translation. New Approaches.** Edited by M. F. Tuite, M. Picard, and M. Bolotin-Fukuhara. 524 pages. 1988.

Vol. 15: **Photosensitisation. Molecular, Cellular and Medical Aspects.** Edited by G. Moreno, R. H. Pottier, and T. G. Truscott. 521 pages. 1988.

Vol. 16: **Membrane Biogenesis.** Edited by J. A. F. Op den Kamp. 477 pages. 1988.

Vol. 17: **Cell to Cell Signals in Plant, Animal and Microbial Symbiosis.** Edited by S. Scannerini, D. Smith, P. Bonfante-Fasolo, and V. Gianinazzi-Pearson. 414 pages. 1988.

Vol. 18: **Plant Cell Biotechnology.** Edited by M. S. S. Pais, F. Mavituna, and J. M. Novais. 500 pages. 1988.

Vol. 19: **Modulation of Synaptic Transmission and Plasticity in Nervous Systems.** Edited by G. Hertting and H.-C. Spatz. 457 pages. 1988.

Vol. 20: **Amino Acid Availability and Brain Function in Health and Disease.** Edited by G. Huether. 487 pages. 1988.

NATO ASI Series H

Vol. 21: **Cellular and Molecular Basis of Synaptic Transmission.**
Edited by H. Zimmermann. 547 pages. 1988.

Vol. 22: **Neural Development and Regeneration. Cellular and Molecular Aspects.**
Edited by A. Gorio, J.R. Perez-Polo, J. de Vellis, and B. Haber. 711 pages. 1988.

Vol. 23: **The Semiotics of Cellular Communication in the Immune System.**
Edited by E.E. Sercarz, F. Celada, N.A. Mitchison, and T. Tada. 326 pages. 1988.

Vol. 24: **Bacteria, Complement and the Phagocytic Cell.**
Edited by F.C. Cabello und C. Pruzzo. 372 pages. 1988.

Vol. 25: **Nicotinic Acetylcholine Receptors in the Nervous System.**
Edited by F. Clementi, C. Gotti, and E. Sher. 424 pages. 1988.

Vol. 26: **Cell to Cell Signals in Mammalian Development.**
Edited by S.W. de Laat, J.G. Bluemink, and C.L. Mummery. 322 pages. 1989.

Vol. 27: **Phytotoxins and Plant Pathogenesis.**
Edited by A. Graniti, R.D. Durbin, and A. Ballio. 508 pages. 1989.

Vol. 28: **Vascular Wilt Diseases of Plants. Basic Studies and Control.**
Edited by E.C. Tjamos and C.H. Beckman. 590 pages. 1989.

Vol. 29: **Receptors, Membrane Transport and Signal Transduction.**
Edited by A.E. Evangelopoulos, J.P. Changeux, L. Packer, T.G. Sotiroudis, and K.W.A. Wirtz. 387 pages. 1989.

Vol. 30: **Effects of Mineral Dusts on Cells.**
Edited by B.T. Mossman and R.O. Bégin. 470 pages. 1989.

Vol. 31: **Neurobiology of the Inner Retina.**
Edited by R. Weiler and N.N. Osborne. 529 pages. 1989.

Vol. 32: **Molecular Biology of Neuroreceptors and Ion Channels.**
Edited by A. Maelicke. 675 pages. 1989.

Vol. 33: **Regulatory Mechanisms of Neuron to Vessel Communication in Brain.**
Edited by F. Battaini, S. Govoni, M.S. Magnoni, and M. Trabucchi. 416 pages. 1989.

Vol. 34: **Vectors as Tools for the Study of Normal and Abnormal Growth and Differentiation.**
Edited by H. Lother, R. Dernick, and W. Ostertag. 477 pages. 1989.

Vol. 35: **Cell Separation in Plants: Physiology, Biochemistry and Molecular Biology.**
Edited by D.J. Osborne and M.B. Jackson. 449 pages. 1989.

Vol. 36: **Signal Molecules in Plants and Plant-Microbe Interactions.**
Edited by B.J.J. Lugtenberg. 425 pages. 1990.

MIX
Papier aus verantwortungsvollen Quellen
Paper from responsible sources
FSC® C105338

If you have any concerns about our products,
you can contact us on
ProductSafety@springernature.com

In case Publisher is established outside the EU,
the EU authorized representative is:
Springer Nature Customer Service Center GmbH
Europaplatz 3, 69115 Heidelberg, Germany

Printed by Libri Plureos GmbH
in Hamburg, Germany